Cytochalasins: Incidence and Biological Activities

The Authors

Dr. S. Kiran, Assistant Professor, Department of Botany, Satavahana University, Karimnagar. He received his M.Sc and Ph.D degrees in Mycology from Kakatiya University. He has a teaching experience of six years and ten years research experience. He published 23 research papers in reputed national and international journals. He was awarded two fellowships form UGC New Delhi for pursuing Ph.D research work.

Dr. M. Surekha received her Ph.D. degree in Mycology and Plant Pathology with the Professor S. M. Reddy, Kakatiya University (1990) and had postdoctoral research experience in fungal toxins. Her primary field of research is mycology and mycotoxins with an emphasis on animal feeds and fodders and has been working in this area for last 25 years. She has guided three students for their Ph.D degrees and presently six are working. She has published 43 research papers in international and national journals of repute. Currently, Dr. Surekha is working as an Associate Professor in the Department of Botany, Kakatiya University and Principal Investigator for the UGC funded project.

Prof. S.M. Reddy after a brilliant academic career joined Osmania University and moved to Kakatiya University. He rose to the position of Head, Department of Botany. Successfully guided 40 research scholars for Ph.D degree and completed 24 research projects. He has published more than 520 papers in National and International journals and authored/edited about two dozen books. He is associated with different scientific societies and occupied coveted positions and organized several national level conferences. He is bestowed with Best Teacher award by Govt. of A.P. and Indian Society of Mycology and Plant Pathology. He is a member of different scientific organization such as DBT, DST, CSIR and UGC etc.

Cytochalasins: Incidence and Biological Activities

S. Kiran

Department of Botany
Satavahana University
Karimnagar, TS, India

M. Surekha and S.M. Reddy

Department of Botany
Kakatiya University
Warangal, TS, India

SCIENTIFIC
PUBLISHERS

Published by

SCIENTIFIC PUBLISHERS (INDIA)
5 A, New Pali Road
P.O. Box 91
Jodhpur - 342 001 INDIA
E-mail: info@scientificpub.com
Website: http://www.scientificpub.com

Print: 2019

ISBN: 978-93-88043-59-5 (Hardbound)
978-93-88043-82-3 (E-book)

Printed in India

PREFACE

Cytochalasins, emerging mycotoxins, are diverse in their structure and biological activities and being elaborated by equally diverse moulds both in taxonomy and ecologically. They live on diverse substrates and in different mode of association. These mycotoxins are unique category by their characteristic of toxicity causing specific inhibition of cytokinesis. They are also reported to cause diverse biological activities related to cell. Because of inhibition of cytokinesis, a multinucleate cells of giant size are formed. Probably due to inhibition of cytokinesis, they act as anticarcinogenic and inhibit tumor formation which is likely to be highly useful as anticancer therapeutic agent. In that cure they are likely to be benguine mycotoxins.

The mycotoxins which are non-antigenic and adversely affect health of animals including man and induce variety of health hazards of chronic to serious nature. Further, under modern living conditions contamination of foods with mycotoxigenic moulds and mycotoxins are investable contaminants of foods and feeds are common rather than exceptions. Inspite of scientific and technological developments, contamination of mycotoxins are unavoidable, inadvertent and cannot be avoided. Therefore, one has to minimize the mycotoxin problem.

Thus management of mycotoxins problem in not only problematic but also intricate in view of its multiple dimensions. Therefore, deeper understanding of conditions of foods and feeds and moulds behaviour is of paramount importance. In view of the emerging nature of cytochalasins their incidence, biological activities are little known. No doubt future studies will reveal more interesting facts which may help us to tackle this problem more effectively.

The present book is an outcome our studies during last fifteen years. We tried to cover different aspects of these toxins based on our observations, analysis of problem and existing literature. However, much more in depth studies are needed to unravel many of their biological activities.

We take this opportunity to thank all those who helped us directly or indirectly. We have liberally borrowed the literature from different sources; we gratefully acknowledge their generous gesture. The encouragement provided by the University authorities are gratefully acknowledged. One of the authors (Dr. S.

Kiran) thanks Prof. M. Komal Reddy, Registrar, Satavahana University, Karimnagar for his generous help and encouragement. We are also thankful to Sri Pawan Kumar Sharma and Tanay Sharma of Scientific Publisher (India), Jodhpur for timely bringing out of this book in nice getup.

Authors

CONTENTS

Chapter-1

Introduction and Classification

The cytochalasins, secondary metabolites of fungi, have attracted the attention of chemists and biologists alike in view of their diverse chemical structures and cytotoxicity. Rothweiler and Tamm (1966) from Switzerland and Aldridge *et al.* (1967) from Great Britain (1967) have isolated phomin and dehydrophomin and cytochalasin A and B from culture filtrates of *Phoma* species and *Helminthosporium dematioideum* respectively. Subsequently it was proved that phomin is cytochalasin A and dehydrophomin is cytochalasin B. Hayakawa *et al.* (1968) have isolated Zygosprorin A from culture filtrates of *Zygosporium masonii* which was identical to cytochalasin D isolated from culture filtrates of *Metarrhizium anisopliae* (Aldridge *et al.*, 1969). Umeda *et al.* (1975) reported cytotoxicity of culture filtrates of *Chaetomium globosum* and *C. cochlioides* towards HeLa cells and named them as chaetoglobosins. A novel group of cytochalasins containing indolyl group instead of phenyl group as in other cytochalasins. Keller-Schierlein and Kupfer (1979) have isolated apochalasins which contained isopropyl group from culture filtrates of *Aspergillus microcysticus. Chaetomium elatum* is reported to produce prochaetoglobosin III (Thohinung *et al.*, (2010). Recently scoparasins (Liu *et al.,* 2014) and xylastriasan (Lei *et al.*, 2018) were added to this group of mycotoxins. Similarly, Chen *et al.* (2015) have reported multirostatin, a new cytochalasin from culture filtrates of *Phoma multirostrata.* All the above compounds exhibited uniformly cytotoxicity, inhibiting cytokinesis without affecting karyokinesis, resulting in the formation of binucleate and in extreme conditions multinucleate cells. Similarly large number of compounds exhibiting cytotoxicity were discovered. All these have been assigned different trivial names, such as zygosporins, phomins, chaetoglobosins, apocytochalasins, paspalins and cytochalasins. The later name proposed by English scientists was adopted for basic skeleton of the class of compounds because of the novel effects on mammalian cells. The name cytochalasin epidemiologically means (cytos=cell; chalasins=relaxation) was suggested to cover all these compounds based on their origin and their biological effects. As a result cytokinesis is efficiently inhibited, while mitosis remains unaffected, thereby generating gaint multinucleated or even at higher concentration denucleated cells. Carter (1967) exploited these properties in molecular and cell biology research especially in cell imaging methods of cytoskeleton (Goldman and Spector, 2004). In nature cytochalasins play a key role in fungal virulence and food spoilage but may also be involved in the maintenance of symbiotic fungus-plant relationship (Lin *et al.,* 2009). Inhibition of macromolecular synthesis is considered to be secondary. Olinsukan *et al.* (1975) were the first to recognize these compounds as mycotoxins. As some of these trivial names overlapped, Binder *et al.* (1973) proposed systematic nomenclature to these compounds based on the following rules

1. The skeleton of the perhydroisoindole unit including macrocycle and the C-substituents of the isoindole system except for the phenyl and the indol-3-yl rings, is named as cytochalasin. The number of atoms in the macrocycle including the atoms common to it and the isoindole ring, is indicated by a number in square brackets preceding the name cytochalasin.
2. The absolute configuration of isoindole system is established and does not require specification. For the substituents in the macrocycle, chirality may be desired by use of the symbols (R) and (S).
3. The clockwise numbering system indicated in the formulae 1-4 is used for designating the positions of the substitute on the macrocycle and additional functional groups attached to the isoindole system.
4. Oxygen atoms which are part of the macrocycle are designated by the prefix oxa-with appropriate locant.
5. Prefixes (E) and (Z), indicate *-trans* and *-cis* geometry of olefinic double bonds respectively according to the IUPAC (1974) recommendations for selection E, fundamental stereochemistry.

This system allows the assignment of the same locants to that part of the molecule which is common to the hitherto known metabolites. The following four basic skeletons are known

1. (11) cytochalasins
2. (13) cytochalasins
3. 24-oxa-(14) cytochalasins
4. 21, 23-dioxa-(13) cytochalasins

Cytochalasins are the compounds where phenylalanine or tryptophan moiety is linked which in turn linked to a C_{16} –K_{18}. Polyketide ring system containing a carbocyclic (cytochalasin C), a lactate (cytochalasin A) or a cyclic carbonate (cytochalasin E) moiety.

Cytochalasins exert a wide range of biological effects. They target microfilament and affect several cellular processes such as cytokinesis, intra-cellular mobility as well as exo-and endocytosis (Hirose *et al.,* 1990, Peterson and Mitchison, 2002 and Foissner and Wasteneys, 2007). Actin which is abundant in eukaryotic cells gets polymerized to form microfilaments and represent the key component of the cytoskeleton. Cytochalasins specifically interact with actin filament network by capping the barbed ends and altering the dynamic properties of microfilaments and inhibit cytokinesis without affecting nuclear division. Cytochalasins at sub toxic level cause multinucleated cells from a proliferating cells and at higher concentrations of these compounds results in denucleation of cells. This property has been exploited in the study of cellular processes.

Cytochalasins can be divided into general groups based on whether they contain a phenylalanine or tryptophan fixed at C_{10} position of the perhydro-isoindole ring system. Compounds which contain a tryptophan moiety generally have been referred as chaetoglobosins. Natori (1977) has classified the cytochalasins into several groups based on the (1) substituents at position 10 (phenyl or indole-3-yl). (2) The type of six membered ring in the perhydroisoindole ring system (four different types). (3) The structure of the macrocyclic ring after $C_{16.}$

1. Structure and classification

More than 120 cytochalasins have been isolated from wide range of fungi including *Phomopsis* (Tomioka, *et al.,* 1987 and Izawa, *et al.,* 1989), *Chalara* (Tomas, 1981), *Hypoxylon* (Edwards and Maitland 1989), *Xylaria* (Espada *et al.,* 1997), *Daladinia* (Buchanan *et al,* 1995, 1996), *Pseudoeurotium* (Feng *et al.,* 2002), *Phoma exigua* (Evidente *et al.,* 2003), *Spicaria elegans* (Zhenjian Lin *et al.,* 2009) and *Aspergillus flavipes* (Zue *et al.,* 2017). Cytochalasins are structurally characterized by the presence of a halogenated and highly substituted bicyclic isoindolone unit fused to a macrocyclic ring. The macrocyclic appendage varies widely within the cytochalasins. Cytochalasin Z_4, Z_5 and Z_6 were elaborated by *Phoma exigua* var. *heteromorpha* on wheat medium, which showed culture root elongation inhibition and exhibited 20% less brine shrimp larva mortality (Evidente *et al.,* 2003). It can either be a lactone, a carbonate or more frequently a carbocycle as exhibited by cytochalasins B, E and D respectively and seems to play an important role in the determination of biological activity. From the structures of the six-membered perthydroisoindole ring the cytochalasins are 23 types (type A, B, C and D)

Type A	- Compounds 1, 13, 15, 18
Type B	- Compounds 2-5, 7, 8, 11, 17, 22, 23
Type C	- Compounds 6, 12
Type D	- Compounds 9,10, 14, 16, 19-21

a. From the group at C (the structure of the macrocyclic ring after C), (10) phenyl (derived from phenylalanine compounds 1-8, 10-13, 21-23, indol-3-yl (derived from tryptophan) compounds-9, 14-20.

b. From the size and inserted oxygen atom of the macrocylic ring (11) cytochalasins-compound-1-9 dioxa-(13) cytochalasin-compound 10 from eight acetate-malonate units, (13) cytochalasins compounds 11-20, oxa (14) cytochalasin-compounds 21-23 from nine acetate malonate units.

c. From the number of incorporated C1 unit: 2 units-compounds 9, 11-13, 21-23; 3 units – compounds 1-8, 10, 14-20.

Cytochalasins are confirmed in the solid state by x-ray analysis, while in a solutions by H-NMR studies. The six membered rings in the perhydroisoindolone is *-cis-* fused to the five membered lactam ring and *trans*-fused to the macrocyclic ring, adopting slightly twisted boat conformation irrespective of the structural type. Recently Ding *et al.* (2012) reported two pentacyclic cytochalasins known as trichoderones A and B produced by an endophytic *Trichoderma gamsii.*

Chen *et al.* (2016) have described nine new cytochalasins belonging to a group of chaetoglobosins. Ding *et al.* (2014) reported production of trichodermone with tetracyclic nucleus by *Trichoderma gamsii.* Further it is interesting to note that a single species could elaborated tri-tetra and pentacyclic cytochalasins.

Zhang *et al.* (2013) isolated periconiasin A, a tricyclic ring system from the endophytic fungus, *Periconia* sps. Tricyclic core structure in which a macrocycle is fused to a bicycle lactam (isoindoline) system (Binder and Tamm, 1973). Cytochalasins which are characterized by highly substituted perhydro-iso indoline moiety to which typically a macrocyclic ring - either a carbocycle, a lactone or a cyclic carbonate is fused. Depending on the amino acid incorporated into a polyketide chain, the hydrogenated isoindolone either bears a benzyl group are known as cytochalasins, while the compound bearing a *P*-methoxybenzyl group are known as pyrichalasins. On the other hand, compounds bearing an indole-3-yl group are known as chaetoglobosins, while the apochalasins bear a 2-methylproyl group. More recently Zhang *et al.* (2008) reported compounds bearing a methyl group are known as alachalasins. Variation of the size and substitution pattern of the macrocycle further increases the chemical diversity. Since the discovery of cytochalasins in 1966, a plethora of structurally intriguing cytochalasins have been isolated from various sources and characterized. Chen *et al.* (2016) feel that chemical diversity of cytochalasins remained untapped and stressed the need to take up more intensive studies. The encapsulation of indocyanine green (ICG) into cytochalasin B induced membrane vesicles led to slowing down the body clearance of ICG and improving the effectiveness of photothermal antitumor therapy *in vivo* (Sheng *et al.,* 2016).

Apart from the identification of number of new cytochalasins, their biological activities and targets, in recent decades much has been learned about their biosynthesis. Therefore, an attempt has been made to update the information in a holistic manner.

■■■■

Chapter-2

Structure of Different Cytochalasins

2.1. Cytochalasins

Cytochalasins are the compounds in which phenylalanine or typtophan moiety linked to perhydroisoindole moiety which intrun is linked to C_{16}-C_{18} polyketide ring system containing a carbocyclic (cytochalasin C), a lactate (cytochalasin A) or a cyclic carbonate (cytochalasin E) moiety. The number of these toxins are being increased day by day. At present there are 60 cytochalasins elaborated by various fungi which are taxonomically quite unrelated (Table 2. 1).

Table 2.1. Cytochalasins produced by different fungi

Toxin	Species	Reference
Cytochalasin A	*Ascochyta heteromorpha*	Ohtsubo *et al.* (1978)
	A. lathyri	
	Curvularia lunata	
	Drechslera dematioideum	
	D. erythrostoma	
	D. biseptata	
	Gnomonia biseptata	
	Hypomyces odoratus	
	Phoma exigua	
	Pyrenophora semeniperda	
Cytochalasin B	*Ascochyta lathyri*	Wells *et al.* (1981)
	Curvularia lunata	Evidente *et al.* (2002)
	Drechslera erythrostoma	
	Drechslera dematioideum	
	Gnomonia biseptata	
	Hypomyces odoratus	
	Hormiscium sp	
	Phoma exigua	

Toxin	Species	Reference
	Phoma exigua var. *heteromorpha*	
	Pyrenophora semeniperda	
	Xylaria psidii	
Cytochalasin C	*Bevaeria bassiana*	El-Kady and Mostafa (1995)
	Bipolaris neergaardii	
	Geneculosporium sp	
	Hypoxylon terricola	
	Metarrhizium anisopliae	
	Phoma herbarum	
	Xylaria arbuscula	
	Xylaria hypoxylon	
Cytochalasin D	*Bipolaris neergaardii*	El-Kady and Mostafa (1995)
	Coriolus vernicipes	
	Drechslera dematioideum	
	Hypocrella bambusae	
	Hypoxylon terricola	
	Metarrhizium anisopliae	
	Microporus afinis	
	Phoma herbarum	
	Xylaria arbuscula	
	Xylaria hypoxylon	
	Zygosporium masonii	
Cytochalasin E	*Aspergillus clavatus*	Steyn *et al.* (1982)
	A. terreus	Udagawa *et al.* (2000)
	Alternaria chlamydospora	Melissa *et al.* (2000)
	Cochliobolus tuberculatus	Hidayat *et al.* (2014)
	Drechslera dematioideum	
	Rhinocladiella sp	
	Rosellinia necatrix	
	Physalospora sp	
Cytochalasin F	*Aspergillu clavatus*	Cooper (1987)
	Drechslera dematioideum	
	Nigrosabulum sp.	
	Phoma exigua var. *heteromorpha*	
	Pyrenophora semeniperda	
	Rhinocladiella sp	

Toxin	Species	Reference
	Spicaria elegans	
Cytochalasin G	*A. clavatus* *Nigrosabulum* sps *Phomopsis paspali* *Pseudeurotium zonatum*	Cameron *et al.* (1974) Hazel *et al.* (1989)
Cytochalasin H	*Endothia gyrosa* *Geniculosporium* sp *Phomopsis paspali* *P. sojae* *Xylaria* sp	Wells *et al.* (1976) Yu Li *et al.* (2012) Zhang *et al.* (2006)
Cytochalasin I	*Daldinia vernicosa*	Stefan *et al,* (2009)
Cytochalasin H_2	*Xylaria* sp	Yu Li *et al.* (2012)
Cytochalasin J	*Endothia gyrosa* *Phomopsis paspali* *P. sojae* *Physalospora* sp	Shu Xu *et al.* (2009) Hidayat *et al.* (2014)
Cytochalasin K	*A. clavatus* *Chalara microspora*	Tomas (1981) Qiao *et al.* (2011)
Cytochalasin L	*Chalara microspora*	Tomas (1981)
Cytochalasin M	*Chalara microspora*	Tomas (1981)
Cytochalasin N,O,P,Q,R	*Hypoxylon terricola* *Phomopsis* sp. *Xylaria hypoxylon*	Edwards *et al.* (1989)
Cytochalasin Q	*Halorosellinia oceanica* *Hirsutella* sp *Xylaria hypoxylon* *Xylaria arbuscula*	Lang *et al.*, (2005)
19,20-epoxycyto-chalasin Q	*Xylaria obovata*	Dagne *et al.*(1994)
Cytochalasin S	*Phomopsis* sp	Izawa *et al,* (1989)
Cytochalasin T	*Pyrenophora semeniperda*	Evidente *et al.* (2002)
Cytochalasin T,U,V	*Phoma exiqua* var. *heteromorpha*	Evidente *et al.* (1992)
Cytochalasin U	*Ascochyta heteromorpha* *Geniculosporium* sp	Zhang *et al.* (2006)

Toxin	Species	Reference
	Pestalotiopsis	
	Phoma exiqua var. *heteromorpha*	
Cytochalasin W	*Ascochyta heteromorpha*	Evidente *et al.* (1996)
	Phoma exiqua var. *Heteromorpha*	Vurro *et al.* (1997)
Cytochalasin X, Y, Z	*Pseudeurotium zonatum*	Feng *et al.* (2002)
Cytochalasin Z_1, Z_2 and Z_3	*Pyrenophora semeniperda*	Evidente *et al.* (2002)
Cytochalasin Z_4	*Aspergillus elegans*	Liu *et al.* (2014)
Cytochalasin Z_4, Z_5 and Z_6	*Phoma exiqua* var. *heteromorpha*	Evidente *et al.* (2003)
Cytochalasin Z_7- Z_{15}	*Spicaria elegans*	Liu *et al.* (2006)
Cytochalasin Z_{10}- Z_{11}	*Endothia gyrosa*	Shu Xu *et al.* (2009)
Cytochalasin Z_{16}	*Aspergillus clavatus*	Hu *et al.* (2014)
Cytochalasin $Z_{16,}$ Z_{17}	*Aspergillus terreus*	Zhang *et al.* (2010)
Cytochalasin Z_{16}- Z_{20}	*Aspergillus flavipes*	Lin *et al.* (2009)
Cytochalasin Z_{21}-Z_{23}	*Spicaria elegans*	Wang *et al.* (2011)
Cytochalasin Z_{24}	*Aspergillus elegans*	Zheng *et al.* (2013)
Cytochalasin $Z_{27,}$ Z_{28}	*Xylaria* sp	Zhang *et al.* (2014)

Cytochalasins A (Fig 2.1): It is elaborated by *Phoma exigua, Drechslera dematioideum* and *Hormiscium* sp. Cytochalasin A is an oxidized analog of cytochalasin B which uniquely inhibits HIV-1 protease (IC_{50} = 3 μM). Cytochalasin A also induces the phosphorylation of the tyrosine phosphatase PTP3 of Dictyostelium, activating STATc (Araki and Williams, 2012). Production and purification methods are similar to those employed for deoxaphomin.

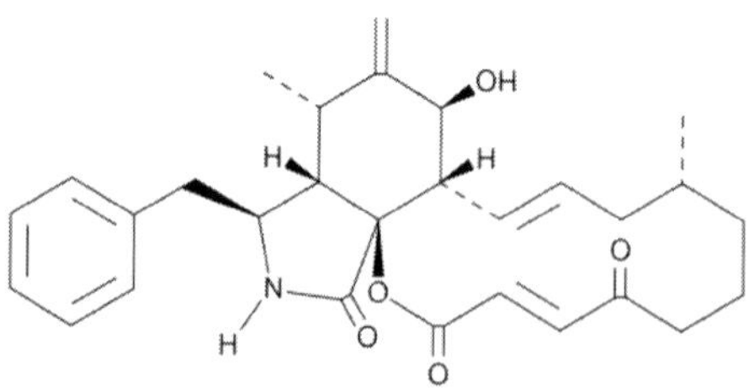

Fig. 2.1: Structure of cytochalasin A

Cytochalasin B (Fig. 2.2): TLC on silica gel G reveals of several derivatives of cytochalasin B was first reported in 1967 from a mould (Smith *et al.,* 1967). Cytochalasin B contains several highly polar keto- and hydroxyl groups and has one peripheric lipophilic benzyl unit. Other producers include *Phoma* spp., *Hormiscium* spp. and *Curvularia lunata.* It can also be synthesized in the laboratory. There are several approaches to do so. Firstly, it is possible to form the six-membered ring of the isoindolone core and the larger macrocyclic ring simultaneously in a late-stage intramolecular Diels-Alder cyclization. Secondly, it is possible to first form the isoindolone core in an intramolecular Diels-Alder reaction and in a second step appends the macrocycle in a stepwise fashion (Haidle and Myers, 2004). Cytochalasin B causes multinucleation in cells and significantly affects cell motility. The multinucleated cells probably arise from the failure of mitotic control, leading to variation in size and shape of interphase nuclei. Many cytochalasins inhibit cell division and are known to induce an apoptotic response, due to the inhibition of polymerization of actin. Cytochalasin B is used as a tool for the study of cell division and cell motility (Rubtsova *et al.*, 1998). Cytochalasins have also been reported to exhibit diverse biological activities such as L-696, 474 inhibits HIV-1, protease that are unrelated to actin binding (Hungate *et al.,* 1996).

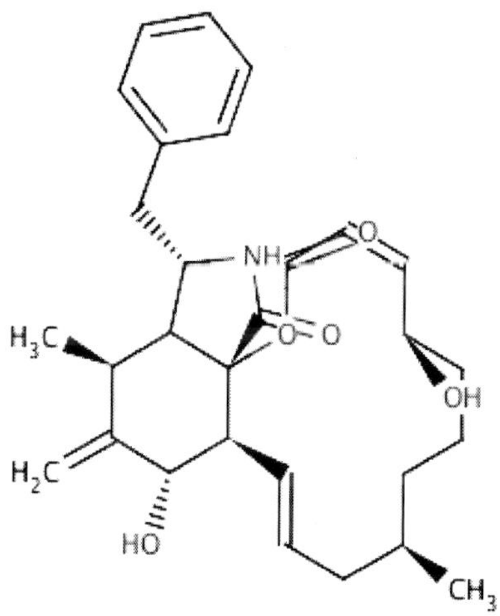

Fig 2.2: Structure of cytochalasin B

Cytochalasins C and D (Fig 2.3 & 2.4): *Metarrhizium anisopliae*, *Zygosporium masonii* and *Z. mycophilum* are the producers of these mycotoxins. Biosynthetic experiments with cytochalasin D using growing cultures of *Z. masonii* revealed that the one phenylalanine, three units of methionine and acetate unit form the building blocks. Cytochalasin C acts as a potent inhibitor of actin filament and contractile microfilaments (Trendowski, 2015). Cytochalasin D is a cell permeable mycotoxin, which causes both the association and dissociation of actin subunits, disrupts actin filaments. The inhibition of actin polymerization 10 times more than cytochalasin B. Unlike cytochalasins A and B, cytochalasins D and E does not inhibit glucose transport.

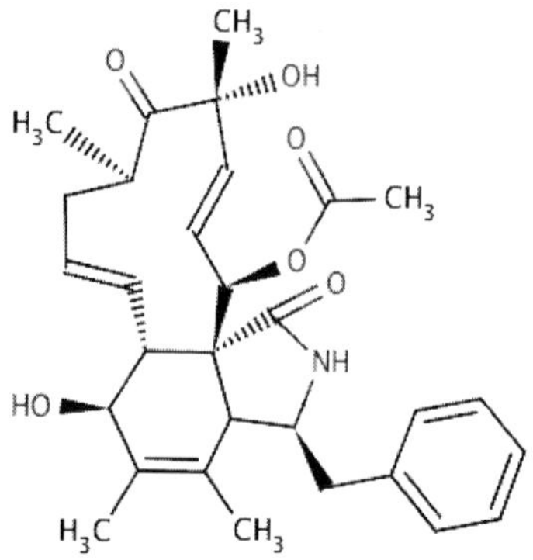

Fig. 2.3. Cytochalasins C

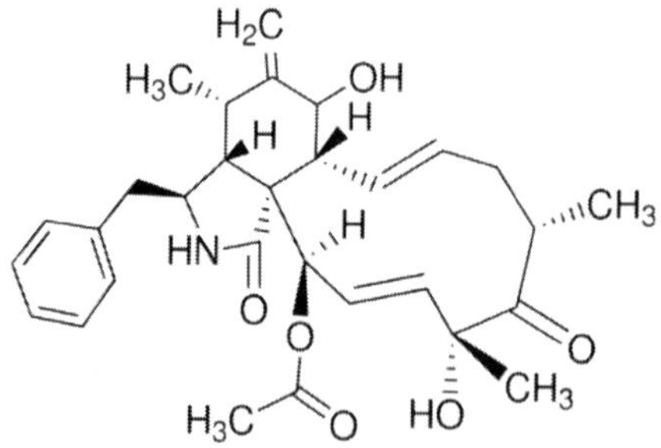

Fig. 2.4. Cytochalasins D

Fig. 2.3 & 2.4: Structures of cytochalasins C and D

Cytochalasin E (Fig. 2.5): 21, 23 Dioxa cytochalasin (cytochalasins E) is produced by *Rosellinia necatrix* and *Aspergillus clavatus.* Cytochalasins E bearing an β,γ-unsaturated ketone moiety. Cytochalasin E contains an arrangement of atoms that span elements of the macrocyclic ring, the epoxide group, and the core ring found in the conserved regions of the angiogenesis inhibitor TNP-470 and its parent compound. This cytochalasin does not inhibit glucose transport or HIV-1 protease activity (Lombardo *et al.*, 1996).

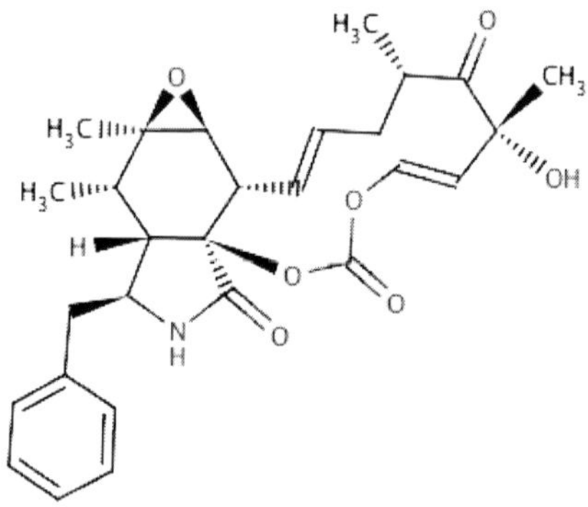

Fig. 2.5: Structure of cytochalasin E

Cytochalasin E contains a substituted perhydroisoindolone scaffold fused with a macrocyclic ring that is the hallmark of cytochalasins, which is proposed to be derived from an intramolecular Diels-Alder reaction of the PKS-NRPS product following its release and formation of the pyrrolinone dienophile (Schumann and Hertweck, 2007). Unlike chaetoglobosin A, cytochalasin E is derived from a shorter polyketide chain (octaketide instead of nonaketide), with a different amino acid building block (phenylalanine instead of tryptophan), and contains a unique vinyl carbonate moeity, which warrants further biosynthetic investigation at the molecular genetics level. Cytochalasin E is toxic against brine shrimp with a LC50 of 2.79 lM, which is comparable to the commercial insecticidal toosendanin (LC50 = 4.03 lM).

Cytochalasin E inhibits angiogenesis induced by basic fibroblast factor by 40-50% in the mouse cornea assay and inhibits the growth of Lewis lung tumors by approximately 72%. Hence, it holds promise for the treatment of cancer and other types of pathogenic angiogenesis (Udagawa *et al.,* 2000). The bovine capillary endothelial cells (BCE) are the most sensitive to cytochalasin E, particularly at lower concentrations (Udagawa *et al.,* 2000). Cytochalasin E at high concentrations was reported to exhibit unique histopathologic effects on the vasculature, including hemorrhage and injury to the vascular walls accompanied by alterations in vascular permeability (Lipski *et al.,* 1987).

Cytochalasin F (Fig. 2.6): It is a minor metabolite of *Drechslera dematioideum* and isomeric with cytochalasin B. Its NMR spectrum shows most of the features of the spectrum of cytochalasin B, the significant differences being (i) the absence of signals due to an exocyclic methylene group, (ii) the presence of a signal due to a vinylic methyl group, and (iii) a change in the pattern of the signal due to H-7. These features of the NMR spectrum of cytochalasin F are consistent with structure and this structure was confirmed by mild treatment of cytochalasin F with acid to give cytochalasin B (Aldridg *et al.,* 1972).

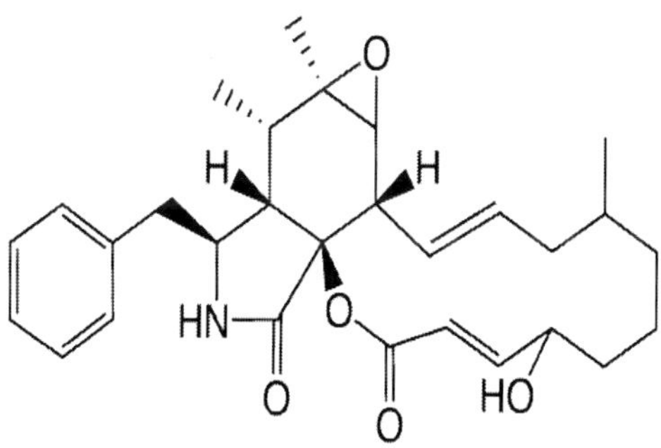

Fig. 2.6. Structure of cytochalasin F

Cytochalasin G (Fig 2.7): This was isolated from an unidentified species of *Nigrosabulum* (Cameron *et al.*, 1974) and *Pseudoeurotium zonatum* (Probst and Tamm, 1981). This compound is indole-substituted perhy-droisoindol-1-one containing an 11-membered macrocycle. The X-ray analysis of cytochalasin G suggested that the six-membered ring adopts a slightly twisted boat conformation, and the macrocyclic ring, which contains an E-alkene, is arranged in a chairlike conformation.

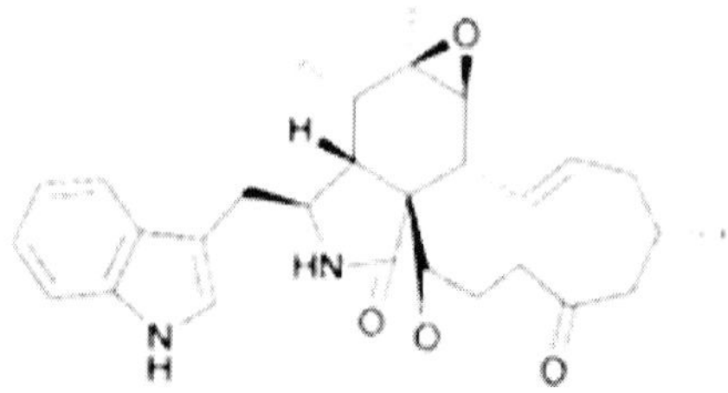

Fig 2.7: Structure of cytochalasin G

Cytochalasins H and J (Fig. 2.8 & 2.9): Cytochalasins H and J were originally named as kodo-cytochalasin-1 and kodo-cytochalasin-2 respectively and reported to be produced by *Phomopsis paspalli* growing on *Paspalum scrobiculatum* Linn., a millet consumed in India (Patwardhan *et al.,* 1974). Cytochalasins H and J have shown to affect central nervous system activity. Cytochalasin J blocks or slows chromosome motion and affects spindle architecture. These compounds were demonstrated to be substantially cytotoxic to the human leukaemia K562 cell line (Shu Xu *et al.,* 2009). Shu Xu *et al.* (2009) have reported epoxycytochalsin H produced by *Endothia gyrosa.*

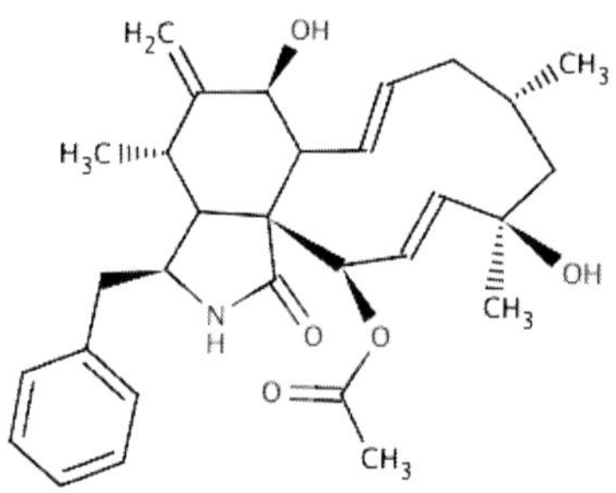

Fig. 2.8. Cytochalasin H

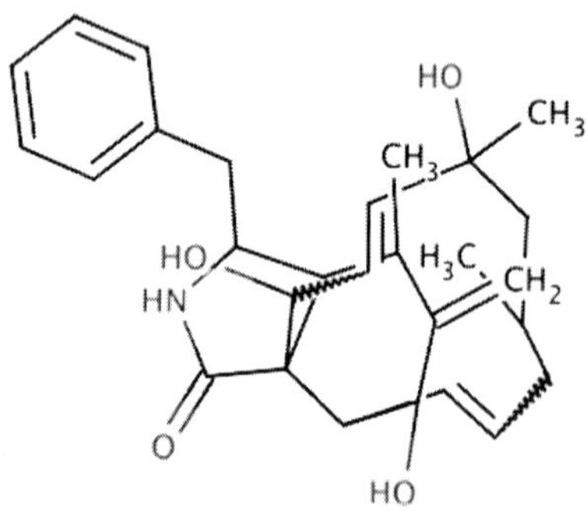

Fig. 2.9. Cytochalasin J

Fig. 2.8 & 2.9: Structures of cytochalasins H and J

Cytochalasins K, L and M (Fig 2.10-12): *Chalara microspora* (Tomas, 1981) grown in a stationary culture elaborated these toxins. The cytochalasin K, L and M were isolated both from the medium and mycelium using reversed-phase chromatography. A comparison of these data with those of chaetoglobosin A and its acetate, indicated that cytochalasin K is identical with the acetate of chaetoglobosin A, except that the indolyl group of the latter is replaced by a phenyl group. Biogenetically, cytochalasin L may be derived from cytochalasin K by a Baeyer-Villiger type reaction, for which there is precedence among other cytochalasins.

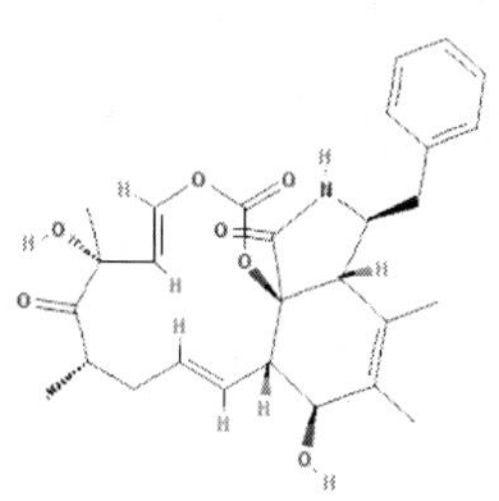

Fig. 2.10. Cytochalasin K

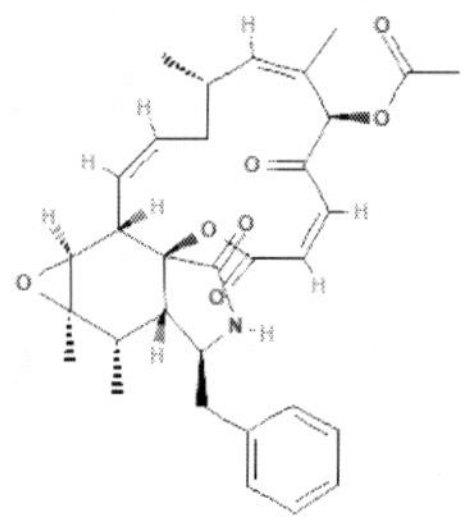

Fig. 2.11. Cytochalasin L

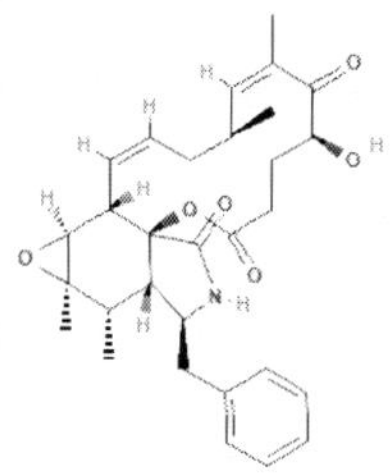

Fig. 2.12. Cytochalasin M

Fig. 2.10 to 2.12. Structures of cytochalasins K, L and M

Cytochalasins N, O, P, Q and R (Fig 2.13 - 17): These are produced by *Hypoxylon terricola* (Edwards and Maitland (1989). Cytochalasin N is the 5,6-epoxide of cytochalasin C, and cytochalasins O and P are two epimeric 6-hydroxy analogues. Cytochalasin Q is a 6,7-epoxide and cytochalasin R is a 6,7,13,14-diepoxide. The epoxides formed by cytochalasin C and D are described and the acid- and BF_3-induced rearrangement products of cytochalasin N are identified.

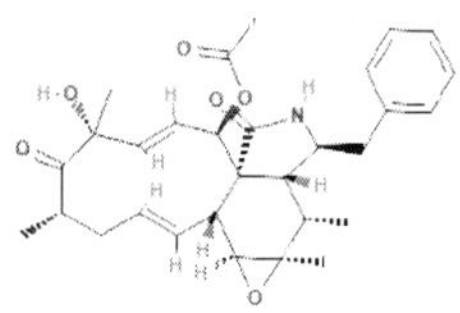

Fig. 2.13. Cytochalasin N

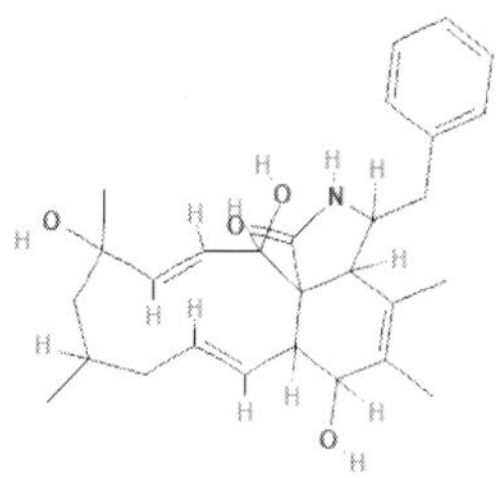

Fig. 2.14. Cytochalasin O

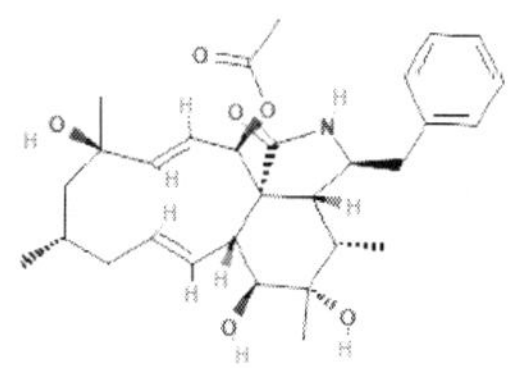

Fig: 2.15. Cytochalasin P

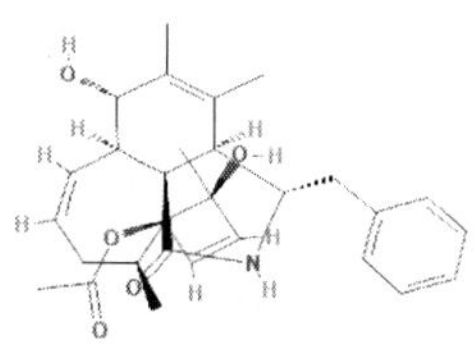

Fig. 2.16. Cytochalasin Q

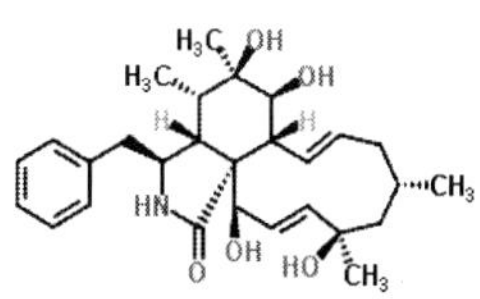

Fig. 2.17. Cytochalasin R

Fig. 2.13 to 2.17 Structures of cytochalasins N, O, P, Q and R

Cytochalasins S (Fig 2.18): This compound 10-pheynl-[11] cytochalasins was isolated from *Phomopsis* sp. (Izawa *et al.,* 1989). Its 5α, 7β- diol-type structure in the cyclohexane part of the molecules. This novel compound exhibited weaker cytotoxicity to mammalian cells and inhibition of capping.

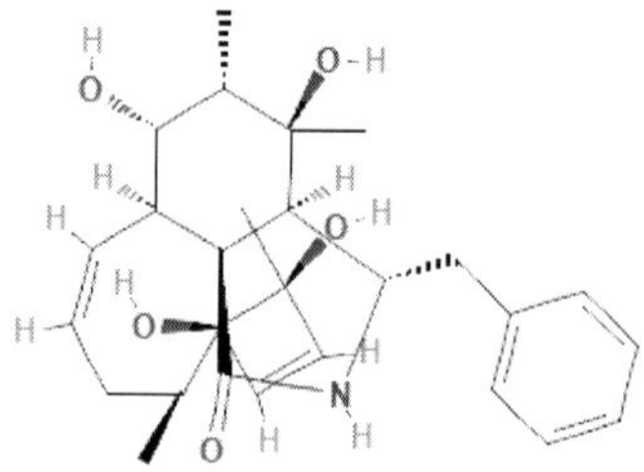

Fig. 2.18. Structure of cytochalasin S

Cytochalasins T, U and V (Fig. 2.19-2.21): These are produced by *Phoma exiqua var. heteromorpha* (Capasso *et al.,* 1991). These are a formyl group on the macrocyclic ring. Cytochalasins U and V were characterized as the 25, 26-dioxa cytochalasin and the first 25-oxa cytochalasin respectively (Evidente *et al.,* 1992). These are compounds show significant and weak activity in the brine shrimp assay.

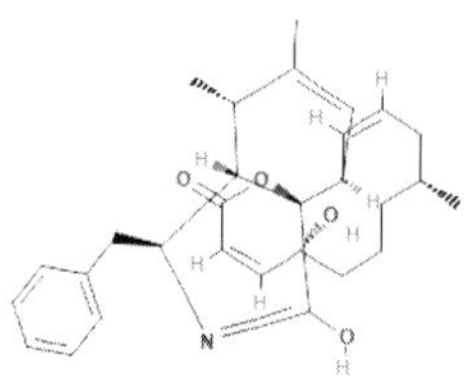

Fig. 2.19. Cytochalasin T

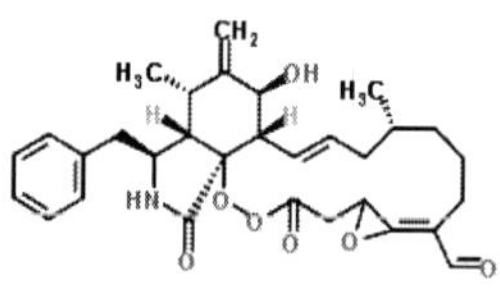

Fig. 2.20. Cytochalasin U

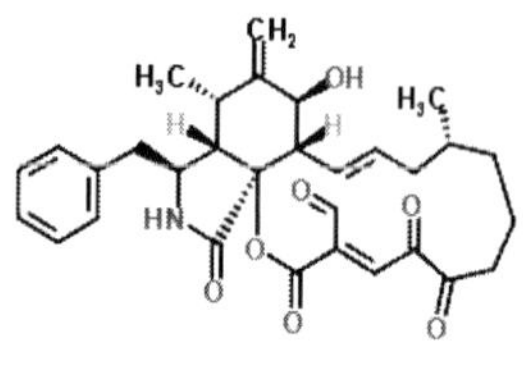

Fig. 2.21. Cytochalasin V

Fig 2.19 to 2.21. Structures of cytochalasins T, U and V

Cytochalasins W (Fig 2.22): This was isolated from *Phoma exigua* var *heteromorpha* (Evidente *et al.,* 1996). It is 24-oxa[14] cytochalasin, bearing a formyl group on the macrocyclic ring. It shows toxic activity in the brine shrimp assay. Cytochalasin W showed the presence of an exocyclyc α, β-unsaturated aldehyde group appearing as a single and as a doublet at the typical chemical shift values of δ 10.0 and 181.2, respectively.

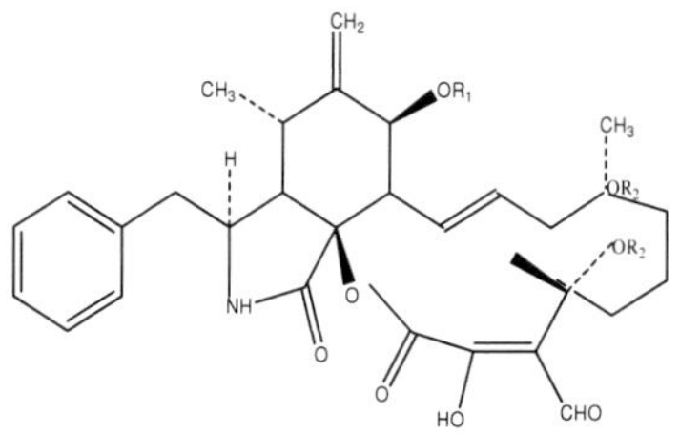

Fig 2.22. Structure of cytochalasin W

Cytochalasins X, Y and Z (Fig 2.23-25): These are produced by *Pseudeurotium zonatum* (Feng *et al.,* 2002). These contain an indole-substituted perhydroisoindol-1-one fused with an 11-membered macrocycle.

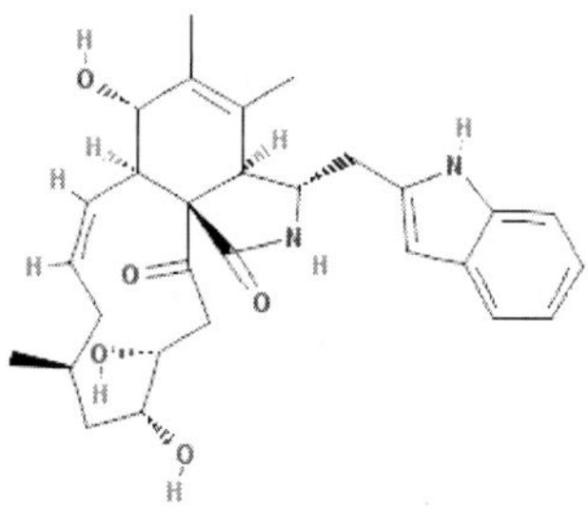

Fig. 2.23. Cytochalasin X

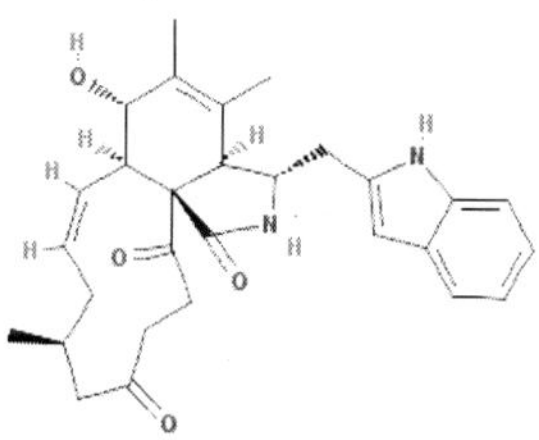

Fig. 2.24. Cytochalasin Y

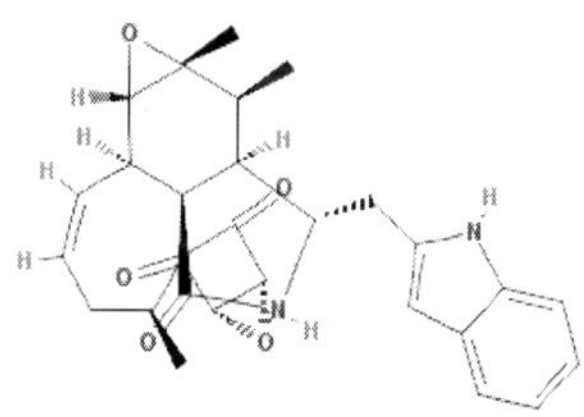

Fig. 2.25. Cytochalasin Z

Fig. 2.23 to 2.25. Structures of cytochalasins X, Y and Z

Cytochalasin Z_1, Z_2 and Z_3 (Fig. 2.26-28): These are produced by *Pyrenophora semeniperda* (Evidente *et al.,* 2002). These cytochalasins were isolated from the wheat culture of *Pyrenophora semeniperda*, a fungus proposed for biological control of grass weeds. Cytochalasins Z_1 and Z_2 proved to be structurally related to cytochalasin T, whereas cytochalasin Z_3 was related to cytochalasin B. When assayed on wheat and tomato seedlings, cytochalasin Z_3, in comparison to cytochalasin B and its 21,22-dihydroderivative is more toxic. All these three cytochalasins were characterized as 24-oxa[14] cytochalasins.

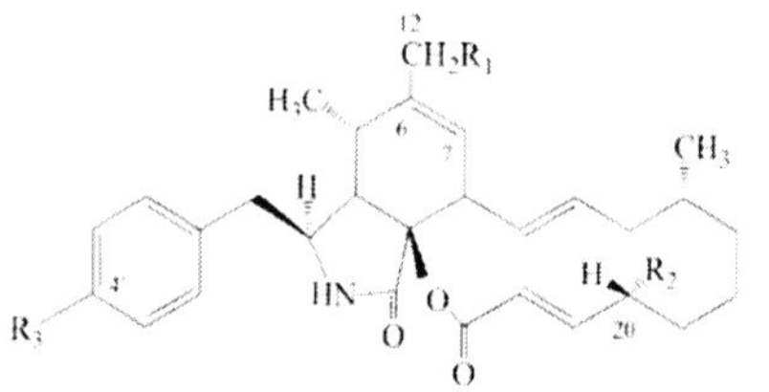

Fig. 2.26. Cytochalasin Z1= 5 R1=R2=H, R3=OH
Fig. 2.27. Cytochalasin Z2= 6 R1=R2=OH, R3=H

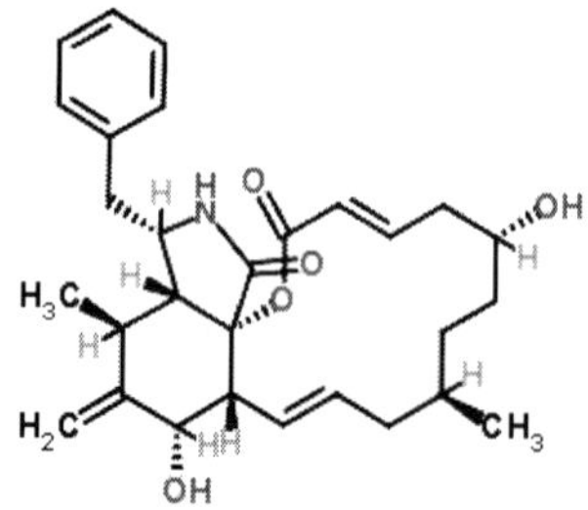

Fig. 2.28. Cytochalasin Z3

Fig. 2.26 to 2.28. Structures of cytochalasins Z_1, Z_2 and Z_3

Cytochalasin Z_4, Z_5 and Z_6 (Fig 2.29-31): These are produced by *Phoma exigua* var. *heteromorpha (*Evidente *et al.,* 2002) on wheat culture medium. Cytochalasins Z_4 and Z_5 proved to be structurally related to cytochalasin B, whereas Z_6 was related to cytochalasin F. These were isolated along with cytochalasins A, B, F,T, Z_2 and Z_3, 7-*O* acetylcytochalasin B, and deoxaphomin from the organic extract of a large amount of *P. exigua* var. *heteromorpha* wheat kernel culture.

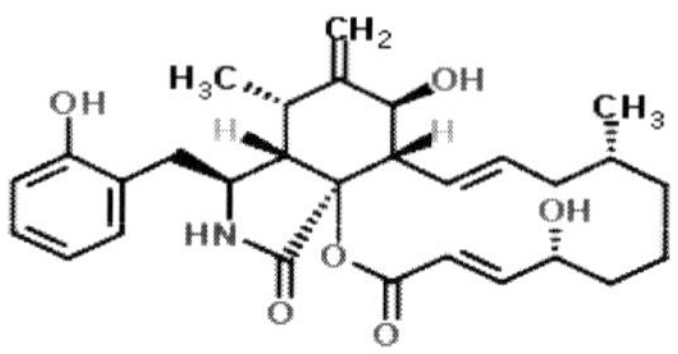

Fig. 2.29. Cytochalasin Z_4

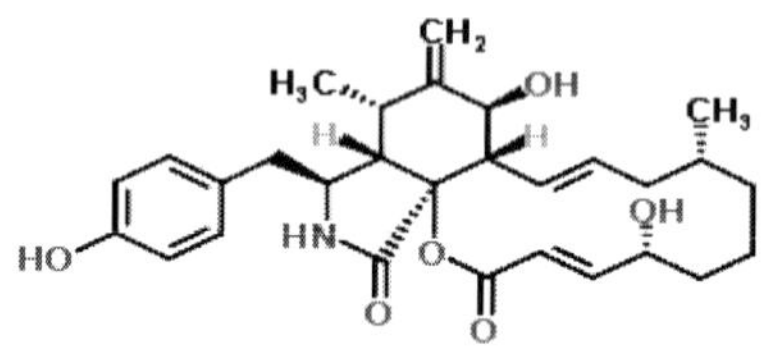

Fig. 2.30. Cytochalasin Z_5

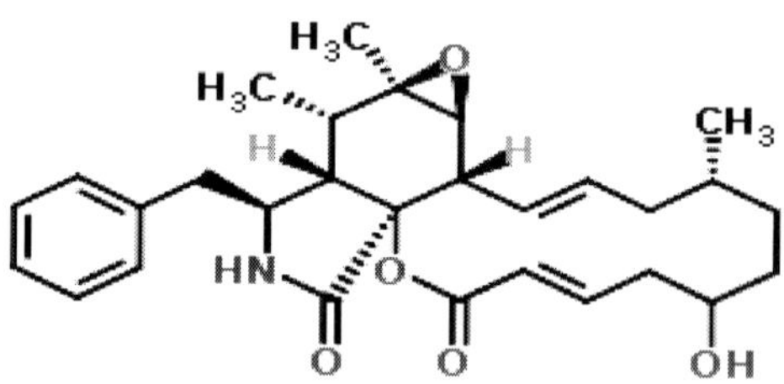

Fig: 2.31. Cytochalasin Z_6

Fig. 2.29 to 2.31. Structures of cytochalasins Z_4, Z_5 and Z_6

Cytochalasin Z_7 -Z_9 (Fig 2.32-34): 10-phenyl-[12]-cytochalasins Z_7, Z_8 and Z_9 were isolated from *Spicaria elegans* (Rui *et al.,* 2006). These compounds contain a 12-membered macrocyclic ring. All three cytochalasins were evaluated for their cytotoxic effects on P388 and A-549 cell lines by the MTT method.

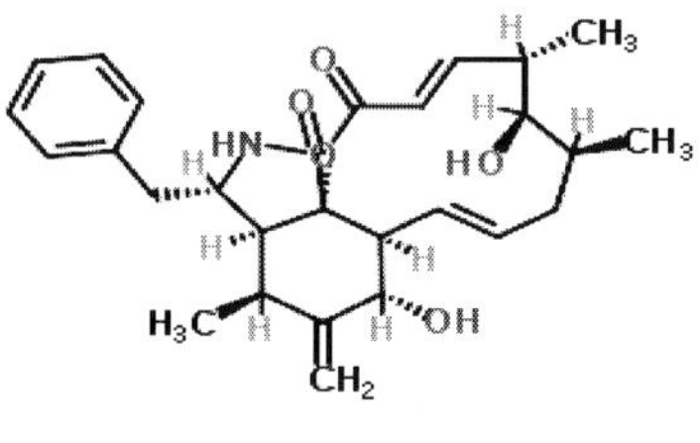

Fig. 2.32. Cytochalasin Z_7

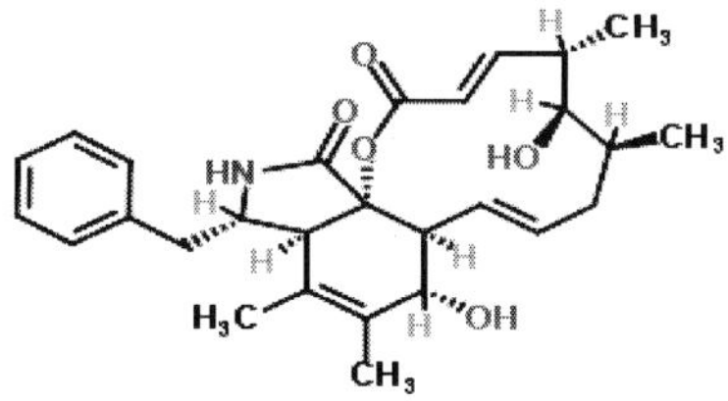

Fig. 2.33. Cytochalasin Z_8

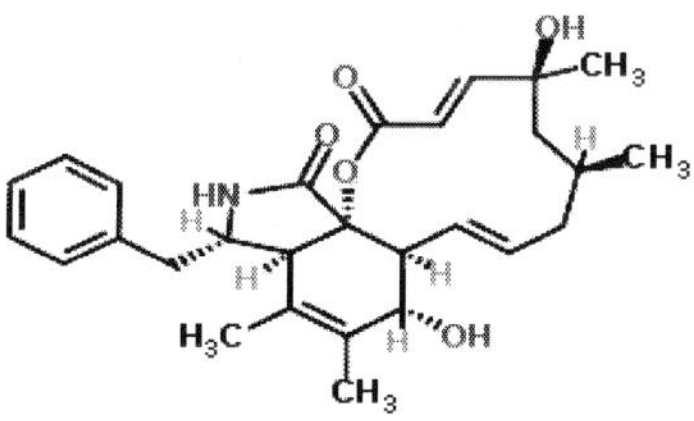

Fig: 2.34. Cytochalasin Z_9

Fig 2.32 to 2.34. Structures of cytochalasins Z_7, Z_8 and Z_9

Cytochalasin Z_{10} – Z_{15} (Fig. 2.35-40): These are produced by *Spicaria elegans* (Rui *et al.,* 2008). These possess a highly substituted perhydroisoindol- 1-one core usually fused with an 11, 13, 12 or 14 membered macrocyclic rings of six novel open-chain. These compounds display a wide range of biological activities. These compounds show cytotoxic effects on P388, A-549, HL-60, and BEL-7402 cell lines of all compounds were evaluated by the MTT method.

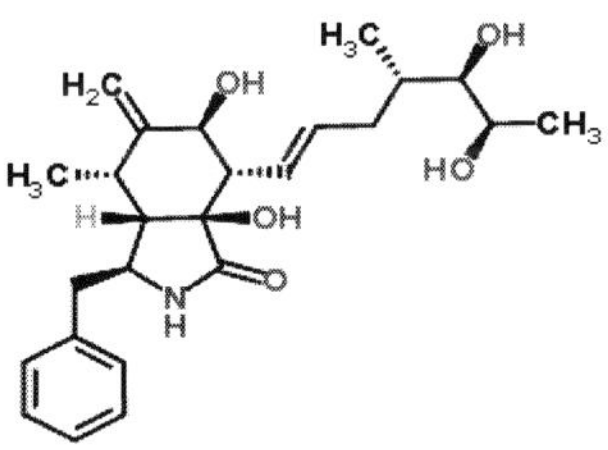

Fig. 2.35. Cytochalasin Z_{10}

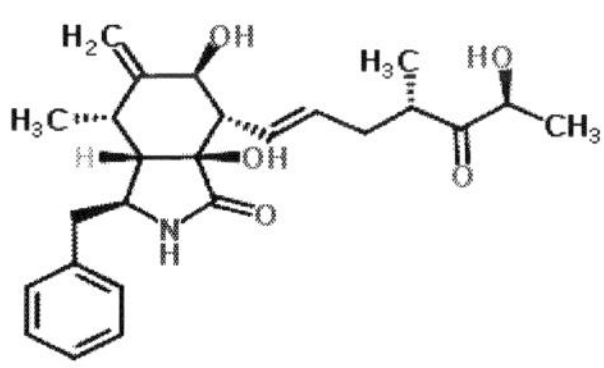

Fig. 2.36. Cytochalasin Z_{11}

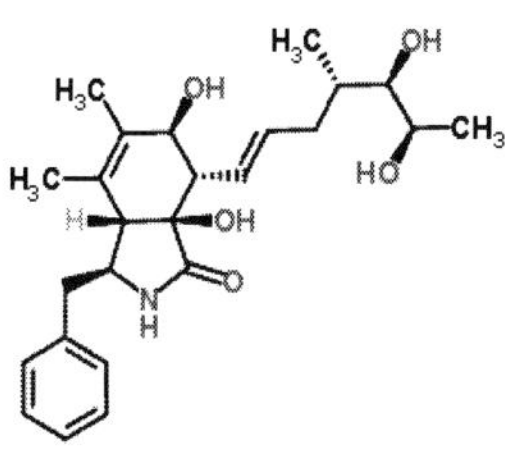

Fig. 2.37. Cytochalasin Z_{12}

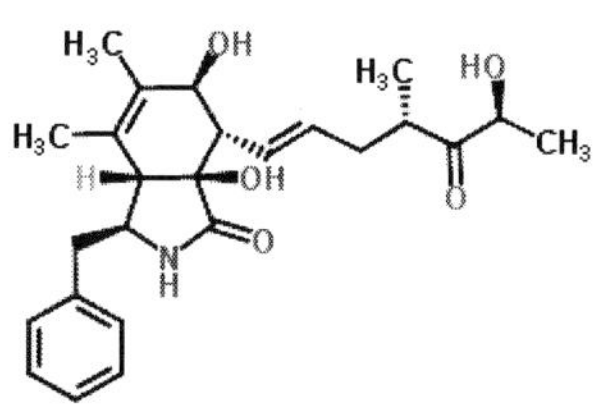

Fig. 2.38. Cytochalasin Z_{13}

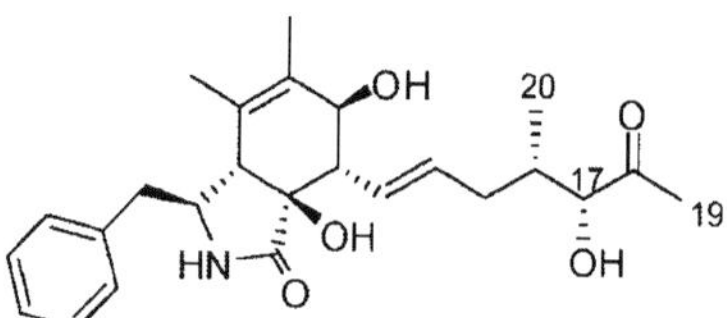

6

Fig. 2.39. Cytochalasin Z_{14}

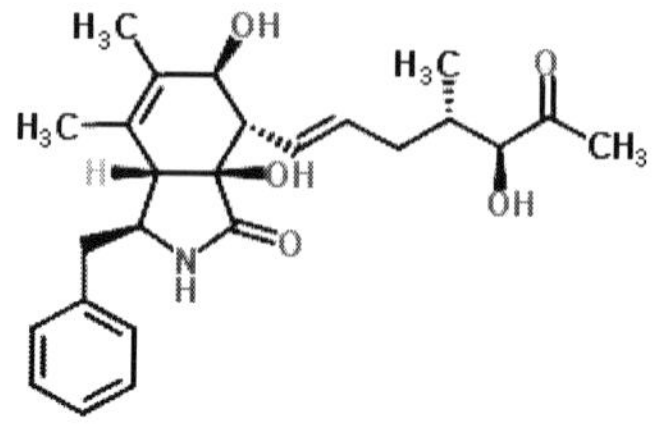

Fig. 2.40. Cytochalasin Z_{15}

Fig 2.35 to 2.40. Structures of cytochalasins Z_{10} to Z_{15}

Cytochalasin Z_{16} – Z_{17} (Fig 2.41 & 42): These are produced by *Aspergillus terreus* (Zhang *et al.,* 2010). These compounds are 15b-dehydro-5- N-acetylardeemin, 10-phenyl-[12]-cytochalasins were characterized from the liquid culture of *Aspergillus terreus.* Cytochalasin Z_{17} had moderate cytotoxicity (IC50: 26.2 μM) against human nasopharyngeal epidermoid tumor KB cell line.

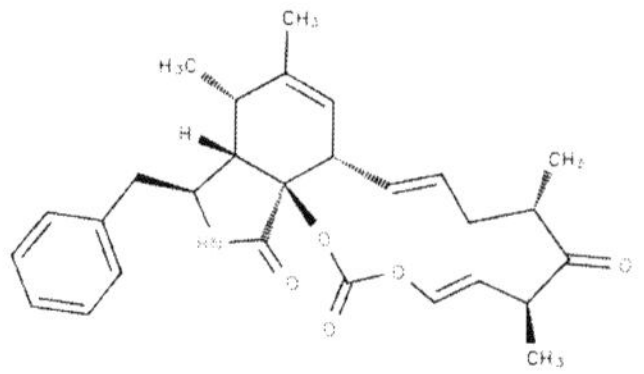

Fig. 2.41. Cytochalasin Z_{16}

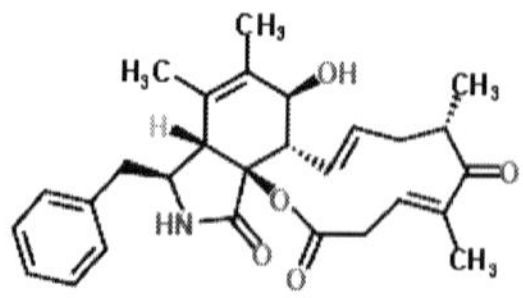

Fig. 2.42. Cytochalasin Z_{17}

Fig 2.41 & 2.42. Structures of cytochalasins Z_{16} to Z_{17}

Cytochalasin Z_{21} – Z_{23} (Fig. 2.43-45): These is compounds were isolated from *Spicaria elegans* (Wang *et al.,* 2011). These are open-chain 10-phenyl-cytochalasin with an 6,7-epoxy group and a C_8 side chain. This configuration of its core skeleton, 10-phenyl-substituted 6,7-epoxyperhydroisoindol-1-one, could be determined by biogenetic origin and comparison of the NMR data with those of cytochalansin E. Cytochalasins Z_{21} and Z_{22} showed cytotoxic activities against A-549 cell lines.

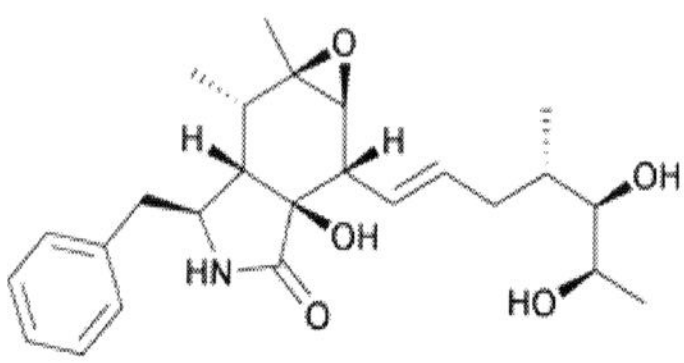

Fig. 2.43. Cytochalasin Z_{21}

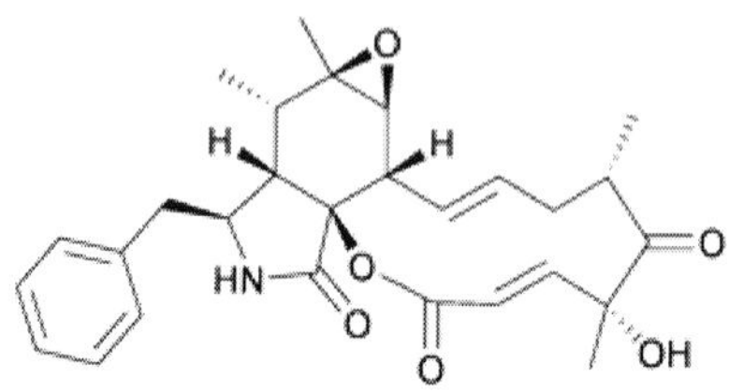

Fig. 2.44. Cytochalasin Z_{22}

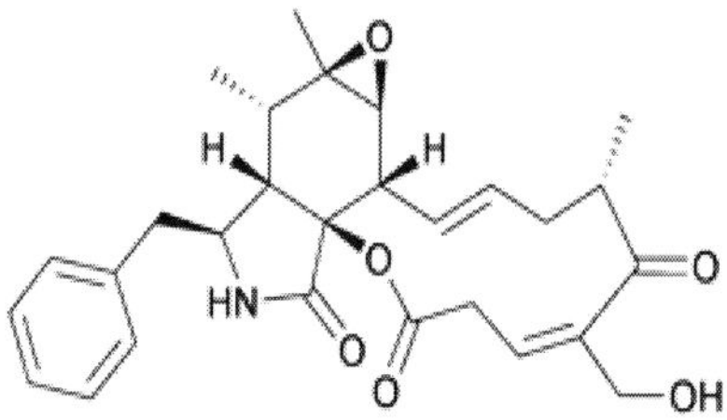

Fig. 2.45. Cytochalasin Z_{23}

Fig 2.43 & 2.45. Structures of cytochalasins Z_{21} to Z_{23}

Cytochalasin Z_{24} (Fig. 2.46): This compound is produced by *Aspergillus elegans* (Zheng *et al.,* 2013). It contains 10-phenyl-substituted 6,7-epoxyperhydroisoindol-1-one skeleton. This compound shows anti tumor activity.

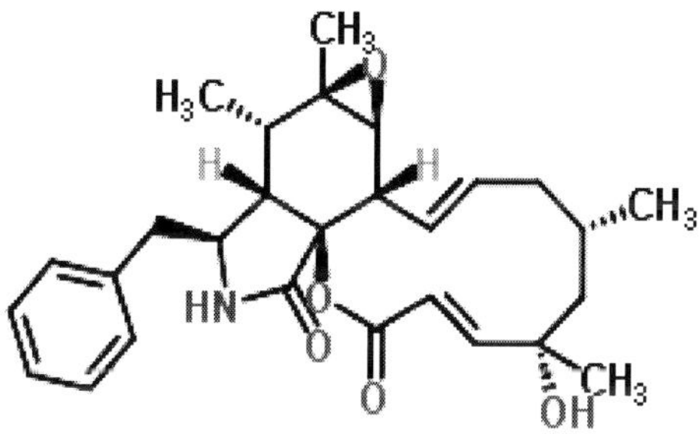

Fig. 2.46. Structure of cytochalasin Z_{24}

Cytochalasin Z_{27} –Z_{28} (Fig 2.47-48): These compounds were isolated from *Xylaria* sp. These constitute a methyl substituted six-membered ring as seen in cytochalasin Z_{16} (Zhang *et al.,* 2014).

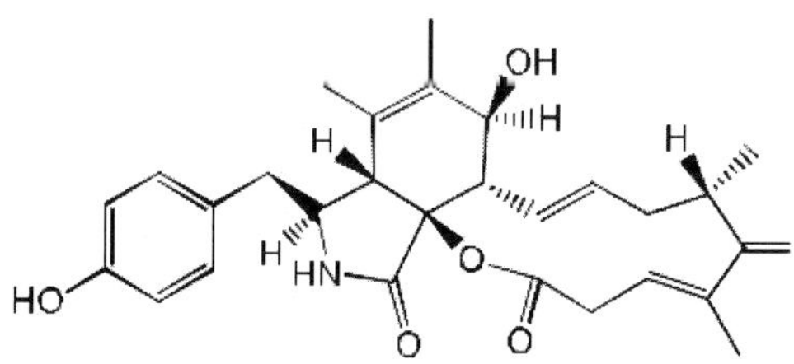

Fig. 2.47. Cytochalasin Z_{27}

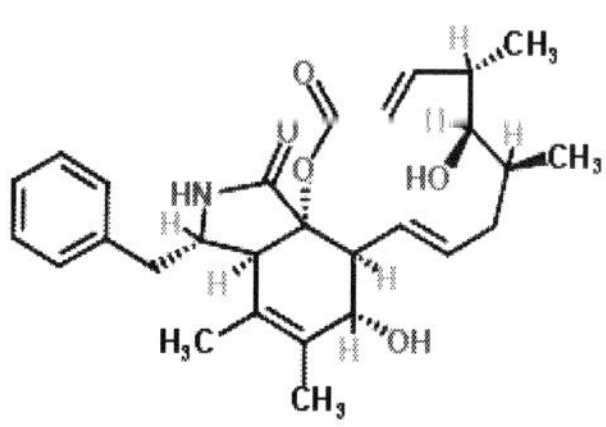

Fig. 2.48. Cytochalasin Z_{28}

Fig 2.47& 2.48. Structures of cytochalasins Z_{27} to Z_{28}

Physico-chemical characteristics of different cytochalasins are given in **Table 2.2**.

Table 2.2. Physico-chemical properties of different cytochalasins

Name of the Cytochalasin	Molecular formula	Molecular weight (Daltons)	Melting point (in °C)	Solubility
Cytochalasin A	$C_{29}H_{35}NO_5$	477.59	193-195	Acetone, DMSO, Ethanol
Cytochalasin B	$C_{29}H_{37}NO_5$	479.61	218-221	Ethanol, DMSO
Cytochalasin C	$C_{29}H_{37}NO_6$	507.62	260	Dichloro Methane
Cytochalasin D	$C_{30}H_{37}NO_6$	507.62	268-271	Chloroform, DMSO
Cytochalasin E	$C_{28}H_{33}NO_7$	495.56	206	Chloroform, Acetonitrile, DMSO, Ethyl acetate
Cytochalasin F	$C_{29}H_{37}O_5N$	479.2	177-178	Chloroform,
Cytochalasin G	$C_{29}H_{34}O_4N_2$	474.2	255-257	Methanol
Cytochalasin H	$C_{30}H_{39}NO_5$	493.63	268-271	-
Cytochalasin J	$C_{28}H_{37}NO_4$	451.61	137-139	-
Cytochalasin K	$C_{32}H_{37}NO_6$	531	Amorphous	Ethanol
Cytochalasin L	$C_{32}H_{37}NO_7$	547	Amorphous	Ethanol
Cytochalasin M	$C_{30}H_{37}NO_6$	507	161-162	Ethanol
Cytochalasin N	$C_{30}H_{37}NO_6$	523	253-254	Methanol
Cytochalasin O	$C_{28}H_{37}NO_4$	451.6	258-268	Methanol
Cytochalasin P	$C_{30}H_{41}NO_6$	511.6	169-173	Methanol
Cytochalasin Q	$C_{30}H_{37}NO_6$	507.6	145-147	Acetone, ethyl acetate
Cytochalasin R	$C_{28}H_{39}NO_5$	469.6	159-167	Acetone
Cytochalasin S	$C_{28}H_{39}NO_5$	469.6	149-151	Acetone
Cytochalasin W	$C_{30}H_{37}\ NO_7$	523.6	-	Methanol
Cytochalasin X	$C_{29}H_{34}\ N_2O_4$	474.6	-	Methanol
Cytochalasin Y	$C_{29}H_{37}N_2O_5$	493.2	-	Methanol
Cytochalasin Z	$C_{29}H_{33}\ N_2O_5$	488.5	-	Methanol
Cytochalasin Z_3	$C_{29}H_{37}NO_5$	479.6	-	Methanol
Cytochalasin Z_4	$C_{29}H_{37}NO_6$	495.6	-	Methanol
Cytochalasin Z_5	$C_{29}H_{37}NO_6$	495.6	-	Methanol
Cytochalasin Z_6	$C_{29}H_{37}NO_5$	479.6	-	-

Name of the Cytochalasin	Molecular formula	Molecular weight (Daltons)	Melting point (in °C)	Solubility
Cytochalasin Z_7	$C_{28}H_{33}NO_6Na+$	502.5	215-217	Methanol
Cytochalasin Z_8	$C_{28}H_{33}NO_6Na$	502.5	252-254	Methanol
Cytochalasin Z_9	$C_{28}H_{35}NO_5Na$	488.2	200-202	Methanol
Cytochalasin Z_{10}	$C_{28}H_{37}NO_5$	490.5	118-120	Methanol
Cytochalasin Z_{11}	$C_{30}H_{39}NO_6$	532.2	196-198	Methanol
Cytochalasin Z_{12}	$C_{25}H_{35}NO_5$	429.5	-	-
Cytochalasin Z_{13}, Z_{14}	$C_{25}H_{33}NO_5$	427.5	-	-
Cytochalasin Z_{16}	$C_{28}H_{33}NO_5Na$	486.2	142-146	Methanol
Cytochalasin Z_{17}	$C_{28}H_{33}NO_5Na$	486.2	181-183	Methanol
Cytochalasin Z_{18}	$C_{31}H_{43}NO_9$	596.2	-	Methanol
Cytochalasin Z_{19}	$C_{29}H_{37}NO_8$	528.2	-	Methanol
Cytochalasin Z_{20}	$C_{25}H_{33}NO_5$	428.2	-	Methanol
Cytochalasin Z_{21}	$C_{25}H_{35}NO_5$	430.2	-	Methanol
Cytochalasin Z_{22}	$C_{28}H_{34}NO6$	480.2	-	-
Cytochalasin Z_{23}	$C_{28}H_{33}NO_6$	480.2	-	-
Cytochalasin Z_{24}	$C_{28}H_{35}NO_5$	466.2	-	CHC13/Methanol
Cytochalasin Z_{27}	$C_{28}H_{33}NO_6Na$	502.2	-	Methanol
Cytochalasin Z_{28}	$C_{28}H_{33}NO_6Na$	502.2	-	Methanol
Dihydro - Cytochalsasin B	$C_{29}H_{39}NO_5$	481.62	198-203	Methanol

2.2. Aspochalasins

Aspochalasine are elaborated by species of *Aspergillus* (Zhou *et al.,* 2004 and Rochfort *et al.,* 2005) and *Spicaria elegans,* isolated from marine sediments (Lin *et al.,* 2009). These compounds are an isoindolone moiety bearing a 2-methylpropyl group at the C3 position and a macrocyclic ring connecting the C8 and C9 positions. Aspochalasins exhibit various bioactivities such as antibiotic (Kohno *et al.,* 1999), cytotoxicity (Tomikawa *et al.,* 2001 and Zhou *et al.,* 2004), anti-HIV (Rochfort *et al.,* 2005), TNF-alpha (Liu *et al.,* 2012) and melanogenesis inhibitors (Choo *et al.,* 2009). All aspochalasins reported are listed along with their producing organisms in **Table 2.3** and physic-chemical properties are presented in **Table 2.4.**

Table 2.3. Aspochalasins produced by different fungi

Toxin	Species	Reference
Aspochalasin A, C, D	*Aspergillus microcysticus*	Keller-Schierlein and Kupfer (1979)
Aspochalasin A1	*Aspergillus elegans*	Ying *et al.* (2014)
Aspochalasins A-D	*Aspergillus niveus*	Gebhardt *et al.* (2004)
Aspochalasin D, J, and I	*Trichoderma gamsii*	Ding *et al.* (2012)
Aspochalasin E	*Aspergillus microcysticus*	Nobuaki *et al.* (1993)
Aspochalasin F and G	*Aspergillus* spp	Fang *et al.* (1997)
Aspochalasin H	*Aspergillus* spp	Tomikawa *et al.* (2002)
Aspochalasins I,J,K	*Aspergillus flavipes*	Zhou *et al.* (2004)
Aspochalasin L	*Aspergillus flavipes* *Aspergillus microcysticus*	Rochfort *et al.* (2005) Gebhardt *et al.* (2004)
Aspochalasin M,N,O,P and Q	*Spicaria elegans*	Lin *et al.* (2009)
Aspochalasin U	*Aspergillus* spp	Junliang *et al.* (2012)
Aspochalasin ,Q,V & W	*Aspergillus* spp	Ying *et al.* (2014)
Aspochalasin Z	*Aspergillus niveus* *Aspergillus microcysticus*	Gebhardt *et al.* (2004)

Table 2.4. Physico-chemical properties of different Aspochalasins

Name of the Cytochalasin	Molecular formula	Molecular weight (Daltons)	Melting point (in °C)	Solubility
Aspochalasin A	$C_{24}H_{33}NO_4$	401.5	-	-
Aspochalasin B	$C_{26}H_{41}NO_5$	447.6	118	Chloroform
Aspochalasin C	$C_{24}H_{35}NO_4$	401.5	126-128	-
Aspochalasin D	$C_{24}H_{35}NO_4$	401.5	148	-
Aspochalasin F	$C_{24}H_{33}NO_4$	399.0	96-97	$CHCl_3$, EtOH, Methanol
Aspochalasin G	$C_{24}H_{33}NO_3$	383.0	95	CHCl3, EtOH, Methanol
Aspochalasin I	$C_{24}H_{35}NO_5$	417.2	136-138	Ethanol, Methanol, DMSO
Aspochalasin J	$C_{24}H_{36}NO_4$	402.2	136-138	Ethanol, Methanol, DMSO

Aspochalasin K	$C_{25}H_{40}NO_5$	434.2	136-138	Ethanol, Methanol, DMSO
Aspochalasin L	$C_{26}H_{41}NO_5$	447.6	-	Methanol
Aspochalasin M	$C_{24}H_{36}NO_4$	402.2	-	Methanol
Aspochalasin N	$C_{27}H_{40}NO_4$	458.2	-	Methanol
Aspochalasin O	$C_{27}H_{40}NO_5$	458.2	-	Methanol
Aspochalasin P	$C_{24}H_{36}NO_4$	402.2	-	Methanol
Aspochalasin R	$C_{24}H_{36}NO_3$	386.2	-	Methanol
Aspochalasin S	$C_{24}H_{35}NO_4$	401.5	-	-
Aspochalasin T	$C_{24}H_{35}NO_5$	417.5	-	-
Aspochalasin U	$C_{24}H_{37}NO_5$	442.5	-	-
Aspochalasin V	$C_{25}H_{37}NO_4S$	432.2	-	-
Aspochalasin W	$C_{25}H_{37}NO_4S$	448.2	-	-

(i) Aspochalasins A, B, C and D: *Aspergillus microcysticus* (Keller-Schiertein and Kupfer, 1979) produces these mycotoxins. Originally, one of these compounds (asposterol), was isolated as an antimicrobial and antibiotic compound. Five years later, new fermentation extracts were chromatographed. The antibiotic (now named as aspochalasin B) and its inactive co-metabolites, aspochalasin A, B, C and D were isolated.

(ii) Aspochalasins F and G: These compounds were isolated from *Aspergillus* spp (Fang *et al.,* 1997). These are with 17,18-epoxy moiety. Aspochalasin F shows methyl, three methylene, eleven methine and five quaternary carbons. Aspochalasin G is similar to Aspochalasin F except for those of C-9 and C-21, in which both are similar to aspochalasin C. So the structure of Aspochalasin G was suggested to be 17,18-anhydroaspochalasin C.

(iii) Aspochalasins I, J and K: *Aspergillus flavipes* occurring in the rhizosphere of *Ericameria laricifolia* elaborated three aspochalasins I, J and K (Zhou *et al.,* 2004). Aspochalasins I and J, having an α,β-unsaturated lactone moiety, while aspochalasins K and E, are with only one carbonyl group, suggesting the importance of an electrophilic α,β-unsaturated carbonyl moiety for the cytotoxic activity of these cytochalasins.

(iv) Aspochalasins Q, V and W: Two novel aspochalasins, 20-β-methylthio-aspochalsin Q, aspochalasin V and aspochalasin W were isolated from culture broth of *Aspergillus* spp. (Ying *et al.,* 2014). Cytotoxicity against the prostate cancer PC3 cell line and HCT116 cell line. Chemical structures of different aspochalasins are illustrated in **Fig 2.49 to 2.72.**

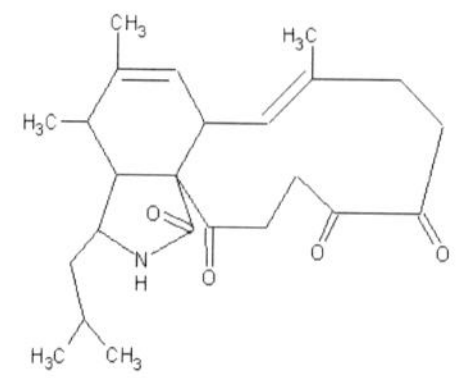

Fig. 2.49. Aspochalasin A

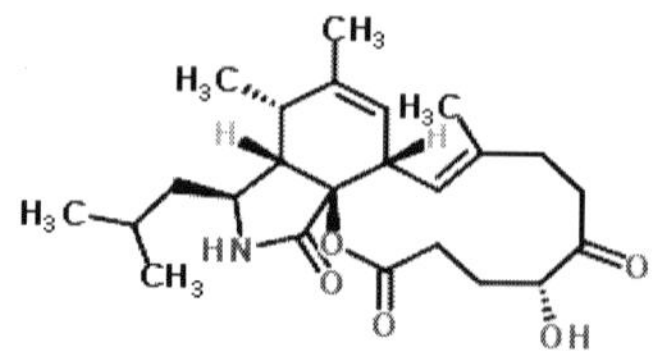

Fig. 2.50. Aspochalasin B

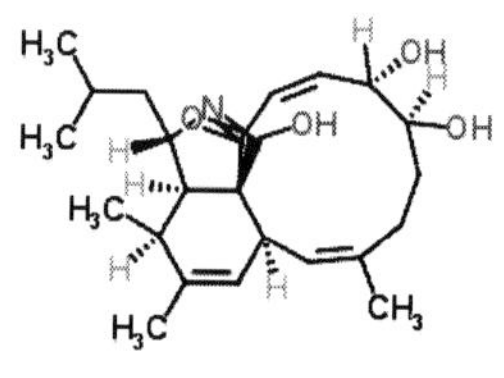

Fig. 2.51. Aspochalasin C

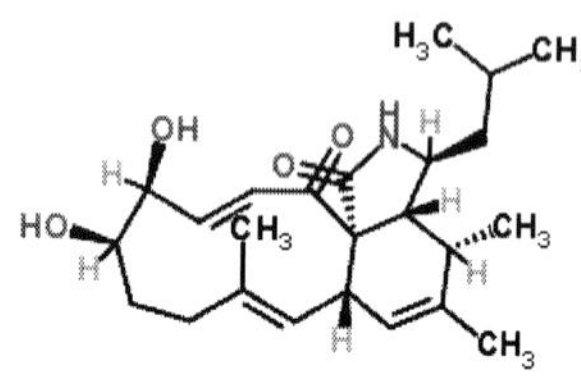

Fig. 2.52. Aspochalasin D

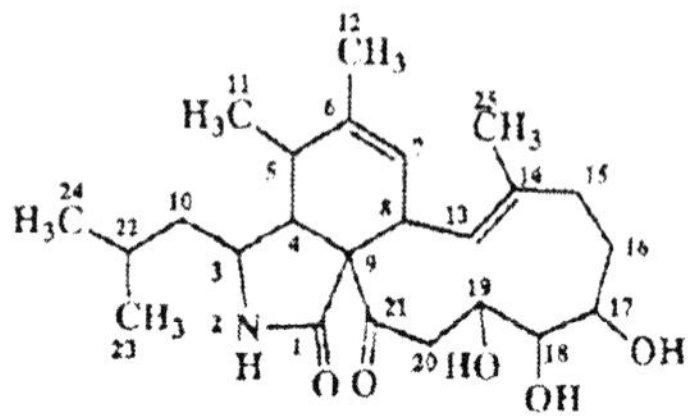

Fig. 2.53. Aspochalasin E

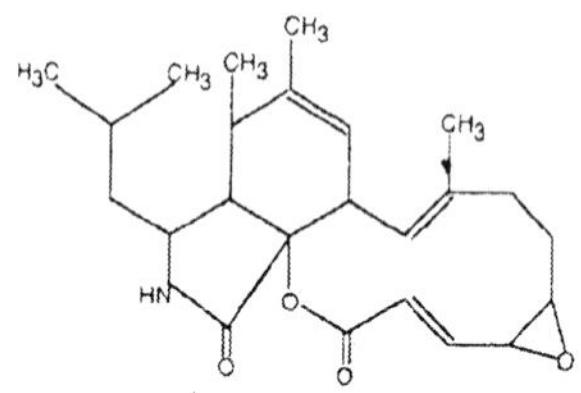

Fig. 2.54. Aspochalasin F

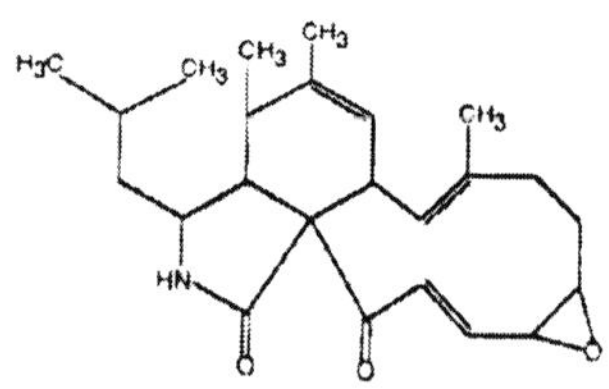

Fig. 2.55. Aspochalasin G

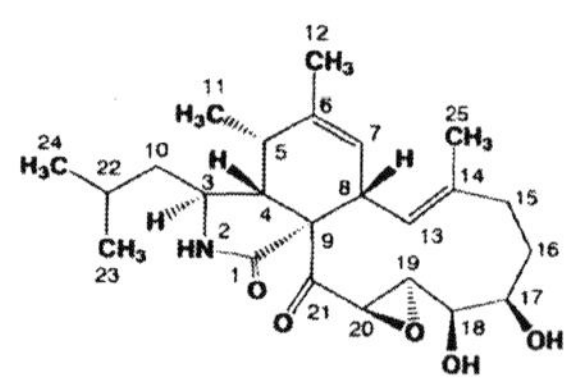

Fig. 2.56. Aspochalasin H

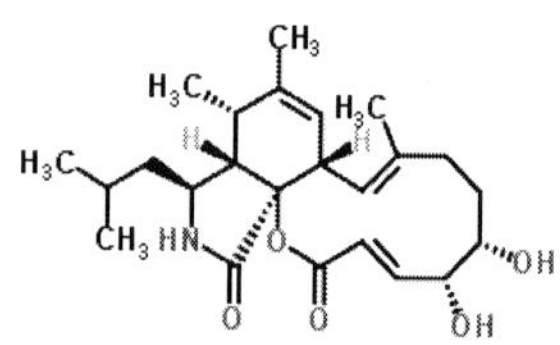

Fig. 2.57. Aspochalasin I

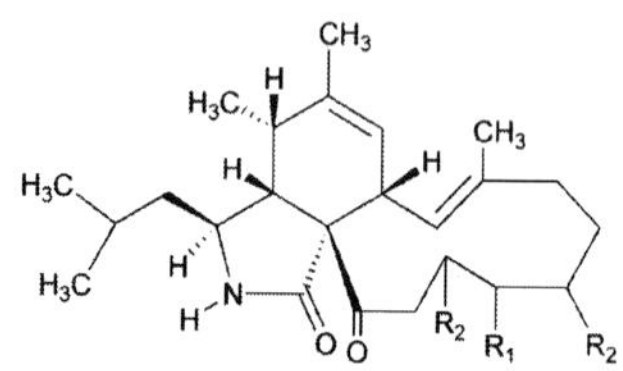

Fig. 2.58. Aspochalasin J

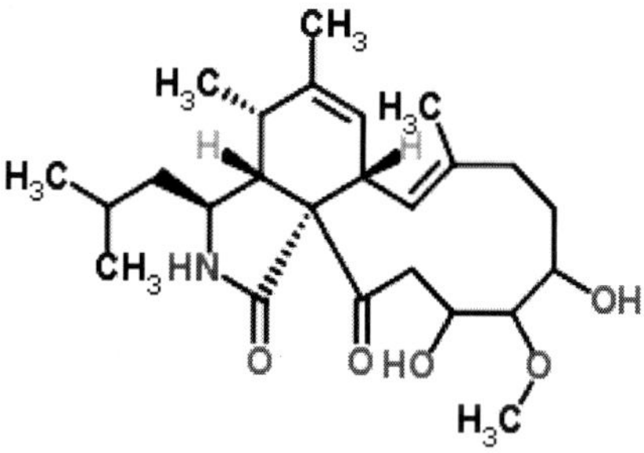

Fig. 2.59. Aspochalasin K

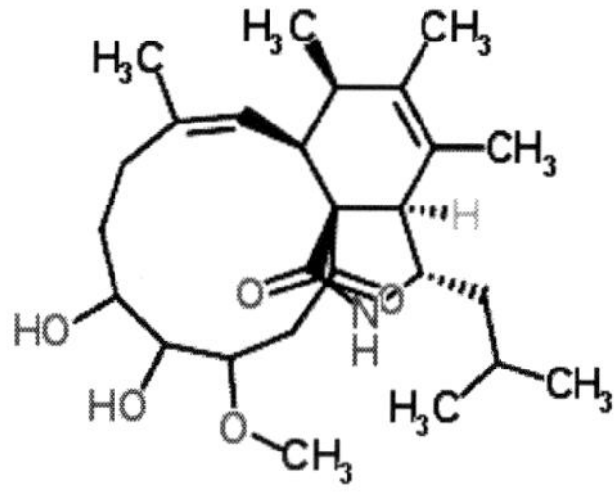

Fig. 2.60. Aspochalasin L

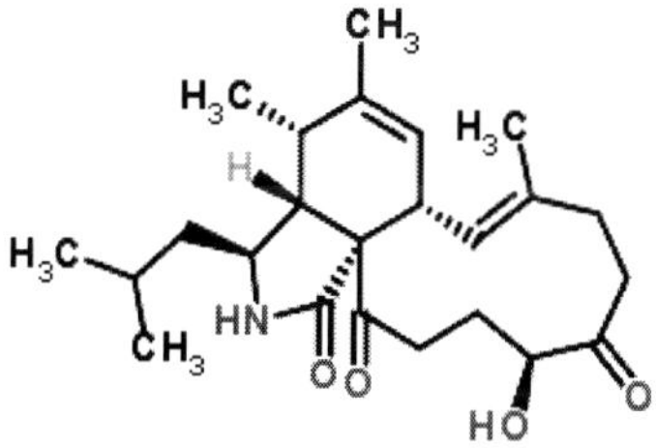

Fig. 2.61. Aspochalasin M

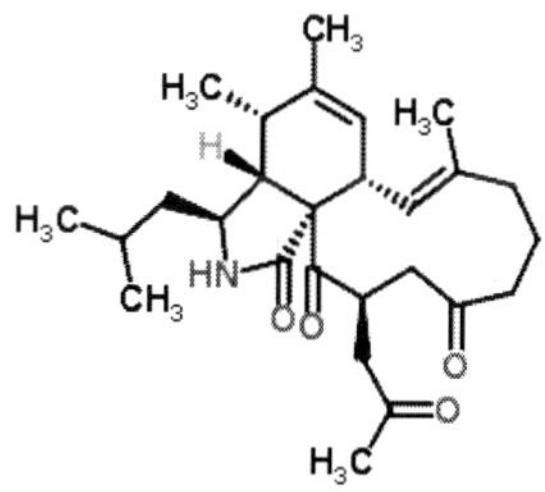

Fig. 2.62. Aspochalasin N

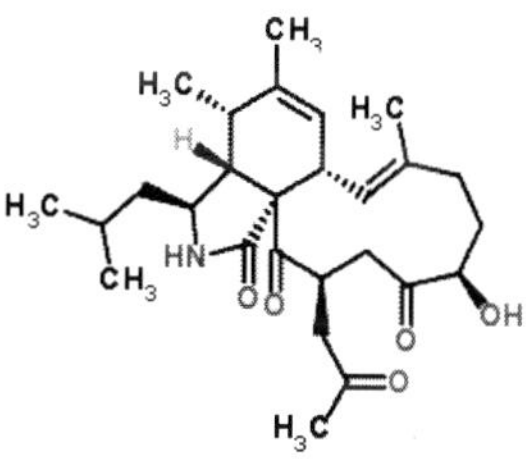

Fig. 2.63. Aspochalasin O

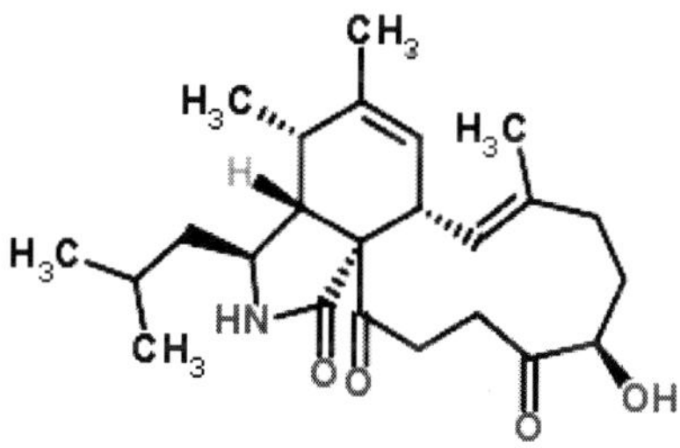

Fig. 2.64. Aspochalasin P

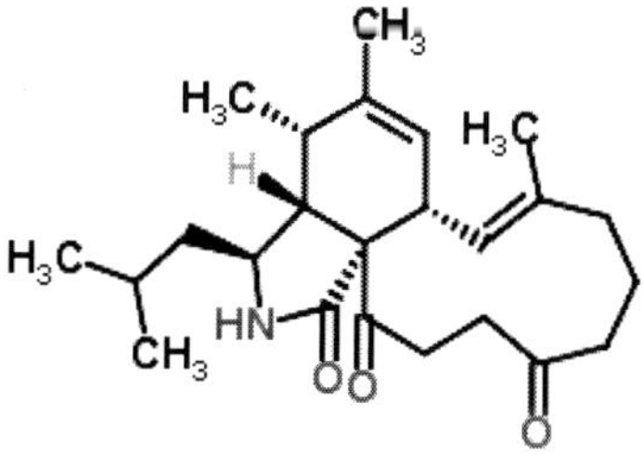

Fig. 2.65. Aspochalasin R

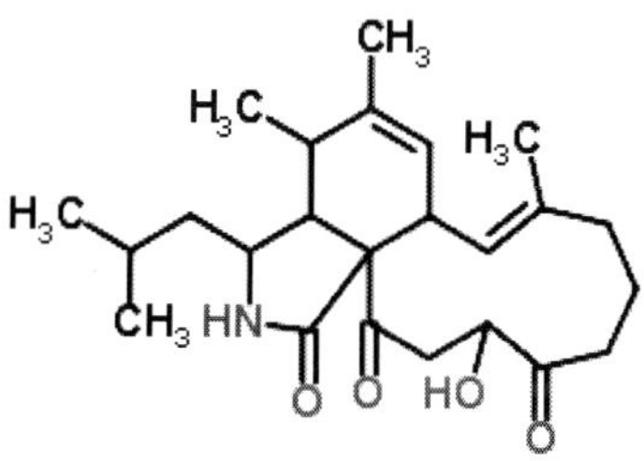

Fig. 2.66. Aspochalasin S

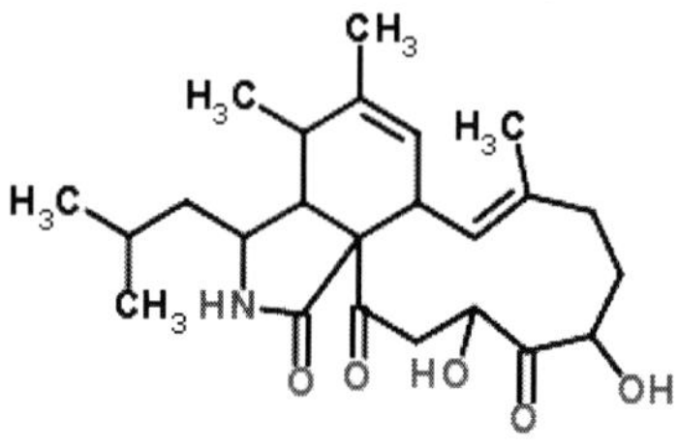

Fig. 2.67. Aspochalasin T

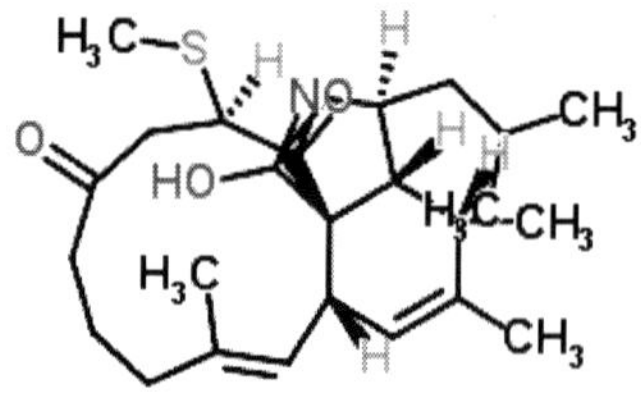

Fig. 2.68. Aspochalasin V

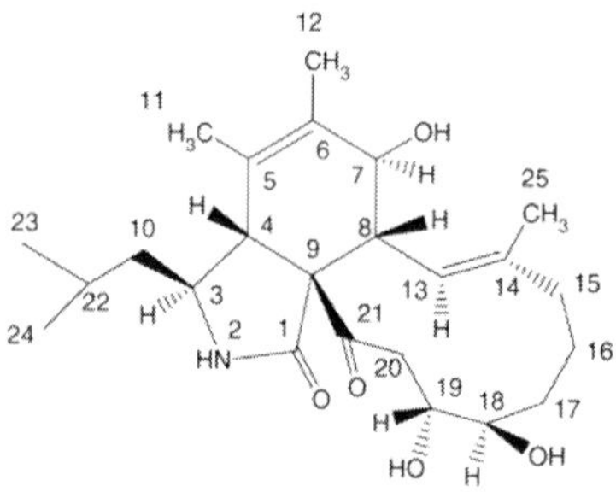

Fig. 2.69. Aspochalasin U

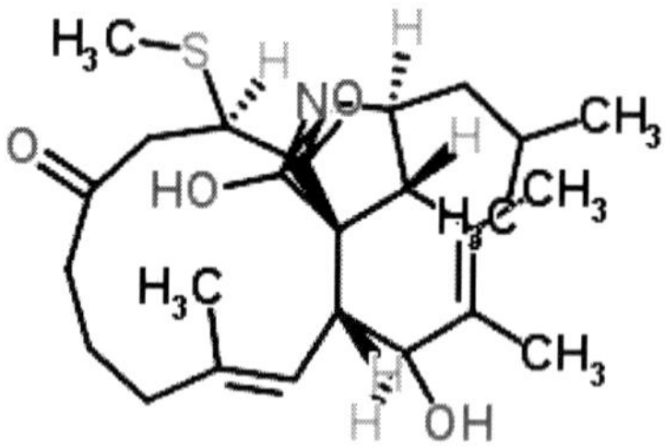

Fig. 2.70. Aspochalasin W

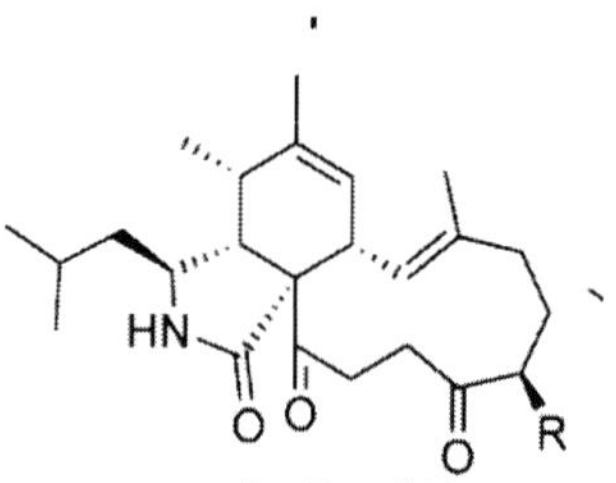

5 R = OH

6 R = H

Fig. 2.71. Aspochalasin Q

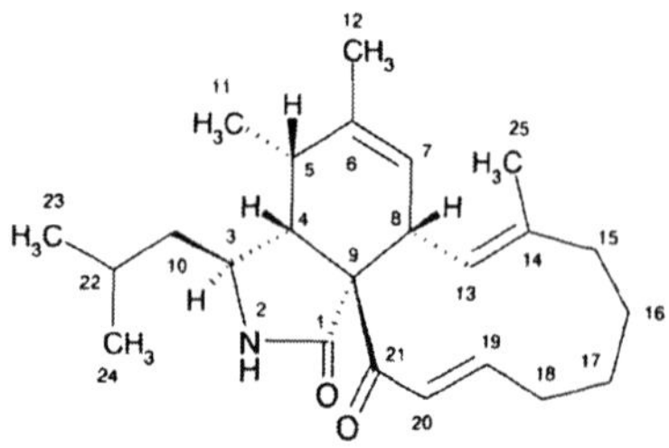

Fig. 2.72. Aspochalasin Z

Fig 2.49-2.72: Structure of different Aspochalasins

2.3. Alachalasins

Zhang *et al.* (2008) reported Alachalasins which are a group of cytochalasins elaborated by *Stachybotrys chartarum.* These are derived from alanine, acetate and three units of methionine. When cytochalasins incorporate phenylalanine, tyrosine, tryptophan, leucine and valine as structural subunits, alachalasins incorporate alanine. Possible biosynthetic pathway is that alanine is attached to an acetate-malonate derived C_{16} polyketide unit that contains two methyl substituents originating from the methyl of methionine (Tomoda *et al.,* 1999 and

Kakeya *et al.,* 1997). These also differ from cytochalasins by virtue of the presence of a C_{17}, C_{18} epoxide moiety rather than an exo-cyclic unit (Fujii *et al.,* 2000 and Namatame *et al.,* 2000). Alachalasin G and H contain adenine units and inhibit HIV replication and have mild antimicrobial activity. When alachalasin A displayed inhibitory effect on HIV-1LAI replication in C8166 cells with an EC50 of 8.01 lM, alachalasins D and G show modest antimicrobial activity (Zhang *et al.* 2008).

Alachalasins A–G are new members of the cytochalasins (Zhang *et al.,* 2005). In all probability, the biosynthesis of these toxins proceeds in a similar manner to that of other cytochalasins (Probst and Tamm, 1981). The molecules may be derived from alanine, acetate and three units of methionine. Most of the known cytochalasins incorporated Phe, Tyr, Try, Leu, and Val as structural subunits, but alachalasins A–G incorporate alanine instead of above-mentioned amino acids. Most noticeably, the presence of an adenine unit in alachalasins G and H is unprecedented in this class of compounds. Alachalasins A–G are the first secondary metabolites to be reported from *S. chartarum.* Physico-chemical properties and chemical structures of these compounds are presented in **Table 2.5** illustrated in **Fig. 2.73-79.**

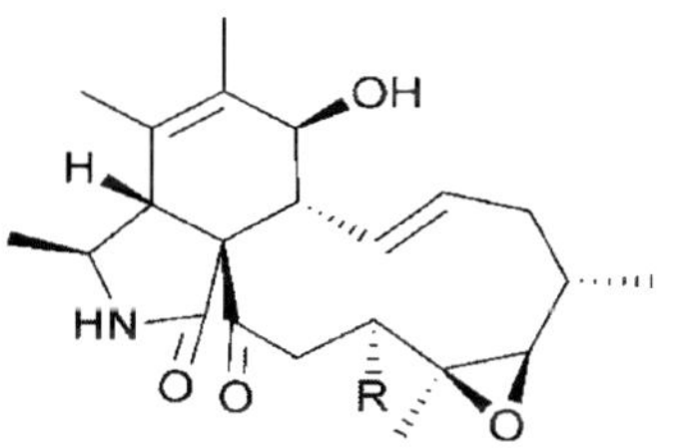

Fig. 2.73. Alachalasin A : R = OH
Fig. 2.74. Alachalasin B : R = OCH_3
Fig. 2.75. Alachalasin C : R = H

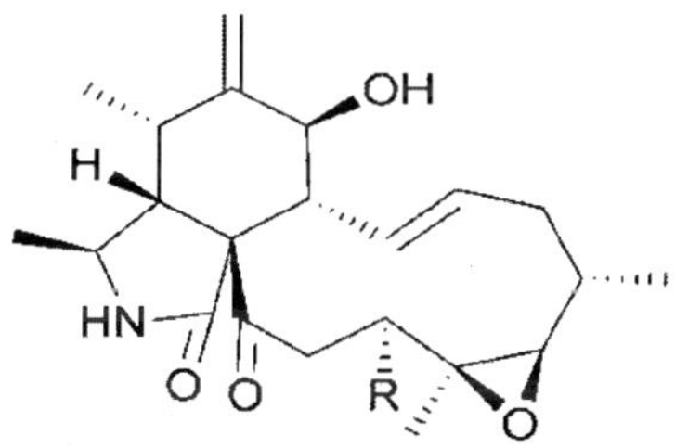

Fig. 2.76. Alachalasin D : R = OH
Fig. 2.77. Alachalasin E : R = OCH_3

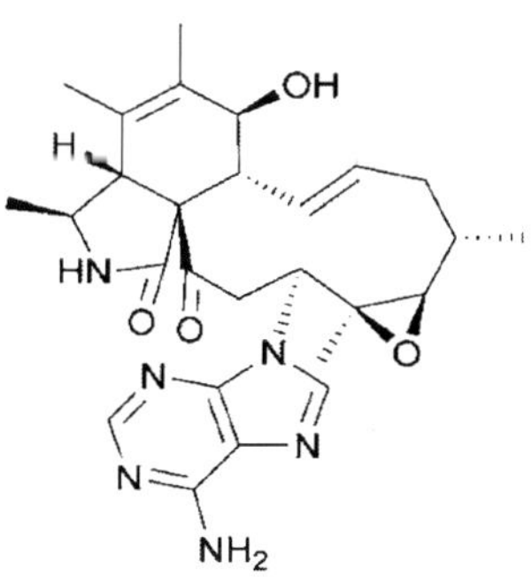

Fig. 2.78. Alachalasin F

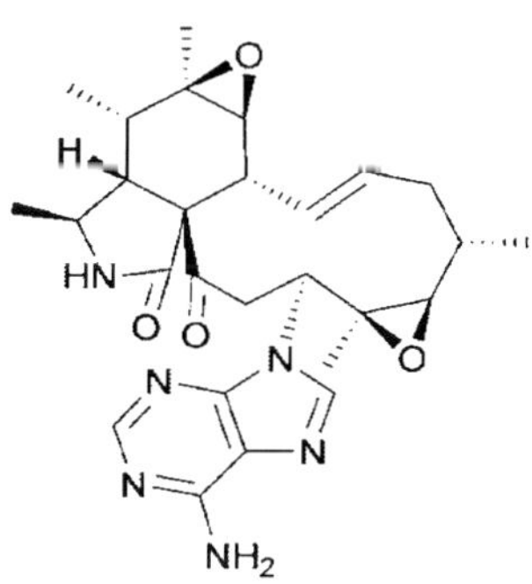

Fig. 2.79. Alachalasin G

Fig. 2.73-79: Structure of Alachalsins

Table 2.5: Physico-chemical properties of different alachalasins

Name of the Cytochalasin	Molecular formula	Molecular weight (Daltons)	Solubility
Alachalasin A	$C_{21}H_{31}NO_5$	412.2	Methanol
Alachalasin B	$C_{23}H_{33}NO_5$	426.2	Methanol
Alachalasin C	$C_{22}H_{31}NO_4$	396.2	Methanol
Alachalasin D	$C_{22}H_{31}NO_4$	412.2	Methanol
Alachalasin E	$C_{23}H_{33}NO_5$	426.2	Methanol
Alachalasin F	$C_{24}H_{35}N_6O_4$	529.2	Methanol
Alachalasin G	$C_{27}H_{34}N_6O_4$	529.2	Methanol

2.4. Chaetoglobosins and acetyl Chaetoglobosins

In this family of cytochalasins, twelve compounds have been identified and characterized. They are chaetoglobosins A-K and three 19-0-acetylchaeto-globosins. Except for chaetoglobosins K and L, the members of this family of mycotoxins have been isolated from strains of *Chaetomium globosum.* Chaetoglobosins A-E were isolated from strains of *C. mollicellum, C. rectum, C. subaffine* and *C. cochliodes*. Chaetoglobosin K and L are isolated from culture of *Diplodia macrospora* (Probit and Tamm, 1982). Chaetoglobosin C was isolated from *Penicillium aurantio-virens* (Springer *et al.*, 1976). Chaetoglobosin A, has a unique inhibitory activity against actin polymerization in mammalian cells (Scherlach *et al.,* 2010). Chaetoglobosins display cytotoxicity on a panel of human cancer cell lines (Zhang *et al.,* 2012). Chaetoglobosins A, C, E and G exhibited acute toxic effects against HeLa cells (IC50 = 3.2–20 μM) (Sekita *et al.,* 1982). Chaetoglobosin A, G, V, C, E and F exhibited moderate cytotoxic activities against four human cancer cell lines (KB, K562, MCF- 7, and HepG2) (Zhang *et al.,* 2010). Chaetoglobosin A and E showed significant growth inhibitory activity against the brine shrimp (Qin *et al.,* 2009) and also had toxic effects to mouse and rat cells and chick embryo (Vesely *et al.,* 1995). Chaeto-globosin A, exhibited significant cytotoxicity against P388 murine leukemia cell line (Jiao *et al.,* 2004). Chaetoglobosins C and G showed cytotoxicity against human breast cancer (HBC-1) cell lines (IC50 = 6.27–20.0 μM) (Thohinung *et al.,* 2010). *Chaetomium globosum* IFB-FO19 growing as endophyte of *Imperata cylindrica* elaborated Chaetoglobosin A along with four other analogue cytochalasins C, F and E and penochalasins A (Ding *et al.,* 2006) and exhibited good antitumor activity.

Chaetoglobosins produced by different organisms and chemical structure are précised in **Table 2.6** and **Fig 2.4.**

Table 2.6: Chaetoglobosins produced by different fungi

Toxin	Species	Reference
Chaetoglobosin A,B,C,D,F,E,G	*Calonectria morganii* *Chaetomium brasiliense* *C. cochlioides* *C. globosum* *C. mollicellum* *C.rectum* *C.subaffine* *Penicillium verrucosum* *P. discolor* *P. expansum*	Wieland and Low (1980) Yang Hu *et al.* (2013) Von Wallbrunn *et al.* (2001)
Chaetoglobosin B,C,D,F and G	*Chaetomium elatum*	Thohinung *et al.* (2010)
Chaetoglobosin C	*C. cochlioides* *C. mollipilium* *Penicillium aurantiovirens* *P. discolor*	Frisvad *et al.* (1997)
Chaetoglobosin J	*Cochliodes globosum* Sub affine	Oikawa *et al.* (1992)
Chaetoglobosin K,L,M	*Diplodia macrospora*	Christoph and Christoph (1988)
Chaetoglobosin M	*Phomopsis leptostromiformis*	Burlot *et al.* (2003)
Chaetoglobosin O	*Cylindrocladium floridanum* *Pencillium expansum* *Rosellinia* sp *Xylaria* sp	Ichihara *et al.* (1996)
Chaetoglobosin P	*Discosia* sp	Donoso *et al.* (1997)
Chaetoglobosin Q, R and T	*Chaetomium globosum*	Jiao *et al.* (2004)
Chaetoglobosin U	*Chaetomium globosum*	Ding *et al.* (2006)
Chaetoglobosin V	*Chaetomium globosum* *C. elatum*	Xue *et al.* (2012) Thohinung *et al.* (2010)
Chaetoglobosin W	*Chaetomium globosum*	Zhang *et al.* (2010)
Chaetoglobosin Y	*Chaetomium globosum*	Zheng *et al.* (2014)
Chaetoglobosin Y	*Chaetomium globosum*	Tao *et al.* (2016)

Cytochalasins that contain tryptophan moiety at position 10 have been referred as chaetoglobosins. The hydroxyl group at C7 and the phenyl group at C10 in Chaetoglobosins are essential for cytotoxicity. The UV and H-NMR spectra of active principle of chromatographic extracts of *C. globosum* were found to be indole derivatives with molecular formula $C_{32}H_{36}O_5N_2$. In H-NMR spectra each proton in the molecules fortunately appears well separated as indole derivative and acetate and the part structures of chaetoglobin A is depicted in **Fig. 2.80** and they are similar as depicted in **Fig. 2.81**. Further, more detailed studies resulted in the isolation of D, E and F. Chaetoglobosin D is an isomer of A-C, while E and F contain two more hydrogen atoms. X-ray analysis of chaetoglobosin A confirmed its structure and established its stereochemistry. The physico-chemical properties of these compounds are presented in **Table 2.7.**

With the help of chemical reactions and spectral data the structures of chaetoglobosins were firmly established (**Fig. 2.80-96**). These compounds bear indol-3-yl groups at the 10-positions instead of the pentyl group that are in other cytochalasins *Penicillium aurautiovireus* produced a compound which is similar to chaetoglobosin C when subjected to x-ray analysis. The molecular frame work of chaetoglobosins appear to be formed from one tryptophane, nine acetate-malonate and three C1 units. *Nigrosabulum* sp is also reported to produce an indole-3-yl cytochalasins, Cytochalasin G.

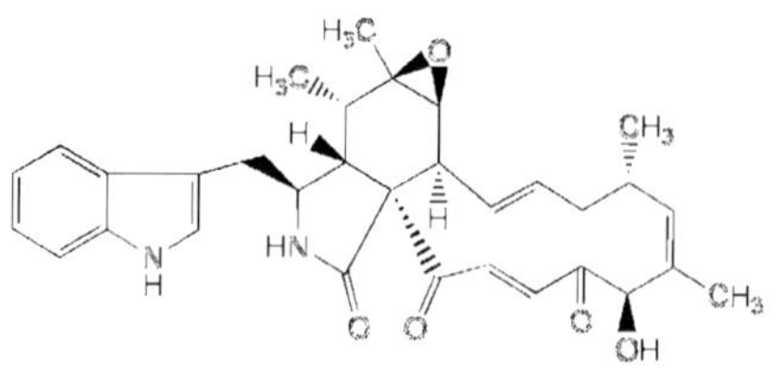

Fig.2.80. Chaetoglobosin A

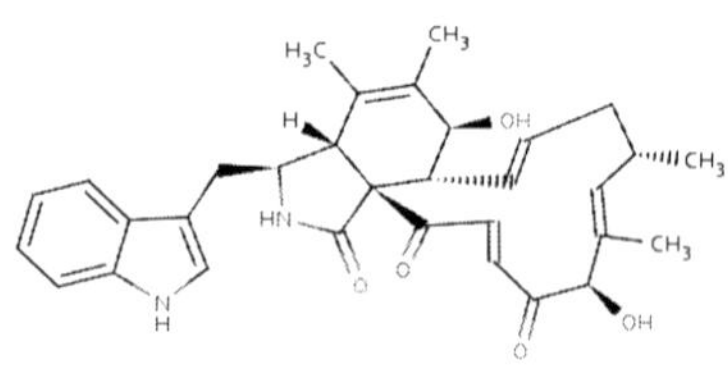

Fig. 2.81. Chaetoglobosin B

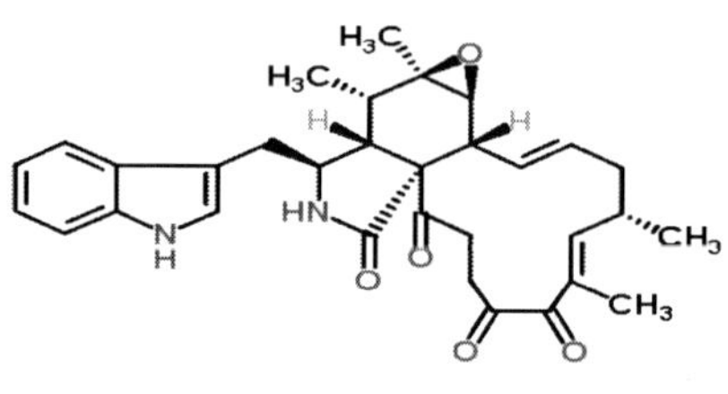

Fig. 2.82. Chaetoglobosin C

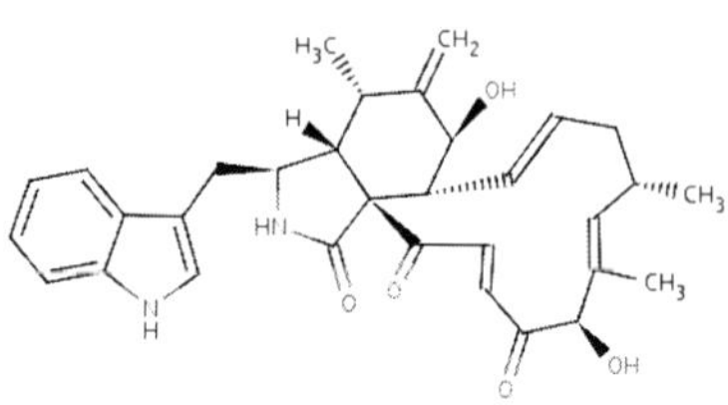

Fig. 2.83. Chaetoglobosin D

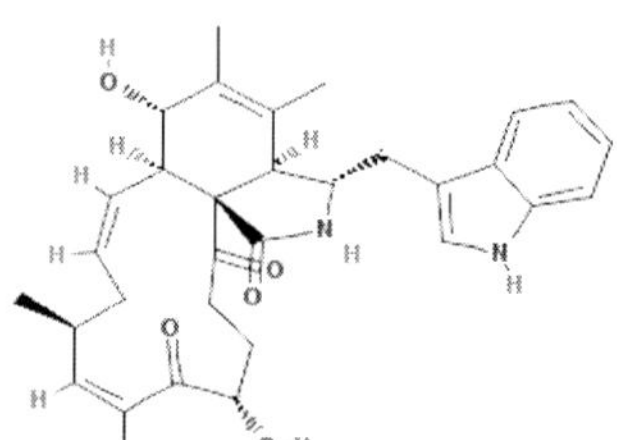

Fig. 2.84. Chaetoglobosin E

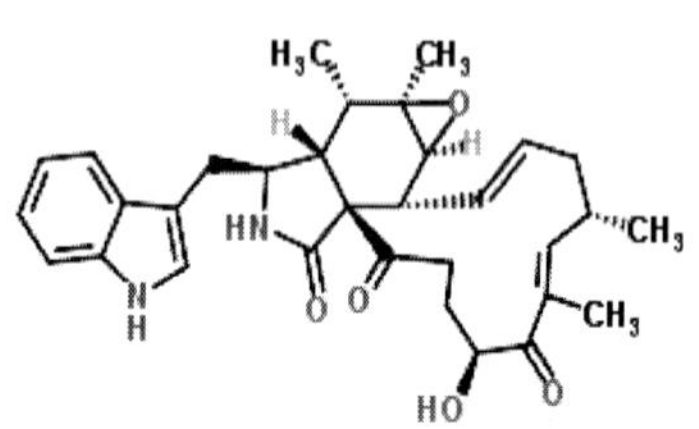

Fig. 2.85. Chaetoglobosin F

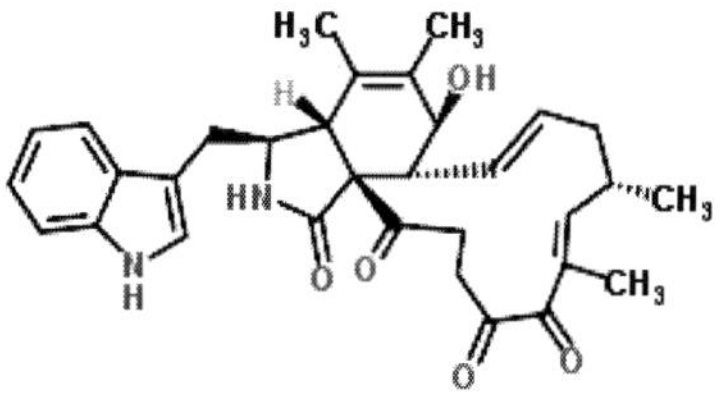

Fig. 2.86. Chaetoglobosin G

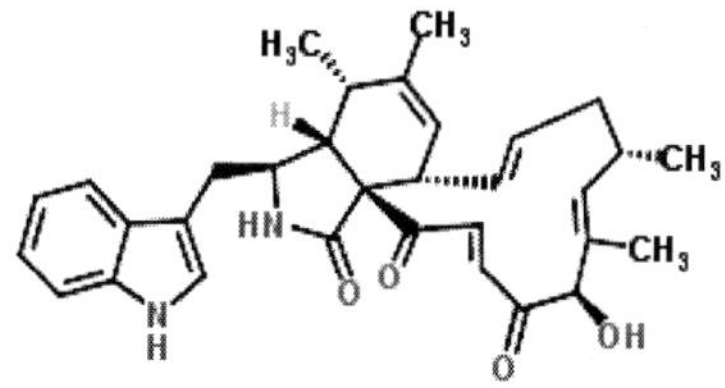

Fig. 2.87. Chaetoglobosin J

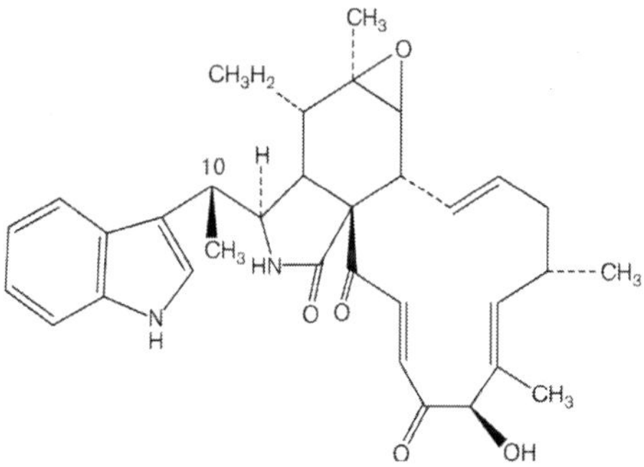

Fig. 2.88. Chaetoglobosin K

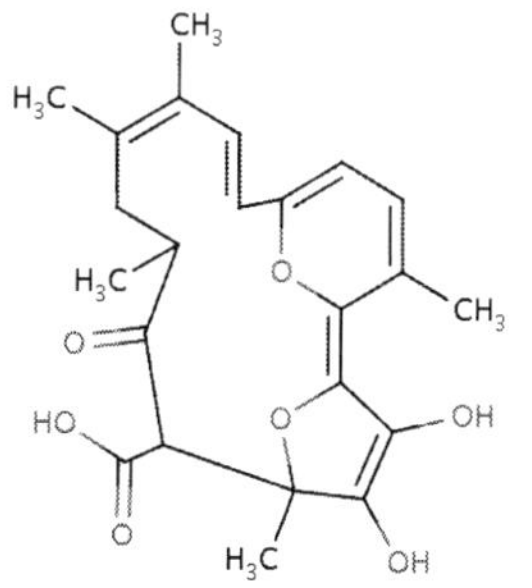

Fig. 2.89. Chaetoglobosin L

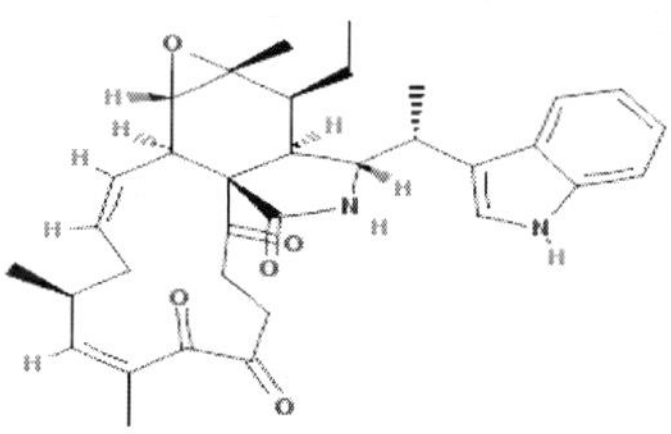

Fig. 2.90. Chaetoglobosin M

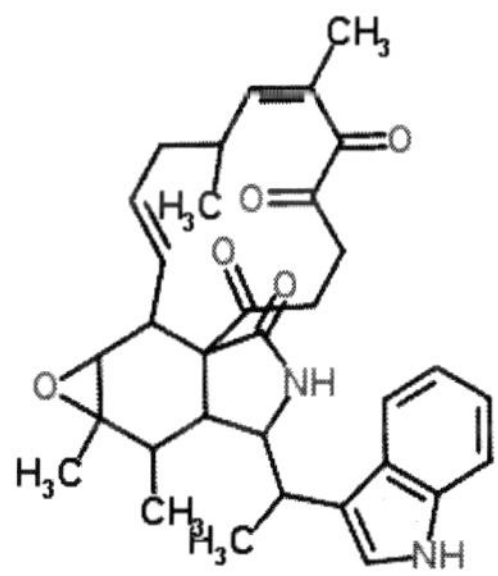

Fig. 2.91. Chaetoglobosin N

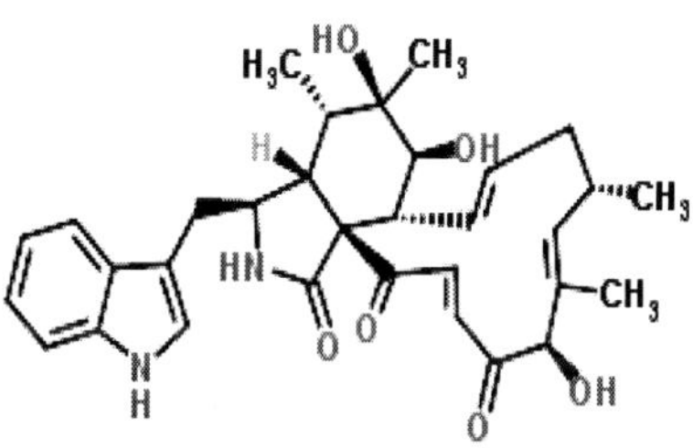

Fig. 2.92. Chaetoglobosin Q

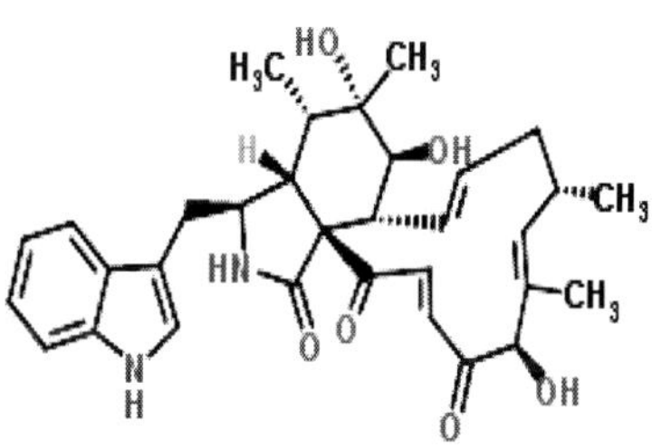

Fig. 2.93. Chaetoglobosin R

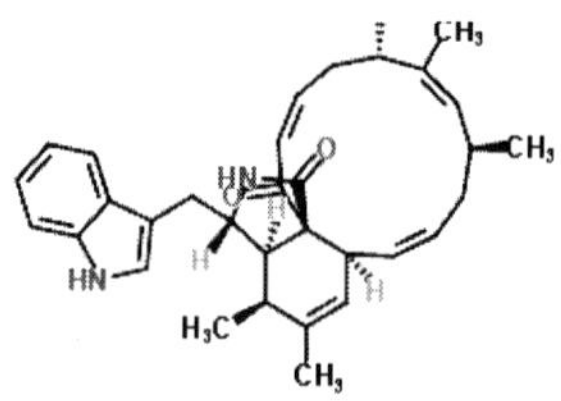

Fig. 2.94. Chaetoglobosin T

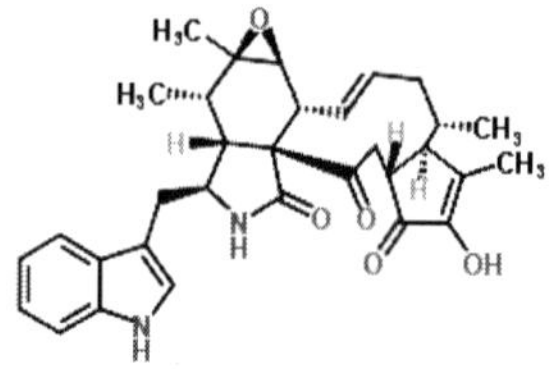

Fig. 2.95. Chaetoglobosin U

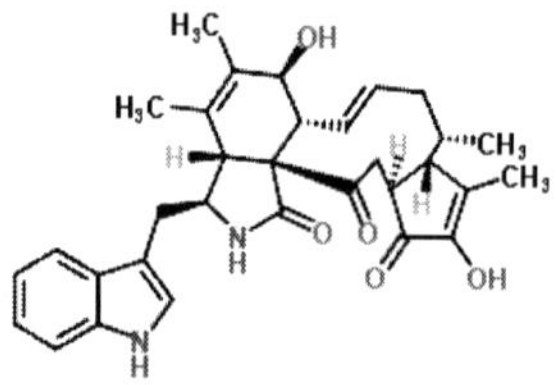

Fig. 2.96. Chaetoglobosin V

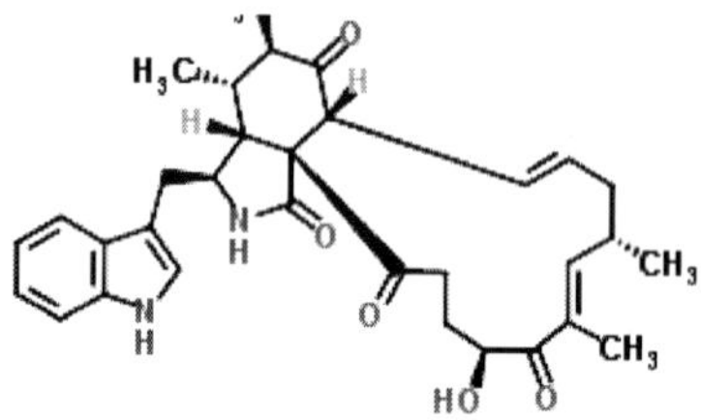

Fig. 2.97. Chaetoglobosin Y

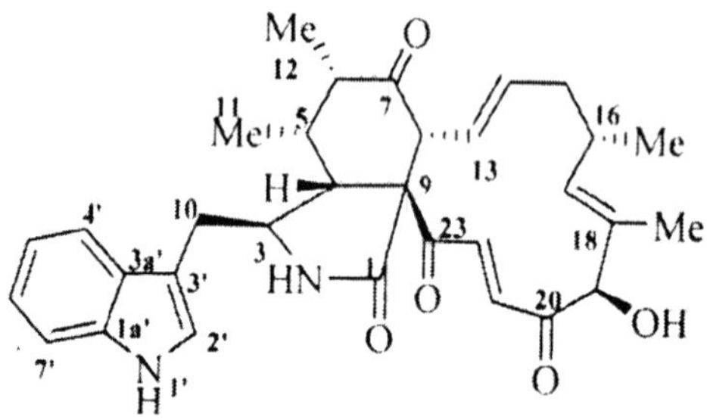

Fig. 2.98. Chaetoglobosin Z

Fig. 2.80-2.98: Structure of different Chaetoglobosins

Table 2.7: Physico-chemical properties of different Chaetoglobosins

Name of the Cytochalasin	Molecular formula	Molecular weight (Daltons)	Melting point (in °C)	Solubility
Chaetoglobosin A	$C_{32}H_{36}NO_5$	528.6	168-170	Methanol, DMSO
Chaetoglobosin B	$C_{32}H_{36}O_5N_2$	528.2	186-187	-
Chaetoglobosin C	$C_{32}H_{36}N_2O_5$	528.6	260-263	-
Chaetoglobosin D	$C_{32}H_{36}\ O_5N_2$	528.2	216	-
Chaetoglobosin E	$C_{32}H_{38}O_5N_2$	530.2	279-280	-
Chaetoglobosin F	$C_{32}H_{38}O_5N_2$	530.6	177-178	-

Name of the Cytochalasin	Molecular formula	Molecular weight (Daltons)	Melting point (in °C)	Solubility
Chaetoglobosin G	$C_{32}H_{36}O_5N_2$	528.2	251-253	Acetone
Chaetoglobosin J	$C_{32}H_{36}O_4N_2$	512.2	149-151	Acetone
Chaetoglobosin K	$C_{34}H_{40}O_5N_2$	556.2	264-266	Chloroform
Chaetoglobosin L	$C_{34}H_{40}\ O_5N_2$	556.7	-	-
Chaetoglobosin M	$C_{34}H_{40}O_5N_2$	556.7	-	MeOH
Chaetoglobosin N	$C_{33}H_{38}N_2O_5$	542.6	205	MeOH
Chaetoglobosin O	$C_{32}H_{38}N_2O_5$	530.2	-	MeOH
Chaetoglobosin P	$C_{35}H_{38}N_2O_5$	543.0	205	MeOH
Chaetoglobosin Q	$C_{32}H_{38}O_6N_2$	546.6	-	MeOH
Chaetoglobosin R	$C_{32}H_{38}N_2\ O_6$	546.6	-	MeOH
Chaetoglobosin T	$C_{32}H_{38}N_2\ O_3$	498.2	-	-
Chaetoglobosin U	$C_{32}H_{36}N_2\ O_5$	528.6	-	-
Chaetoglobosin V	$C_{32}H_{38}N_2\ O_3$	521.6	240-241	Methanol
Chaetoglobosin Y	$C_{32}H_{38}N_2\ O_5$	530.6	-	-
Chaetoglobosin Z	$C_{32}H_6O_5N_2$	529.2	-	Methanol

2.5. Zygosporins

Zygosporins are a group of cytochalasins elaborated by species of *Zygosporium* characterized by cytotoxicity *in vitro* producing a blocked cytoplasmic cleavage, the nuclear division continuing with the production of multinucleated cells (Vedejs and Wittenberger, 1990). Some of the physic-chemical characteristics are given in **Table 2.8**. Chemical structures are précised in **Fig 2.99-102.**

Table 2.8: Physico-chemical properties of different Zygosporins

Name of the Cytochalasin	Molecular formula	Molecular weight (Daltons)	Melting point (in °C)	Solubility
Zygosporin A	$C_{30}H_{40}O_7NBr$	751	242-245	Methanol
Zygosporin D	$C_{28}H_{37}O_6N$	465.2	180-190	Methanol
Zygosporin E	$C_{30}H_{37}O_5N$	491.2	218-223.5	Benzene
Zygosporin F	$C_{32}H_{39}O_7N$	549.2	126-129	Acetone
Zygosporin G	$C_{30}H_{37}O_5N$	491.2	115-125	-
Zygosporin J	$C_{32}H_{36}N_2O_4$	512.6	149-151	MeOH

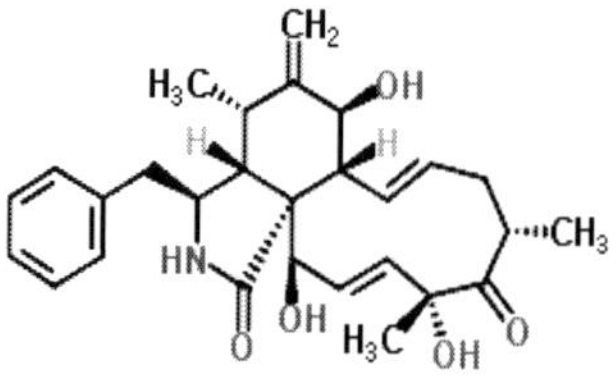

Fig. 2.99. Zygosporin D

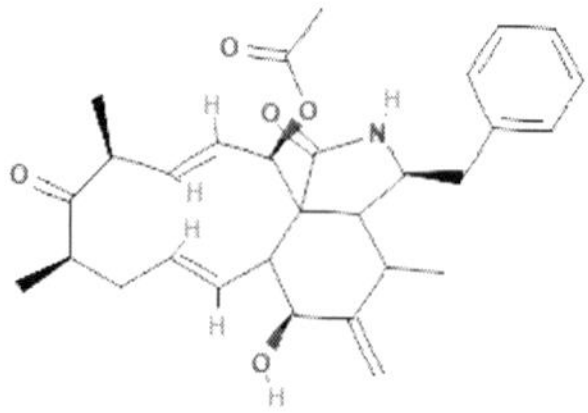

Fig. 2.100. Zygosporin E

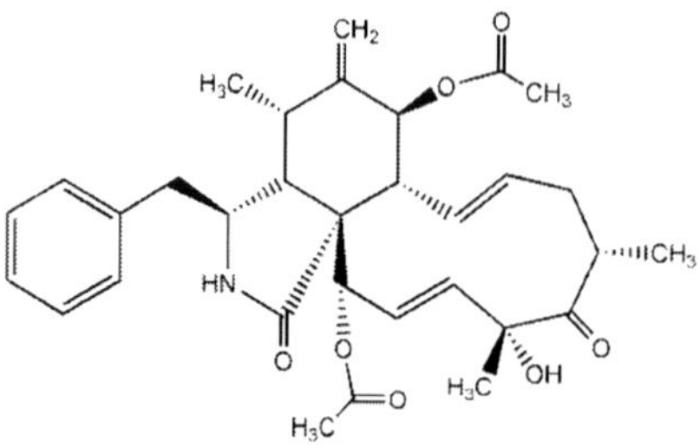

Fig. 2.101. Zygosporin F

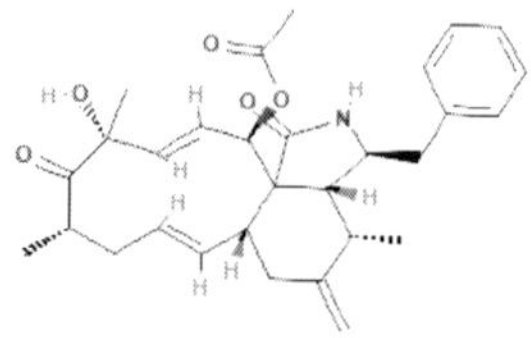

Fig. 2.102. Zygosporin G

Fig. 2.99-102: Structure of zygosporins

2.6. Deoxaphomin, proxiphomin and protophomin

These toxins are produced by *Phoma* sp. (strain S 298) and *Phoma exigua.* These compounds are considered to be a carbocyclic (13) cytochalasin that generate the macrolide system of cytochalasin B by an enzymatic Baeyer-villiger-type oxygen between C9 and C23 (Binder and Tamm, 1973). Deoxaphomin, proxiphomin and protophomin produced by different organisms and Physico-chemical properties of these toxins arc précised in **Table 2.9 and 2.10**. The structures of different deoxaphomin, proxiphomin and protophomin toxins are illustrated in **Fig 2.103-107**.

Table 2.9: Deoxaphomin, proxiphomin and protophomin produced by different fungi

Toxin	**Species**	**Reference**
Deoxaphomin	*Phoma* sp.	Binder and Tamm (1973)
	Pyrenophora semeniperda	Evidente *et al.* (2002)
Proxiphomin	*Phoma* sp.	Binder and Tamm (1973)
Protophomin	*Phoma* sp.	Binder and Tamm (1973)

Table 2.10: Physico-chemical properties of different deoxaphomins, Protophomin and Proxiphomin

Name of the Cytochalasin	Molecular formula	Molecular weight (Daltons)	Melting point (in °C)	Solubility
Deoxaphomin A	$C_{29}H_{37}NO_4$	463.2	-	Chloroforme
Deoxaphomin B	$C_{29}H_{37}NO_4$	464.2	-	
Deoxaphomin C	$C_{29}H_{37}NO_3$	448.2	-	Chloroforme
Protophomin	$C_{29}H_{35}O_3N$	445.2	252-256	Methanol
Proxiphomin	$C_{29}H_{37}NO_2$	431.6	Amorphous	-

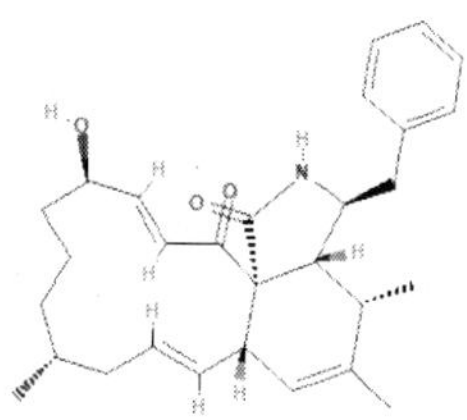

Fig. 2.103. Deoxaphomin A

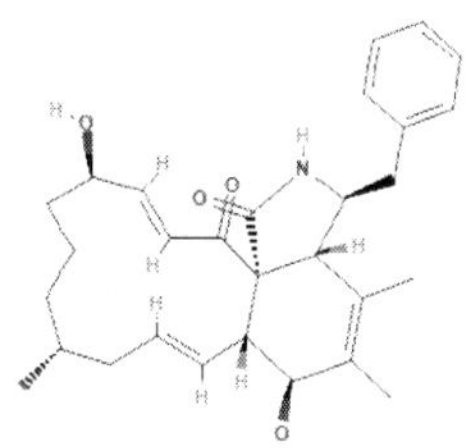

Fig. 2.104. Deoxaphomin B

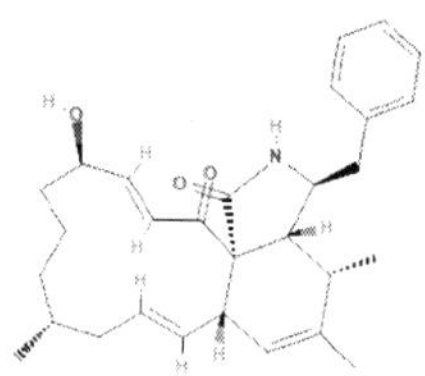

Fig 2.105. Deoxaphomin C

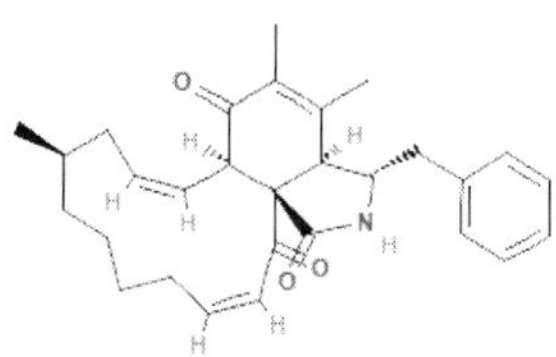

Fig. 2.106. Protophomin

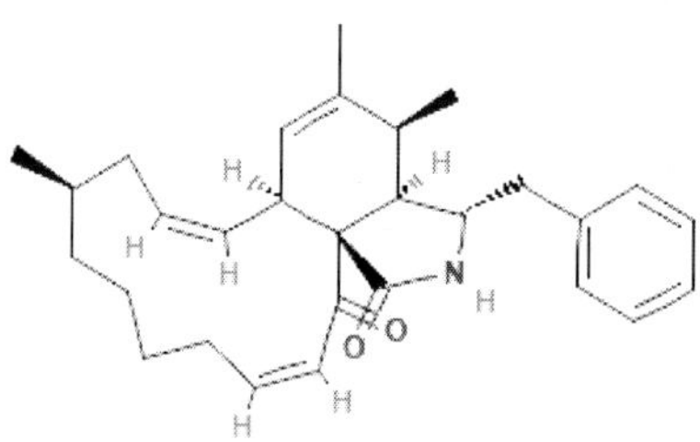

Fig. 2.107. Proxiphomin

Fig. 2.103-107: Structure of Deoxaphomins, Protophomin and Proxiphomin

2.7. Flavichalasines

Flavichalasin A and B, three pentacyclic cytochalasins, flavichalasins C-E, eight tricyclic cytochalasins, flavichalasins F-M, together with eight known analogues were isolated from the solid culture of *Aspergillus flavipes* (Wei *et al.,* 2017). Flavichalasins have cytotoxic effect on seven human cancer cell lines (HL-60 and NB4; human promyelocytic leukemia cell lines; human T- lymphocyte cell line; human breast cancer cell line; HEP- human liver cancer cell line; human colon cancer cell lines) (Wei *et al.,* 2017). The structure of different flavichalasins are given in **Fig. 2.108-120** and physic-chemical characteristics are précised in **Table 2.11**

Table 2.11: Physico-chemical properties of different Flavichalasins

Name of the Cytochalasin	Molecular formula	Molecular weight (Daltons)	Solubility
Flavichalasin A	$C_{24}H_{33}NO_5$	416.2	Methanol
Flavichalasin B	$C_{24}H_{37}NO_4$	426.2	Methanol
Flavichalasin C	$C_{24}H_{35}NO_5Na$	440.2	Methanol
Flavichalasin D	$C_{24}H_{34}NO_5$	416.2	Methanol
Flavichalasin E	$C_{24}H_{35}NO_6Na$	456.2	Methanol
Flavichalasin F	$C_{24}H_{39}NO_5Na$	456.2	Methanol
Flavichalasin G	$C_{24}H_{37}NO_5$ Na	426.2	Methanol
Flavichalasin H	$C_{24}H_{37}NO_3$ Na	410.2	Methanol
Flavichalasin I	$C_{24}H_{36}NO_4$ Na	425.2	Methanol
Flavichalasin J	$C_{24}H_{35}NO_5$ Na	440.2	Methanol
Flavichalasin K	$C_{24}H_{35}NO_5$ Na	440.2	Methanol
Flavichalasin L	$C_{24}H_{38}NO_5$	432.2	Methanol
Flavichalasin M	$C_{24}H_{35}NO_5$ Na	440.2	Methanol

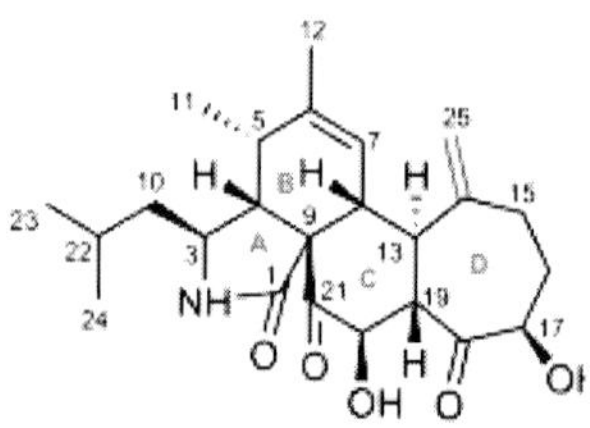

Fig: 2.108. Flavichalasine A

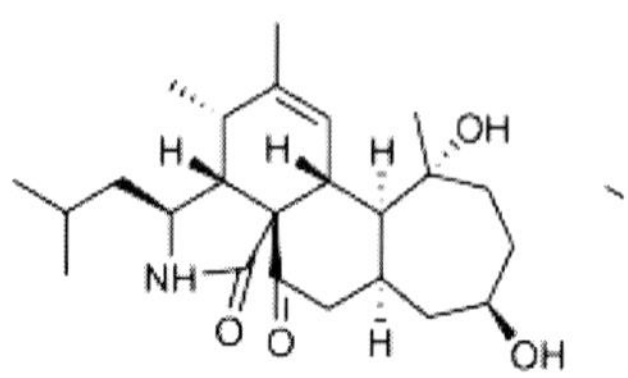

Fig: 2.109. Flavichalasine B

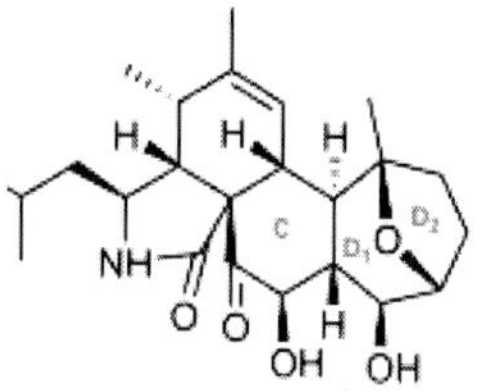

Fig: 2.110. Flavichalasine C

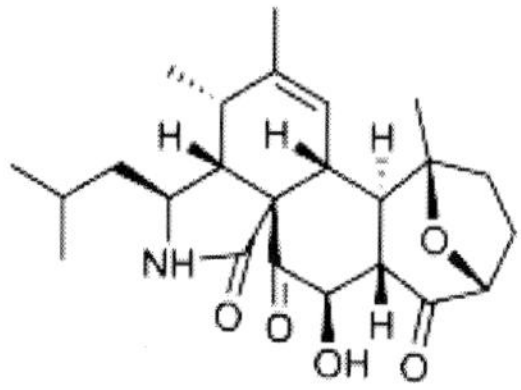

Fig: 2.111. Flavichalasine D

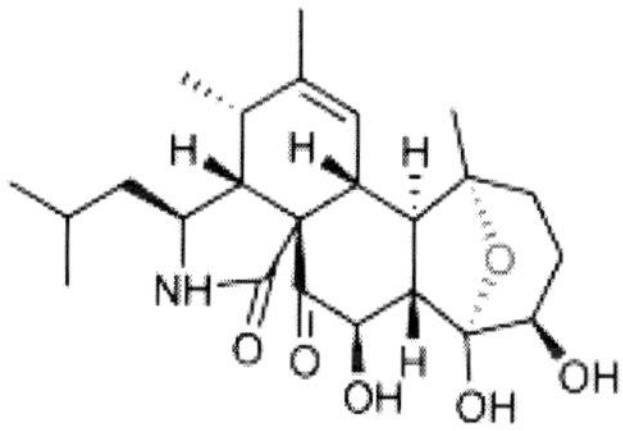

Fig: 2.112. Flavichalasine E

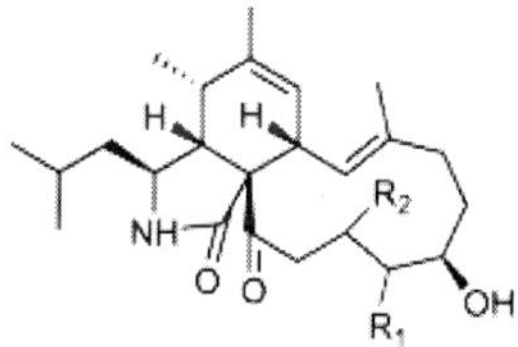

6 $R_1 = \beta OH$ $R_2 = \beta OCH_3$
7 $R_1 = \beta OH$ $R_2 = H$
8 $R_1 = H$ $R_2 = H$
14 $R_1 = \beta OH$ $R_2 = \beta OH$

Fig: 2.113. Flavichalasine F
Fig: 2.114. Flavichalasine G
Fig: 2.115. Flavichalasine M

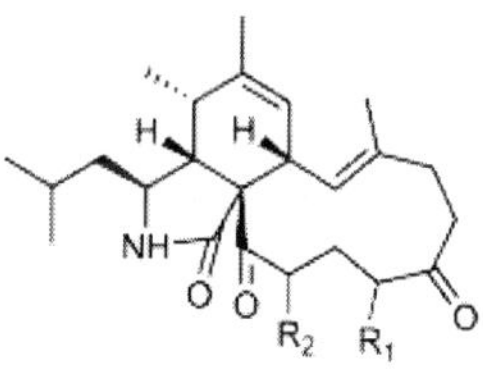

9 $R_1 = \alpha OH$ $R_2 = H$
10 $R_1 = \alpha OH$ $R_2 = \beta OH$
11 $R_1 = \beta OH$ $R_2 = \beta OH$

Fig: 2.116. Flavichalasine H
Fig: 2.117. Flavichalasine I
Fig: 2.118. Flavichalasine J

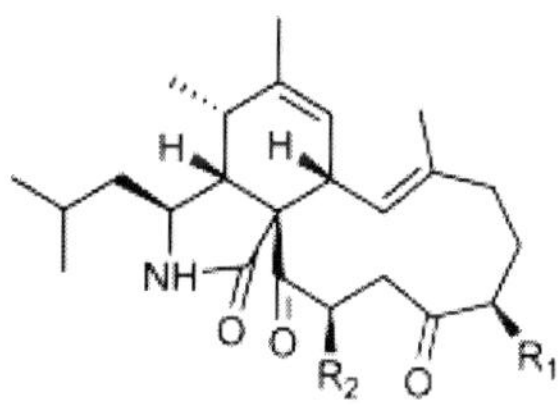

12 $R_1 = OH$ $R_2 = OCH_3$
13 $R_1 = OH$ $R_2 = OH$

Fig: 2.119. Flavichalasine K
Fig: 2.120. Flavichalasine L

Fig. 2.108-120: Structure of different Flavichalasins

2.8. Penochalasins

Penochalasins are a group of macrocyclic structure containing pyrdole moiety. Penochalasins A, B, C and D have been isolated from a strain of *Penicillium* sp. These were originally isolated from the marine alga, *Enteromorpha intestinalis* (Numata *et al.,* 1996). Penochalasins A, B and C, together with known chaetoglobosins A and F, exhibit potent cytotoxicity against cultured P388 cells.

The organisms producing penochalasins are listed in **Table 2.12,** while their physic-chemical characteristics are given in **Table 2.13.** The structure of different penochalasins is illustrated in **Fig. 2.121-129.**

Table 2.12: Penochalasins produced by different fungi

Toxin	Species	Reference
Penochalasin A, B and C	*Penicillium* spp	Numata *et al.* (1996)
Penochalasin D-H	*Penicillium* spp	Iwamoto *et al.* (2001)
Penochalasin K	*Penicillium chrysogenum*	Zhu *et al.* (2017)

Table 2.13: Physico-chemical properties of different Penochalasins

Name of the Cytochalasin	Molecular formula	Molecular weight (Daltons)	Melting point (in °C)	Solubility
Penochalasin A	$C_{32}H_{38}N_3O_3$	509.6	222-224°C	Acetone
Penochalasin B	$C_{32}H_{35}N_3O_3$	509.6	177-179	Acetone
Penochalasin C	$C_{32}H_{35}N_3O_3$	509.6	173-178	Acetone
Penochalasin D	$C_{32}H_{37}N_3O_3$	511.6	-	-
Penochalasin E	$C_{32}H_{38}N_2O_5$	530.6	-	-
Penochalasin F	$C_{32}H_{38}N_2O_5$	530.6	-	-
Penochalasin G	$C_{32}H_{38}N_2O_4$	514.6	-	-
Penochalasin H	$C_{32}H_{38}N_2O_4$	530.2	-	-
Penochalasin K	$C_{32}H_{34}N_2Na_4O$	533.2	218.5-220.3	-

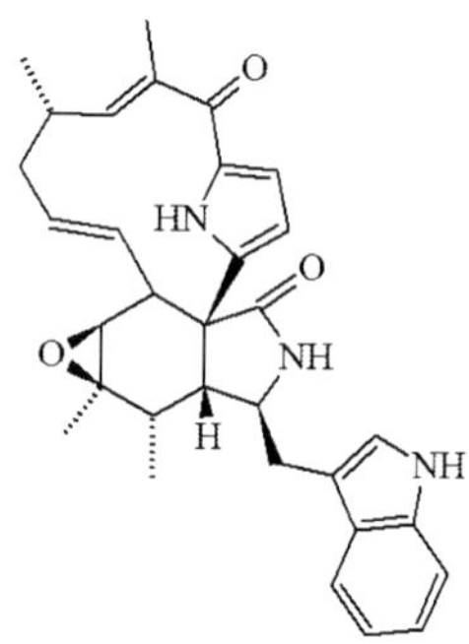

Fig. 2.121. Penochalasin A

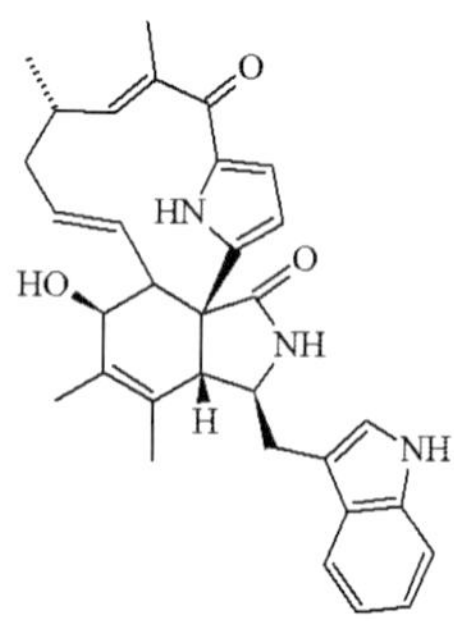

Fig 2.122. Penochalasin B

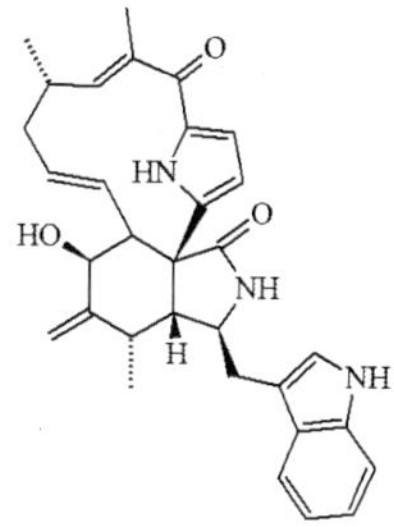

Fig 2.123. Penochalasin C

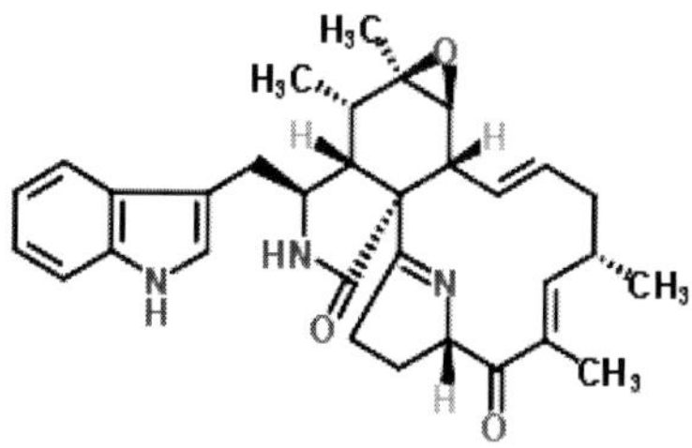

Fig 2.124. Penochalasin D

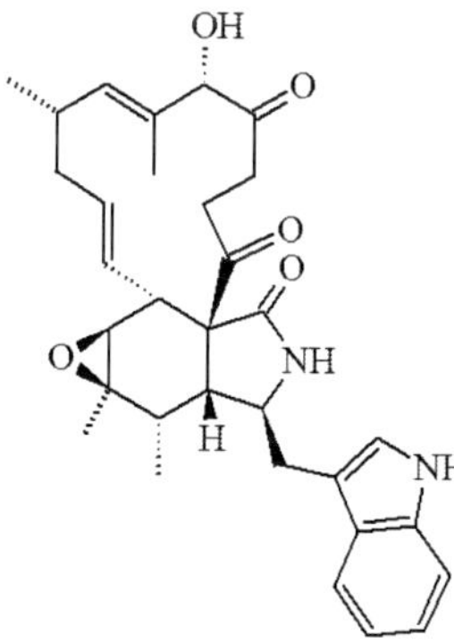

Fig 2.125. Penochalasin E

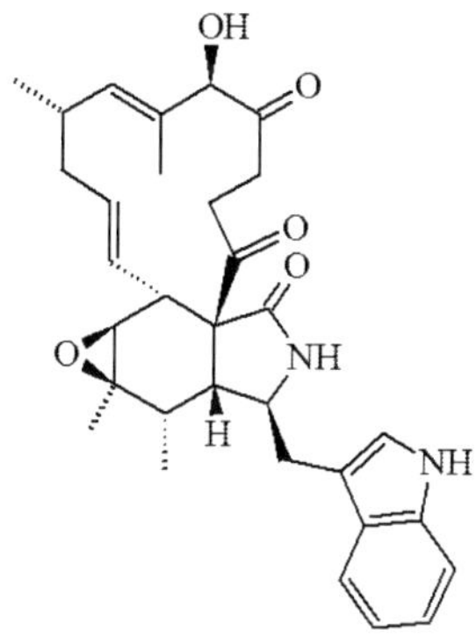

Fig 2.126. Penochalasin F

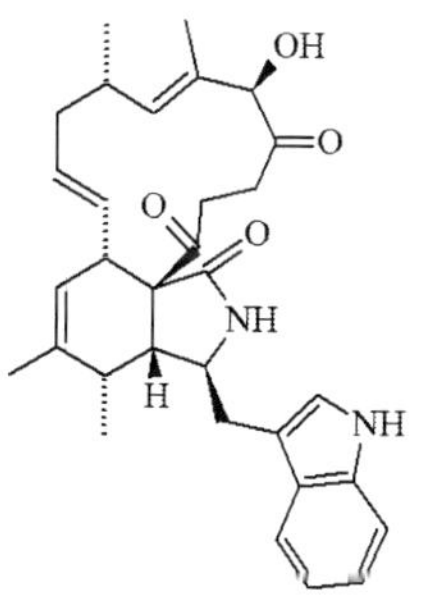

Fig. 2.127. Penochalasin G

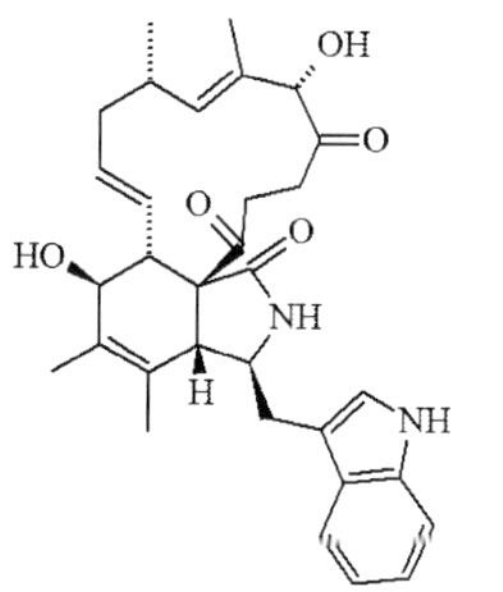

Fig. 2.128. Penochalasin H

Fig. 2.129. Penochalasin K

Fig. 2.121-129: Structure of different Penochalasins

2.9. Epicochalasins

Epicochalasins A and B are a class of cytochalasins bearing unexpected scaffolds consisting of fused aspochalasin and epicoccine dimer moieties. These were isolated from the culture broth of *Aspergillus flavipes* (Zhu *et al.,* 2016). Epicochalasins A and B, two merocytochalasins possessing unprecedented hendecacyclic 5/6/11/5/6/5/6/5/6/6/5 ring systems with an adamantyl cage and G2/M-phase cell-cycle-arrest activity. Molecular formula is $C_{42}H_{49}NO_{12}$. Molecular weight of Epicochalasins A is 782.3. The structures of epicochalasins are given in **Fig. 2.130-131.**

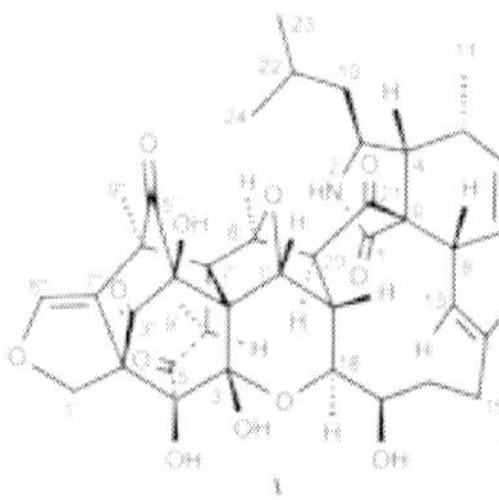

1

Fig. 2.130. Epicochalasins A

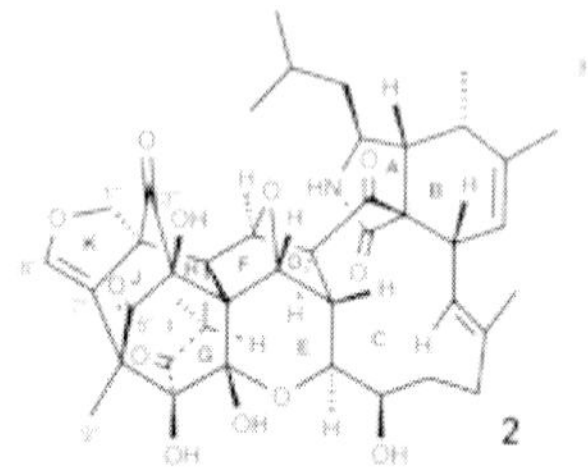

2

Fig. 2.131. Epicochalasins B

Fig 2.130-131: Structure of Epicochalasins

2.10. Asperchalasin

Asperchalasin A, the first cytochalasin dimer featuring a unique decacyclic 5/6/11/5/5/6/5/11/6/5 ring system consisting of 20 chiral centers, was isolated from the culture broth of *Aspergillus flavipes* (Zhu *et al.,* 2015). This compound possesses an unprecedented 13-oxatetracyclo [7.2.1.12,5.01,6] tridec-8,12-dione core structure, is the first example of a dimeric cytochalasan alkaloid. This compound induced significant G1-phase cell cycle arrest by selectively inhibiting cyclin A, CDK2 and CDK6 in cancerous, but not normal cells, highlighting it as a potentially selective cell cycle regulator against cancer cells. The structure of epicochalasin is given below (**Fig. 2.132**). Its molecular formula is $C_{57}H_{72}N_2O_{12}$ and Molecular weight is 977.5.

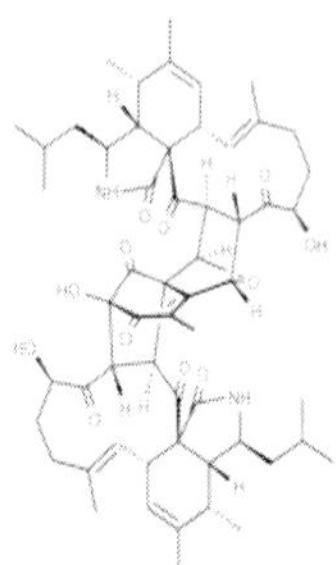

Fig 2.132: Structure of asperchalasin A

2.11. Spicochalasin

This compound was isolated from *Spicaria elegans* (Zhenjian *et al.,* 2009). Spicochalasin A has a unique pentacyclic system and was found to be moderately cytotoxic towards human leukemic HL-60 cells with an *IC*50 value of 19.9 μM. Spicochalasins are growth inhibitors against the A-549, BEL-7402, P388 and HL-60 cancer cell lines (Zhenjian *et al.,* 2009). Molecular formula is $C_{24}H_{33}NO_5$ and Molecular weight is 416.2. The structure of spicochalasin is given in **Fig. 2.133.**

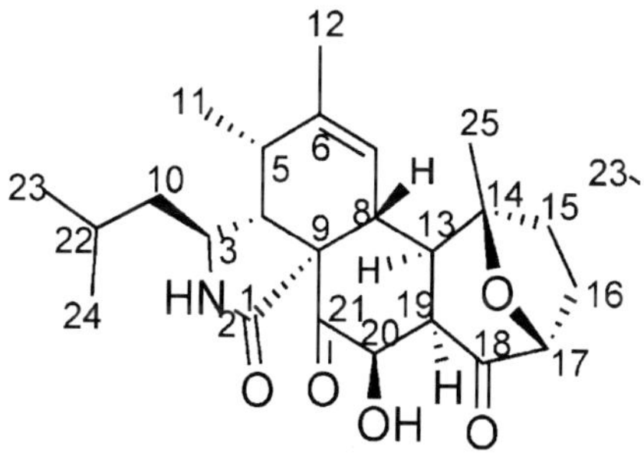

Fig. 2.133: Structure of Spicochalasin

2.12. Rosellichalasin

Rosellinia necatrix causing white root-rot of potato, soybean, tea, peanut and apple produces cytochalasin E (Kimura *et al.,* 1989) and Rosellichalasin. This compound contains the lactam ring and the benzyl group. In a bioassay using lettuce seeds, at a dose of 100 mg/l inhibited the germination by 50% after 24 hr. The root and hypocotyl growth were also inhibited by 70% and 60% at the same concentration, respectively. Molecular formula is $C_{28}H_{33}NO_7$, Molecular weight is 495.2 and Melting point is 121-123°C. The structure of rosellichalasin is illustrated in **Fig. 2.134.**

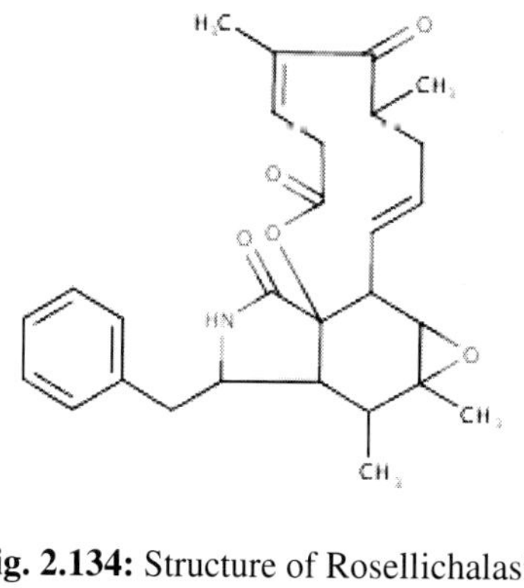

Fig. 2.134: Structure of Rosellichalasin

2.13. Ascochalasins

Ascochyta heteromorpha isolated from oleander (*Nerium oleander* L.), produced known cytochalasins A and B (Capasso *et al.,* 1988). This compound contains (13)-carbocyclic subgroup of cytochalasin. Molecular formula is $C_{29}H_{39}NO_4$. Molecular weight is 465.2. The structure of ascochalasin is illustrated in **Fig. 2.135.**

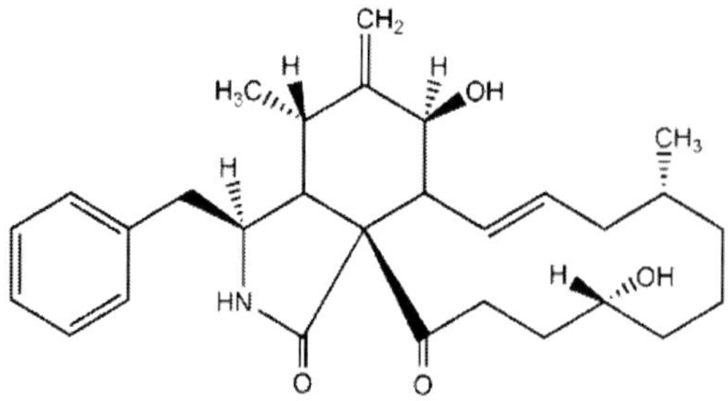

Fig. 2.135: Structure of Ascochalasins

2.14. Pyrichalasin H

Pyrichalasin H, a phytotoxic metabolite was isolated from *Pyricularia grisea* (Manabu and Tsuneo, 1991). It is a dihydro derivative of pyriculol, 2-(3,4-dihydroxyhepta- 1,5-dienyl)-6-hydroxybenzyl alcohol and shows various unique biological properties. This is the first reported case of a *p*-methoxyphenyl group being substituted at C-10 belonging to the cytochalasins containing either one phenyl, indolyl, or isopropyl group. The comparison of the proton-proton coupling constants for cytochalasin H and pyrichalasin H indicates that the relative stereochemistry of both compounds should be the same. Similarly the comparison with the cytochalasin D curves also shows the absolute stereochemistry of both compounds are the same. Pyrichalasin H was found to strongly inhibit the growth of rice seedlings at 1 μg/ml. Molecular formula is $C_{31}H_{41}NO_6$. Molecular weight is 523.2. The structure of pyrichalasin H is given in **Fig. 2.136.**

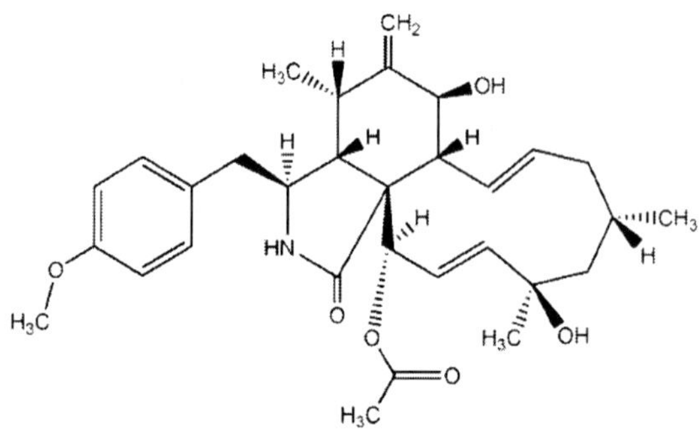

Fig 2.136: Structure of Pyrichalasin H

2.15. Periconiasins A-C

Periconiasins A-C, new cytochalasins with an unprecedented 9/6/5 tricyclic ring system, were isolated from the endophytic fungus *Periconia* sp (Zhang *et al.*, 2013). Their molecular structures together with the relative configuration were unambiguously established on the basis of extensive spectroscopic data analysis as Periconiasin A, B and C with molecular formulas $C_{22}H_{33}NO_3$, $C_{22}H_{33}NO_3$ and $C_{22}H_{31}NO_3$ and molecular weight 360.2, 360.2 and 358.2 respectively. They exhibited significant cytotoxicity against human HCT-8, Bel-7402 and BGC-823 cell lines. The structure of Periconiasins A-C is given in **Fig. 2.137.**

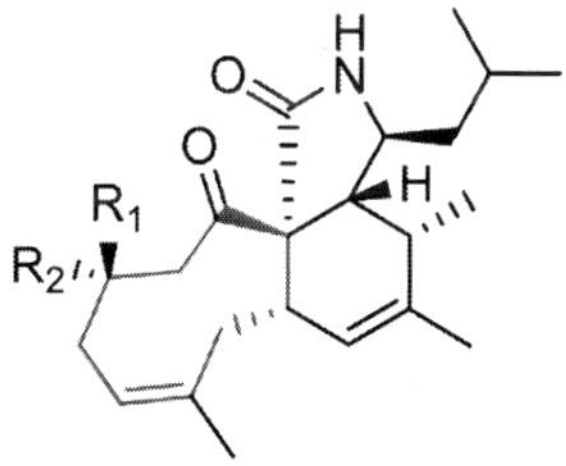

1 R_1 = H, R_2 = OH **1a** R_1 = H, R_2 = 4-ClPhCO$_2$
2 R_1 = OH, R_2 = H **3** R_1 = R_2 = O

Fig. 2.137: Structure of Periconiasins

2.16. Scoparasins

Cytochalasin Z_{24}, Z_{25}, Z_{26} and other new cytochalasin compounds, scoparasin A, B and C was isolated from the fungus *Eutypella* sp. (Liu *et al.*, 2014). These compounds were evaluated for cytotoxic activities against several human tumor cell lines. This compound was ketone and lactam carbonyl functional groups. Scoparasin A, B and C with molecular formulas $C_{28}H_{33}NO_8$, $C_{29}H_{37}NO_9$ and $C_{28}H_{32}NO_8$ and molecular weight 512.2, 543.6 and 407.6 respectively. They show antimicrobial activities against *Staphylococcus aureus* ATCC 25923 and *Microsporum gypseum* SH-MU-4 (Pongcharoen *et al.*, 2006).

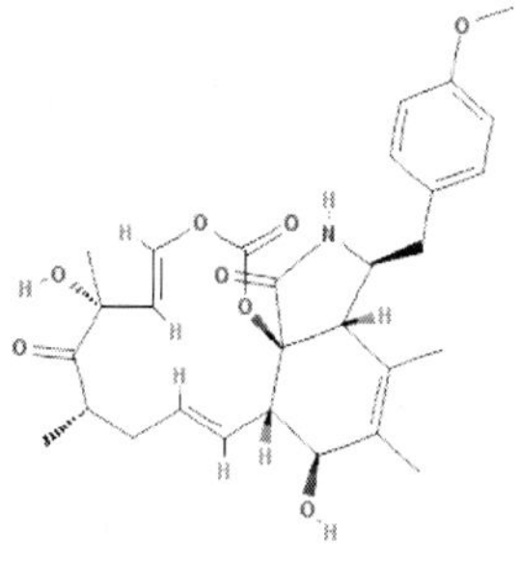

Fig. 2.138. Epicochalasins A

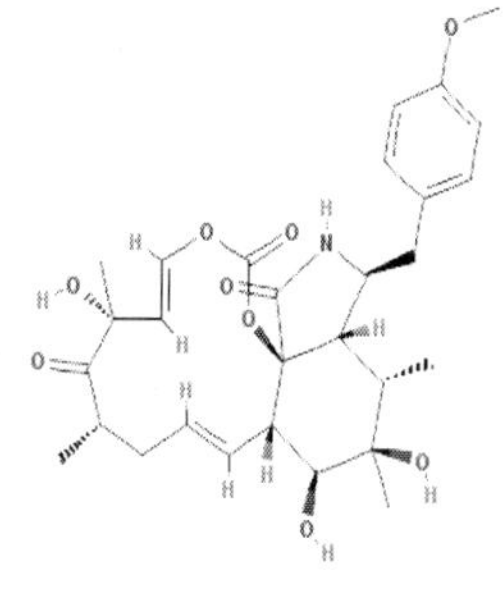

Fig. 2.139. Epicochalasins B

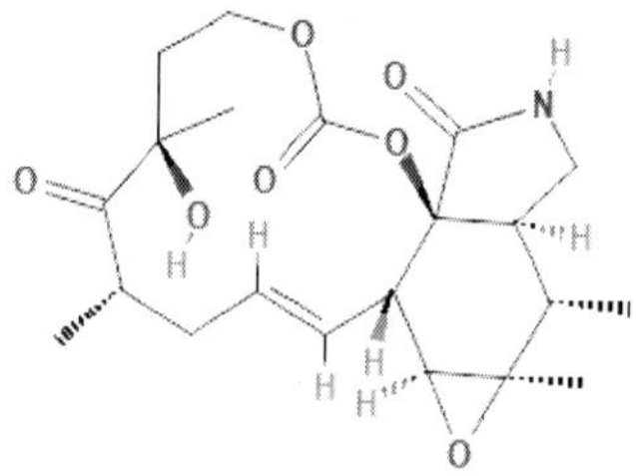

Fig. 2.140. Epicochalasins A

2.17. Xylastriasan

Xylastriasan A, a new cytochalasin an was isolated from the ethanol extract of fruiting bodies of the fungus *Xylaria striata*. (Lei *et al.,* 2018). This compound is alkaloid with a rare 5/6/6/5/6 pentacyclic system, featuring a cyclohexanone (ring C) and an 8 oxabicyclo[3.2.1]octane with one oxygen bridge between C-14 and C-18 (rings D and E). The structure of Xylastriasan A is given in **Fig. 2.141.**

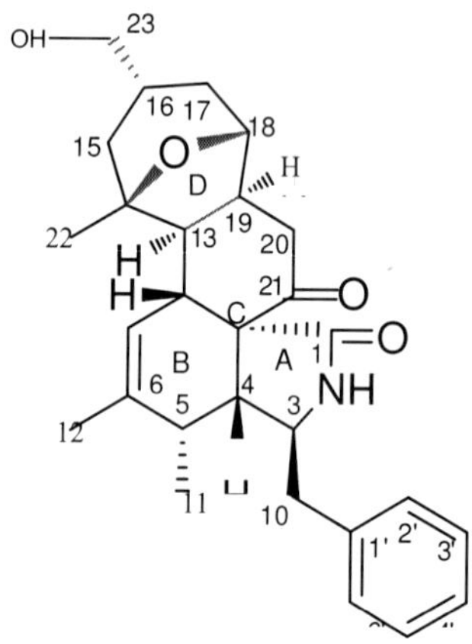

Fig. 2.141. Xylastriasan

This compound is active against five cancer cell lines, including human liver hepatocellular cells (HEPG2), mouse melanoma cells (B16), human colon cancer cells (SW620), human lung adenocarcinoma cells (A549) and mouse breast cancer cells (4T1), and has weak cytotoxic activity against HEPG2, B16 and A549 cell lines with IC50 values of 93.61, 85.61 and 91.58 μM, respectively. Its molecular formula is $C_{28}H_{35}NO_4$ and molecular weight is 472.2.

■■■■

Chapter-3

Taxonomy of Cytochalasins Producing Fungi

In all seventy mould species colonizing variety of substrates were found to elaborate large array of cytochalasins whose identification sometimes makes it difficult. Therefore, a simple synoptic key is developed for its proper identification. A detailed description of each fungus is given to facilitate the investigator working with cytochalasins.

Key to the fungi producing Cytochalasins

1. Fungi reproduce sexually 2

1a. Fungi either do not produce sexual spores or produce rarely 18

2. Fungi reproduce by producing ascocarps and ascospores 3

2a. Fungi reproduce by basidiocarp and basidiospores 17

3. Fungi produce Cleistothecium type of ascomata 4

3a. Fungi produce perithecium type of ascomata 5

4. Ascocarp peridium one layer thick, ascospores hyaline then brown, ascospores globose or elliptical, smooth .... *Pseudeurotium*(1)

4a. Ascocarp peridium several cell layered, ascospores hyaline, globose and smooth wall .... *Nigrosubulum*(2)

5. Asci unitunicate 6

5a. Asci bitunicate 16

6. Asci evanescent, Perithecium glaberous, spherical, hemispherical, flask shaped, ostiolate, hairs of different types arise on the surface of perithecium .... *Chaetomium*

6a. Asci with persistent wall 7

7. Asci apical structure not amyloid, asci without refractive apical caps or rings, ascospores without germ pores or germ slits, ascomata fleshy, bright often white .... *Calonectria*(4)

7a. Asci with apical ring, often loosing from the ascogenous cells and embedded in slimy mass 8

8. Ascomata immersed in a stroma or in a pseudostroma, stroma fleshy, light or mostly yellow, ascospores with a septum in or near the middle .... *Endothia* (5)

8a. Ascomata mostly large and aggregated often connected by stromatic structures, asci mostly small, including a simple amyloid ring in the apex, ascospores one celled, allantoid 9

9. Ascomata numerous formed in pseudostroma which is erumpent often pustulate ostiole long ....*Eutypella* (6)

9a. Ascomata fleshy, asci with an apical cap or body perforated by narrow canal 10

10. Asci with an apical cap or body perforated by a narrow canal.................................. 11

10a. Asci mostly large, amyloid, portion with a simple ring or a system of rings often with a chitinoid pulvillus, ascospores 1-celled to septate not allantoid 12

11. Ascomata medium-sized with short papillate ostiole, asci cylindrical, ascomatal stroma head like or clavate, stalked, arising from endosclerotia .... *Hyphomyces* (7)

11a. Ascospores filiform, arsing from endosclerotial structures in insects or fungus fructification ....*Cordyceps* (8)

12. Ascomata, superficial, single or aggregated or free or surrounded by a hyphal subiculum or immersed in the host tissue and carried by a small stromatic clypeus, ascospores sometimes provided with an apical gelatinous appendage ....*Rosellinia* (10)

12a. Ascomata several to many, embedded in a stoma or sitting in a conspicuous erect or wirelike stroma consisting of entirely of fungal tissue............ 13

13. Stromata composed of light-colored parenchyma, occasionally covered by a dark, crust, ascomata monostichous often developing near the periphery of the stromata, stroma more or less erect or stipitate stromata on plants or debris or originating from termitaria, erect, clavate, branched or with head like fertile apex ....*Xylaria* (10)

13a. Ascomata polystichous, completely embedded in the stoma often of at the base of the stromata, flask of the stoma white 14

14. Stromata with reticulate fertile portion enclosing a sterile portion ascospores comparatively large .... *Engleromyces* (11)

14a. Reticulate fertile portion absent in the stroma 15

15. Stromata subglobose or globose showing concentric zones......................*Daldinia* (12)

15a. Stromata without concentric zones, stromata crust like to hemispherical or globose, ascomata embedded in the periphery of the stroma provided with a short neck, semiglobose to angular not separating.............................*Hypoxylon* (13)

16. Asci interspersed with pseudoparaphyses or with paraphysidal hyphae which branch at the tips to form an epithecium, ascospores spirally coiled with in the ascus, conidial state .... *Bipolaris* (*Helminthosporium*) *Cochliobolus* (14)

16a. Ascospores large, not flattened, hyaline to pale-brown, circular in cross section, 3,5 or 7 septate with a single longitudinal septum, pseudothecia large, usually setose bearing *Drechslera* conidial state .... *Pyrenophora* (15)

17. Basidiomata laterally stipitate, pileus, light-brown, stipe short, pore surface cream to light- brown, basidiospores 1-celld, hyaline, cylindrical .... *Coriolus* (16)

17a. Basidiomata solitary or in groups, laterally stipitate, flabelliform, spathulate, pileus glaberous strongly zonate, basidiospores small cylindrical or oblong elliptical, hyaline, non-amyloid and non-cyanophilous. .... *Microporus* (17)

18.Conidia or conidiophores are produced nakedly, no fruit body is formed 24

18a.Conidia and conidiophores produced either in a pycnidium or acervulus 19

19. Conidia produced in acervulus and conidia 5-celled,pedicellate with setulae .... *Pestalotiopsis* (18)

19a. Conidia produced in a pycnidium 20

20. Pycnidiospores one-celled, hyaline 23

20a. Pycnidiospores 2 to several celled 21

21. Pycnidiospores more than one cell with one setae at each end .... *Discosia* (19)

21a. Pycnidiospores two celled 22

22. Pycnidiospores hyaline .... *Ascochyta* (20)

22a. Pycnidiospores brown to dark-brown .... *Diplodia* (21)

23. Pycnidiospores are only one type, globose or spherical .... *Phoma* (22)

23a. Pycnidiospores two types α-conidia and β-conidia .... *Phomopsis* (23)

24. Conidiophore free and borne on individual hyphae 25

24a. Conidiophore form a distinct synnemata, insecti colous .... *Metarrhizium* (24)

25. Conidia borne singly and not in dry slimy head 26

25a. Conidia aggregated in a slimy mass at the tip of the conidiophore, globose, oval to short cylindric .... *Acremonium* (25)

26. Conidiogenous cells somewhat swollen at the base borne directly on the hyphae tapering to a long slender tube. Conidia in chains .... *Paceilomyces* (26)

26a. Conidiophore not swollen at the base 27

27. Conidiophore terminated by a swelling, sterigmata arise in one or two series on the surface of the vesicle .... *Aspergillus* (27)

27a. Conidiophore apex thread like 28

28. Conidiophore terminated by thread like zigzag apex, entomogenous .... *Beauveria* (28)

28a. Conidiophore growth indefinite 29

29. Conidia acropleurogenous, polyblastic, 1-3 septa, geniculate or denticulate .... *Pyricularia* (29)

29a. Conidia cylindrical, smooth, pigmented in light shades of yellow, green or brown or combination of three 30

30. Conidia cylindrical germinating laterally, perfect stage in *Pyrenophora* .... *Drechslera* (30)

30a. Conidia not cylindric, perfect state not in *Pyrenophora* 31

31. Conidiophore elongated, bearing conidia acrogenously, Conidia in chains 32

31a. Conidia produced singly and acrogenous 34

32. Conidia ovoid, obclavate, solitary or in chains with a beak, pale-yellow to dark olive-brown conidia with both transverse and vertical septa .... *Alternaria* (31)

32a. Conidiophore, brush like, composed of a stipe terminated by a few or many sterile branches 33

33. Conidiogenous cells or phialides bear conidia in a long dry chains .... *Penicillium* (32)

33a. Conidiogenous cells swollen at the base tapering to a long slender, widely divergent tube, conidia hyaline in chains .... *Paecilomyces* (33)

34. Conidia not cylindric, perfect stage not in *Pyrenophora* 36

34a. Conidia dark, globose, rough developing successing in chains from globose, Conidiogenous cells 35

35. Conidia fusiform or ellipsoid, straight or curved germination only from end cells, *Cochliobolus* teleomorph .... *Bipolaris* (34)

35a. Conidia fusiform, curved, dark coloured with light coloured end cells .... *Curvularia* (35)

36. Conidiophore branched only at the apex .... *Periconia* (36)

36a. Conidiophore branched profusely 37

37. Conidiogenous cells bottle shaped disposed irregularly .... *Trichoderma* (37)

37a. Each branch terminated by phialides 38

38. Phialides produced in cluster 39

38a. Phialides enlarged at the base, conidia not in heads, entomogenous .... *Hirsutella* (38)

39. Phialides divergent, loose tapering to a tube, conidia in chains 40

39a. Phialides pigmented phialides. Conidia one celled formed in slimy mass around the phialides .... *Stachybotrys* (39)

40. Prophialide dark and phialides subhyaline .... *Zygosporium* (40)

40a. Phialides are borne on branched conidiophore 41

41. Conidia septate .... *Cylindrocladium* (41)

41a. Conidia not septate 42

42. Conidiogenous cells repeatedly geniculate .... *Geniculosporium* (42)

42a. Conidiogenous cells not geniculate 43

43. Conidiogenous cells flask shaped producing endogenous conidia............*Chalara* (43)

43a. Conidiogenous cells undifferentiated 44

44. Conidia not separating easily chains mostly horizinate...................... *Hormiscium* (44)

44a. Ramo-conidia frequently present, scars small, conidiophores mostly rather slender.. *Rhinocladiella* (44)

1. *Alternaria chlamydospora* Mouchacca

Colony: Effuse, floccose, dark-olive to blackish-brown

Chlamydospora: Many celled, very variable in size and shape, golden-brown.

Conidiophores: Up to 150μ long, 3-6 μ thick, rather pale-brown.

Conidia: At first obclavate or obpyriform, later swelling and becoming very variable in shape, golden brown, smooth or occasionally minutely verruculose, 20-60 x 6-40 μ, with a short, pale beak 2-6 μ thick.

Cytochalasins: Cytochalasin E

Diagnostic features: Conidiophores many celled, golden-brown, conidia obclavate later swelling becoming variable, golden- brown.

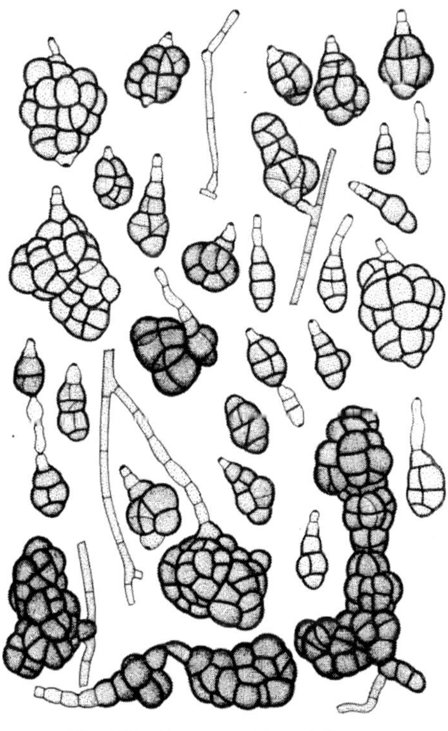

Conidiophores and conidia

2. *Ascochyta* Lib.

Colony: Spreading, hyaline to pale- brown.

Conidiomata: Pycnidia, amphigenous, globose, brown, immersed, uniloculate, thin walled texture angularis, 1-3 cell thick, ostiolate, slightly papillate.

Conidiophore: Absent

Conidiogenous cells: Enterobalstic, phialidic, determinate, discrete, doliform to lageniform, hyaline, smooth, formed from the inner cells of the pycnidial wall.

Conidia: Hyaline, 1-septate, continuous or constricted, thin walled, smooth, guttulate, ovoid to oblong, cylindrical to irregular.

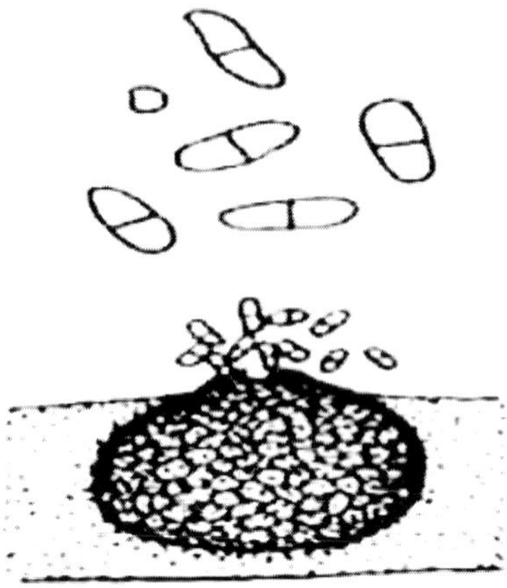

Pycnidium and bicelled conidia

Teleomorph= *Didymella* Sacc

Ascomata: A unioculate perithecioid pseudothecium, scattered, immersed in substrate or erumpent; pseudothecia darkbrown to black, globose to subglobose, ostiolate.

Ascomatal wall: Pseudo-parenchymatous, sometimes thickened around ostiole to form a clypeus. Centrum containing filamentous, often branched, pseudoparaphyses.

Asci: Bitunicate, parallel, clavate to cylindrical, short-stalked, 8-spored.

Ascospores: 2-celled, septum median, constricted or not, oval to fusiform, hyaline.

Cytochalasins: Cytochalasin-A, B and Ascochalasin

Diagnostic features: Pycnidia glaberans, uniloculate, enteroblastic, phialidic determinate, conidia two-celled, hyaline, Teleomorph *Didymella*

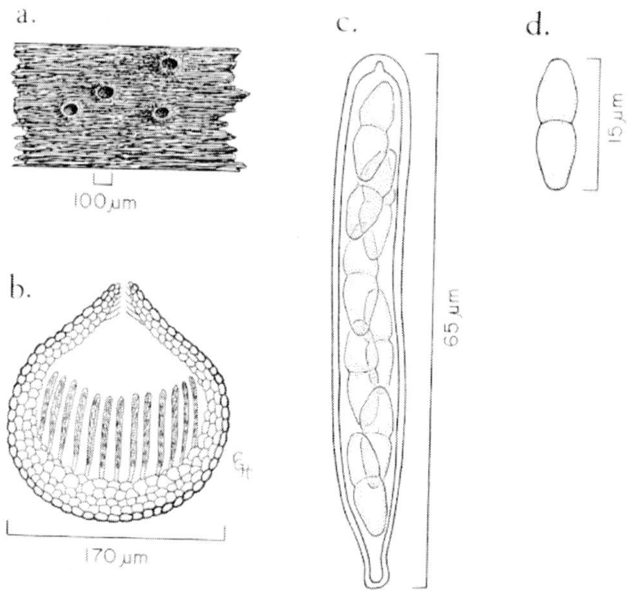

(a) Perithecia in natural habitat, (b) cross section of perithecium showing arrangement of asci and ostiole, (c) Ascus with ascospores (d) bicelled ascospores

3. ***Aspergillus elegans*** Gasperini

Colonies: Colonies restricted, vegetative mycelium forming tough basal felt, thinning to a flat non-sporulating margin, entire but becoming fimbriate due to uneven development of limited white surface mycelium, conidial structures intermixed with aerial hyphae in a cushion-like pattern light-buff to pale-ochraceous buff turning to grey color to tawny olive with age. Reverse dull pinkish brown shades; exudate limited, inconspicuous, conidiophores long-stalked mature ochraceous buff conidial heads.

Conidiophores: Variable, in length, heavy walled, yellow to clear brown shades, varying from smooth or nearly so to coarsely. vesicles often somewhat elongate in small heads but consistently subglobose to globose in large heads, bearing sterigmata over their entire surface; sterigmata in two series, crowded, but occasionally smaller secondaries.

Conidia: Ovate to elliptical, smooth or delicately roughened, 1-celled.

Sclerotia: Sparingly produced in some strains, white to yellowish-cream, ovate to discoid.

Cytochalasin: Cytochalasin Z4 and Aspochalasin A1.

Diagnostic features: Light-buff to pale-ochraceous buff, reverse dull pinkish brown, conidiophore elongate, vesicle sub-globose to globose, sterigmata in two series, conidia ovate to elongate, smooth.

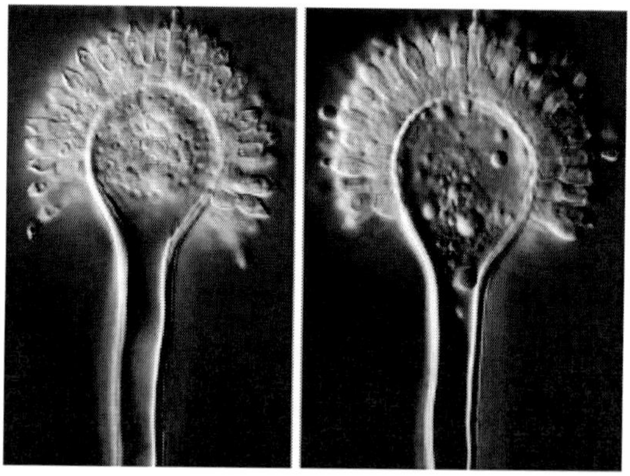

Conidiophores, vesicle, uniseriate phialides and conidia

4. ***Aspergillus microcysticus*** Sappa

Colonies: On Czapek's agar growing slowly, attaining a diameter of 1.5 to 1.7 cm. in 8 days, velvety, raised, radially sulcate, obscurely zonate, margins abrupt and sublobulate, at first yellowish-white or citrine, then gray-blue; exudate limited as minute, dilute amber droplets; reverse gray-brown. Conidial heads crowded in marginal areas, globose, small, 20 to 40 μm or sometimes fragmentary, 10 to 15 μm. Felt compact.

Conidiophores: Erect, or slightly sinuous, 180 to 200 μm by 4.5 to 4.8 μm, smooth; walls thick, 0.5 to 0.8 μm, dilute, brownish. Vesicles small, sometimes regularly hemispherical, 8.0 μm by 6.5-7.0 μm, lower third nude, bearing sterigmata above, sometimes very small, 2.8 to 3.2 μm, as slightly inflated conidiophores and then bearing similar but unequal sterigmata.

Sterigmata: Sterigmata arranged in two series; primaries sparse, only at the apex, 5.6 to 6.4 μm by 2.5 μm, secondaries 5.6 to 6.0 μm by 2.0 to 2.2 μm.

Conidia: globose, dilute-green, smooth or very finely aperulate, measuring 3.0 to 3.2 μm.

Cytochalasins: Aspochalasins A, B, C, and E

Diagnostic features: Colonies velvety, raised, radially sulcate, gray-blue; exudate limited as minute, two series sterigmata, globose, very finely aperulate.

5. ***Aspergillus clavatus*** Desm

Synonyms = *A. apicalis* Mehrotra & Basu 1976

= *A. pallidus* Kamyschko 1963

Subgenus: Clavati

Section: Clavati

Teleomorph: Absent

Colony: Spreading

Colonies: Blue-grey of 3-4 cm diameter in 10 days.

Mycelium: Felt thin, plane, conidial heads abundant, large, clavate (club shaped), blue-green when young, later slate-olive, 1-4 mm in length.

Reverse: Colony colorless at first, light-brown with age.

Exudate: Absent.

Conidial heads: Clavate, 250-400 × 135-175 µm, often splitting into divergent columns at maturity.

Conidiophores: Thin walled, gradually enlarging at the apex into a clavate vesicle, vesicle 180-25 × 38-60 µm, fertile over the entire surface.

Phialides: Single series, basel ones smaller, 2.5-3.4 × 1.8-3 µm, those at the apex of the vesicle larger, 6.5-10 × 2.6-33 µm.

Conidia: Elliptical, smooth, somewhat heavy walled, 3-4.8 x 2.5-3.5 µm, light-green in mass.

Cytochalasins: Cytochalasin E, F, G, K and Z_{16}

Diagnostic features: Conidiophore vesicle is clavate and bears single row of phialides, conidia elliptical and heavy walled light green in mass.

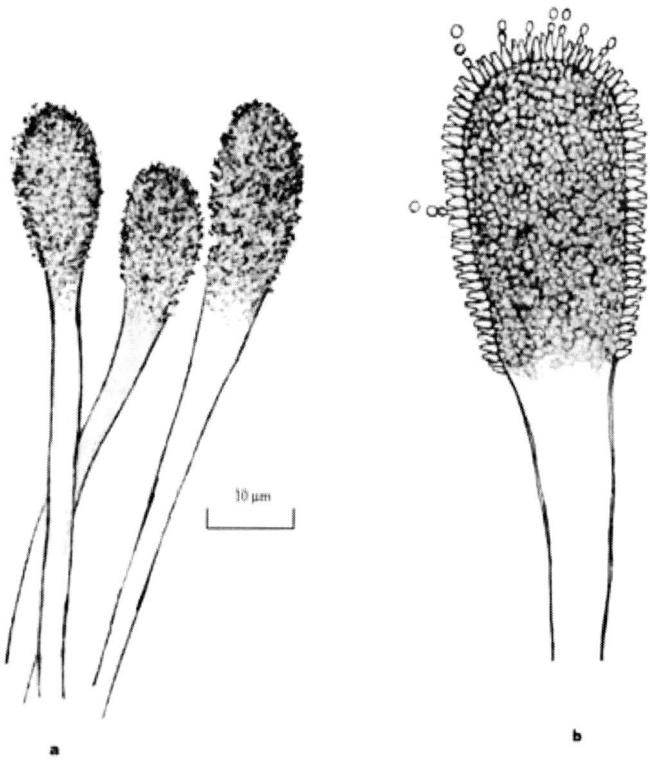

Conidiophores with clavate apex with single row of phialides

6. *A. flavipes* (Baineer & Sartory) Thom & Church

Synonyms = *Fernelia flavipes*

= *Sterigmatocystis flavipes* Baineer & Sartory (1911)

Subgenus: Nidulantes

Section: Flavipedes

Teleomorph: Absent

Colony: Growing moderately, white at first turning to bright-wheat colour later velvety

Reverse: Usually yellow-brown to red-brown, periphery white.

Conidial heads: Columnar, pale-avellaneus shades.

Conidiophores: Smooth or short

Vesicle: Subglobose to vertically elongate

Phialides: Biseriate

Metulae: Always present, curved upwards

Conidia: Subglobose, nearly colourless, smooth

Hulle cells: Sometimes present

Cytochalasins: Cytochalasin Z16 to Z20, Aspochalasins I,J,K and L, Flavichalasins, Epichochalasin A and B, Asperchalasins

Diagnostic features: Colonies white turning to bright-wheat colour and velvety, subglobose to vertically elongate vesicle with biseriate phialides, subglobose hyaline smooth conidia.

Conidiophore with phialides, chain of conidia

7. *A. niveus* Blochwitz

Synonymy= *A. eburneus* Biourge

= *Emericella nivea* Wiley & Simmons

Colonies: Growing slowly, attaining 2.5-5.3 cm diam in two weeks, white to light pink,

Reverse: Dark yellow brown to greenish black, exudates pale to brown.

Conidial heads: White to light pink, loosely columnar, up to 250-300 x 20-30 µm, larger heads up to 6- µm in diam, abnormal proliferating heads and occasional branched

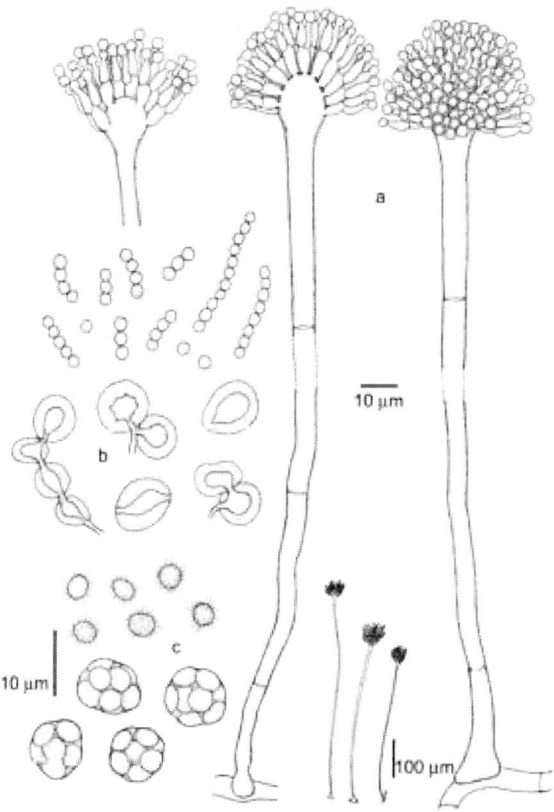

(a) Conidiophores with phialides, (b) conidial chains, (c) Hulle cells, (d) asci and ascospores

Conidiophores: Sometimes observed, conidiophores smooth, hyaline to faint yellow, slender, variable in length, 300-6 x 4-5.6 µm, rarely longer, vesicles hemi-spherical, hyaline, 8-15 µm in diam, with hyaline

Metulae and phialides: metulae 5-8 x 2.5-3 µm, phialides, 5-7 x2-2.5 µm,

Conidia: Globose, smooth, colorless, thin walled, 2-2.5 µm.

Teleomorph: ***Fennellia nivea*** B.J. Wiley & E.G. Simmons

Ascomata: Cleistothecia embedded in basal felt, 80-130 μm in diam, surrounded by yellow hulle cells, wall 1-several layered thick consisting of thin walled hyphae, light-yellow.

Asci: globose, evanescent, 8-spored.

Ascospores: lenticular with inconspicuous equatorial grooves and two very low equatorial crests, valves spiny.

Cytochalasins: Aspochalasin A to D and Z.

Diagnostic features: Colonies white to light-pink. Reverse of the colony dark-yellow-brown to green black, exudates pale to brown, conidiophore light yellow of variable length with hemispherical head. Metulae are present. Conidia globose and smooth.

8. ***A. terreus*** Thom

Synonyms = *A.boedijinii* Ann. Mycol. 32.83

= *A. terreus v. boedijinii* (B1935).

Subgenus: Nidularites

Section: Terrei

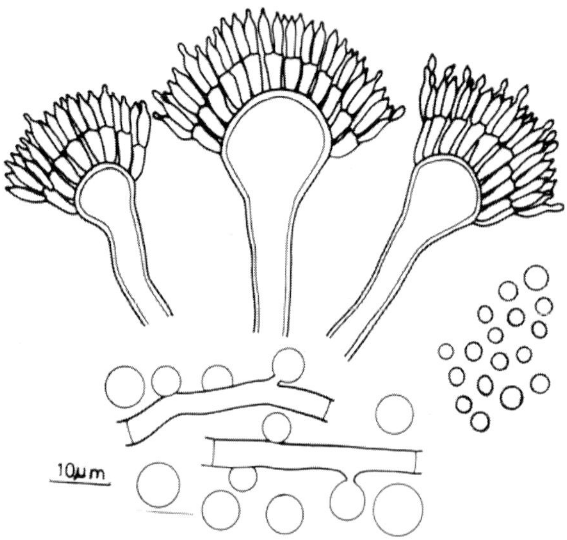

Conidiophores with biseriate phialides, conidia

Teleomorph: Absent

Colony: Spreading, plain or with radial furrows, velvety, floccose.

Colour of the colony: Cinnamon-buff to wood-brown

Reverse of the colour: Dull-brown.

Extrolites: Amber coloured

Vesicles: Long, columnar, compact diam.

Colour: Cinnamon -brown

Vesicles: Hemispherical

Metulae: Crowded, parallel

Phialides: Biseriate, closely packed

Conidia: Globose to subglobose, smooth

Sclerotia: Absent

Cytochalasins: Cytochalasin E, Z16 and Z17

Diagnostic features: Colonies cinnamon-buff to wood-brown, two series phialides, globose to sub globose, smooth conidia.

9. Bipolaris neergaardii Danquah

= *Drechslera neergaardii* Danquah

= *Curvularia neergaardii* Danquah

Colonies: Effuse, pale to mid brown.

Conidiophores: Solitary or in groups of 3-4, straight or flexuous, often geniculate towards the apex, cylindrical, sometimes swollen at the base, pale to mid-brown, up to 150 μ long, 4-8 μ thick.

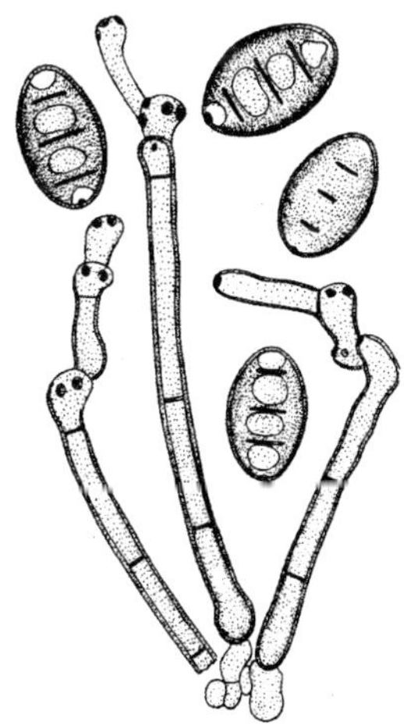

Conidiophores and conidia

Conidia: Straight, broadly ellipsoidal, mid to dark-golden-brown, smooth, 2-3 (mostly 3) pseudoseptate, 19-30 (20-26) x 13.5-20 (15-18) μ thick in the broadest part, cell at each end small, hilum inconspicuous.

Cytochalasins: Cytochalasin C and D

Diagnostic features: Conidiophores solitary straight or flexuous, often geniculate towards the apex, **conidia** straight, broadly ellipsoidal, mid to dark golden brown, 2-3 pseudoseptate.

10. ***Beaveria bassiana*** (Bals. Criv.) Vuvill

Synonyms = *Botrytis bassiana* Bals. – Criv
= *Stradylidium bassianum* Bals. – Criv (Mont)
= *Spicarai bassiana* (Bals. – Criv) Vuvill
= *Sporotrichum globuliferum* Speq
= *Beaveria globulifera* (Speq) Picafrd
= *Botrytis acrediorum* (Brongn & Delacr
= *Spicaria acrediorum* (Brongn & Delacr) Vuvill
= *Botrytis delacroiseic* Sacc
= *Isaria vexans* Peitt
= *Beaveria vexans* Peitt (Petch)
= *Botrytis effuse* Beau
= *B. necans* Mars
= *B. stephanoderris* Balty
= *Beaveria laxa* Petch
= *B. doryphora* Poisson

Colony: White spreading fluffy to powdery parasitic on insect, mycelia strands cylindrical margin, a septate.

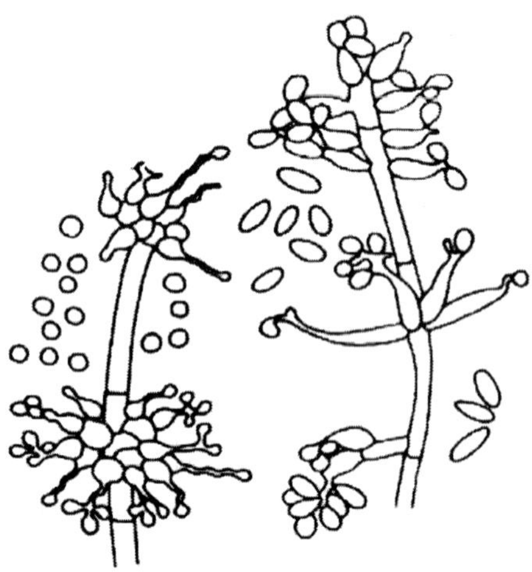

Conidiophores with verticells of phialides and conidia

Conidiophore: Single, irregularly grouped or in verticellate clusters in some species inflated at the base, tapering to a slender spore bearing portion which appears zigzag after spores are produced more or less comparable to cymose with scorpiod, helicoid dichotomy, sense, globose.

Conidia: Hyaline, rounded to ovoid, 1-celled, borne singly on small sterigmata in a several bar fungal spore bearing portion of conidiophore becoming zigzag after bearing several spores.

Cytochalasins: Cytochalasin C

Diagnostic features: Insecticolous, conidiophores in verticellate clusters

11. ***Calonectria morganii*** Crous

Colonies: White, spreading and effuse.

Ascomata: An ostiolate perithecium, solitary to aggregated, superficial, bright colored, turning purple in KOH, ascomata wall composed of two regions, the outer region with angular to globose, thick-walled, pigmented cells, the cells of the inner region elongate, hyaline and thin walled. Centrum containing apical paraphyses.

Asci: Unitunicate, broadly clavate to obovate, undifferentiated at apex.

Ascospores: Elliptic to fusiform, often somewhat curved, 1-several-septate, hyaline.

Cytochalasins: Chaetoglobosin A, B, C, D, F, E and G

Anamorph: ***Cylindrocladium scoparium*** Morgan

Colonies: Reddish-brown from the bottom of the plate with an irregular white margin

Conidiophores: Macronematous, branched, two or more bifurnicate lateral branches to the main stipe. The primary branches give rise to secondary and sometimes tertiary branches with progressively smaller cells. Each branch ending in two or more ovoid to doliform phialides conidiophores ends in swollen apex.

Conidia: Straight, cylindrical or slightly swollen in the upper cell, 1 septate, measuring 50-60 x 4.5-6 mm.

Cytochalasins: Chaetoglobosin A, B, C, D, F, E, G

Diagnostic features: Perithecium globose, bright coloured, asciunitunicate, clavae to obovate, Ascospores 1-several cells, hyaline elliptic to fusiform. Conidiophores branched two to three lines and each branch ends in two or more doliform phialides, conidiophores ends in swollen apex. Conidia cylindrical two celled, microsclerotia are present.

Chlamydospores : Abundant in older cultures, forming in chains or clumps, causing the culture to appear almost black on the reverse of the colony. These chlamydospores are known to as microsclerotia.

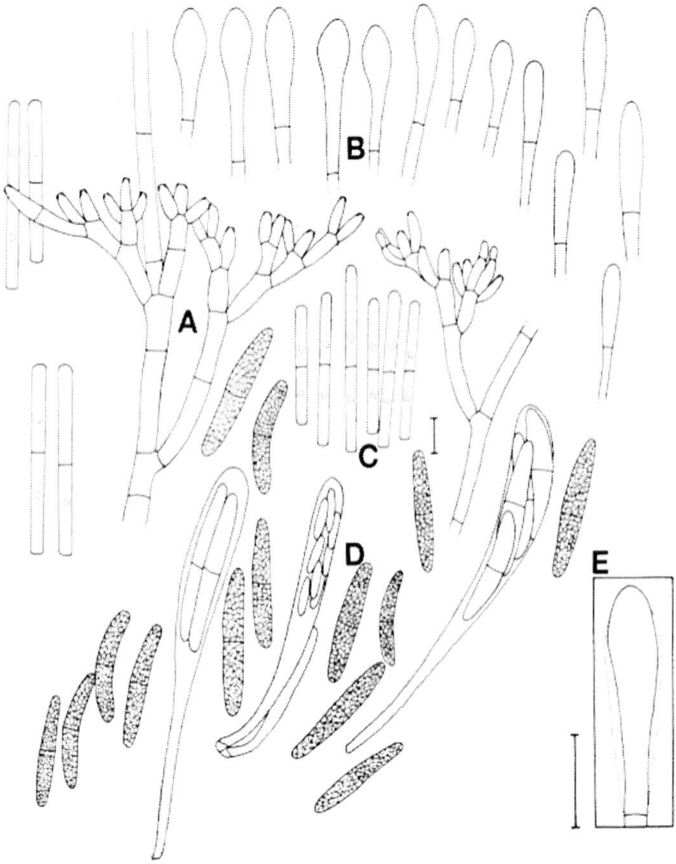

(A) Conidiophores with elongated stipe, (B) cylindrical, (C) cylindrical septate conidia (D) asci and Ascospores (E) swollen apex

12. *Chaetomium brasiliense* Bat. & Pontual

= *Chaetomium hamatum* Bat. & Pontual, Bol. Agr. Com. Pernambuco 15: 70 1948.

≡ *Chaetomium repandum* Bat. & Pontual, Bol. Agr. Com. Pernambuco 15: 70 1948.

≡ *Chaetomium velutinum* L.M. Ames, Mycologia 41: 637 1949.

≡ *Chaetomium alboarenulum* Ames, U. S. Army Res. Develop., ser. 2: 16 1963.

≡ *Chaetomium congoense* Ames, U. S. Army Res. Develop., ser. 2: 18 1963.

≡ *Chaetomium leucophora* Ames, U. S. Army Res. Develop., ser. 2: 28 1963.

≡ *Chaetomium perpulchrum* Ames, U. S. Army Res. Develop., ser. 2: 33 1963

Colonies: With a daily growth rate of 3-4 mm, becoming dark due to pigmented, immersed and superficial hyphae, with a floccose, grey or nearly white aerial mycelium; reverse black.

Ascomata: Maturing within 20-24 days, superficial and surrounded by the aerial mycelium, spherical or nearly so, ostiolate, 80-200µm with a dark wall of angular, 6-10µm cells; ascomatal hairs flexuous, undulate or spirally coiled, septate, dark brown, verrucose or warty, 3-4µm broad.

Asci: Fasciculate, cylindrical, with a short stalk, 8-spored, rather persistent, 45-55 × 6.5-8 µm.

Ascospores: Ovate, bilaterally flattened, dark brown when mature, 7-8.5 × 6-7 × 5-6 µm, with a germ pore at the attenuated end.

Paraphyses: Not observed.

Cytochalasins: Chaetoglobosin A.

Diagnostic features: Colonies becoming dark with age, floccose reverse black. Ascomatal hairs flexuous, undulate or spirally coiled. Asci cylindrical, wall persistent, ascospores are ovate, bilaterally flattened, dark brown.

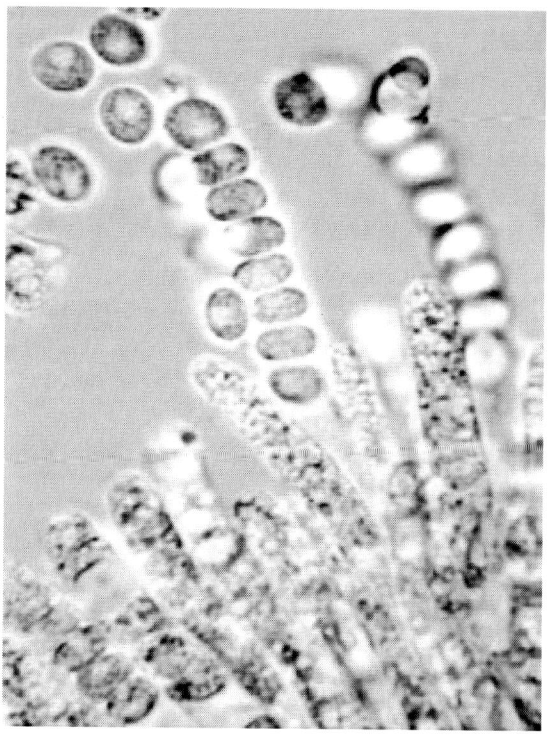

Asci and Ascospores

13. *Chaetomium cochliodes* Pallise

Synonymy= *C. flexmosum* Palliser

Colonies: Broadly spreading, dark olive-green to dark-olive-grey.

Perithecium: Dark-olive-brown to black, sub-globose, ostiolate.

Terminal hairs: Dark-olive-brown at first, straight below and sinuous above at maturity two types.

(a) Straight below spirally coiled above,

(b) Slender varied from irregularly twisted or undulated to loosely coiled rather light-colored, minutely roughened.

Lateral hairs: Dark olive-brown, straight or flexed, gradually interwoven with the terminal hairs, roughened, septate.

Ascus: Club- shaped, 8 spored

Ascospores: Single celled, olive-brown, lemon shaped, sub- apiculate, at both ends. **Cytochalasins:** Chaetoglobosin A, B, C, D, E, F and G

Diagnostic features: Colonies dark- olive brown - olive- brown, dark-olibegrey, dark olive brown terminal hairs two types (1) Straight below and spirally coiled above and slender varied from irregular twisted or (2) undulated to loosely coiled and minutely roughed.

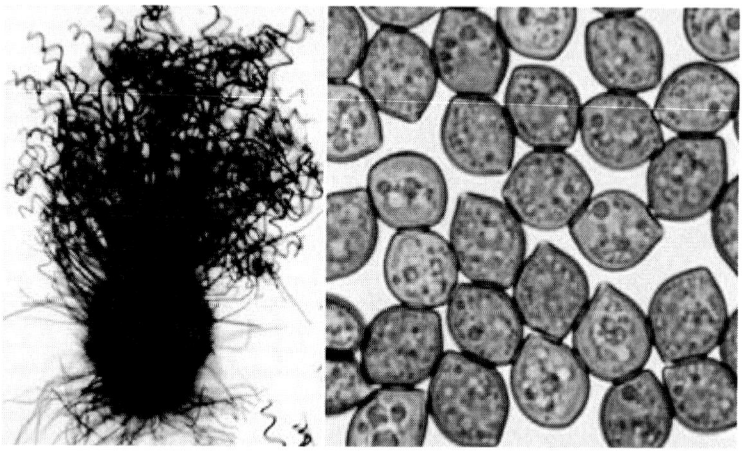

Perithecium with two types of terminal hairs, ascospores

14. *Chaetomium elatum* Kunze ex. Fries

Synonyms = *C. comatum* (Tode) Fr.

= *Psilona cylindrical* (Pers.)

= *C. pannosum* Waller.

= *Sphaeria chaetomium* Lib.

= *Chlorodium ovoideum* Corda

= *C. lageniforme* Corda

= *C. affine* Corda

= *C. cylindricum* (Pers.) Rabenh

= *C. graminis* Rabenh

= *C. graminicolum* Fuckel

= *C. fieberic* Fuckel

= *C. libertii* Roum. et Pat.

= *C. velutinum* Ell. et Ev.

= *C. atrum* var. *distinctum* Roum

= *C. glabrescens* Ell. et Ev.

Colonies: Spreading broadly, dark-greenish-grey to greenish-black, consisting of a matted basal felt with aerial hyphae, white to pale-yellow.

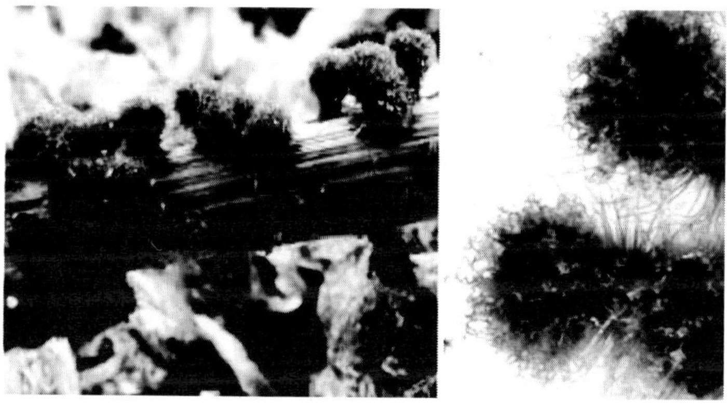

Perithecium with hairs

Perithecium: Black, ovate, ostiolate

(i) **Terminal hairs:** Dark-olive-brown, unbranched hairs obscure, branched hairs straight below, several times dichotomously branched above with branches at right angles and often reflexed, coarsely roughened throughout, stiff with pointed tips, later formed hairs similar but with branches at acute angles, more narrow and sinuous than those of the former

(ii) **Lateral hairs:** Dark olive-brown, straight or flexed, unbranched with rounded tips roughened, obscurely septate

Ascus: Club shaped, 8-spored, very evanescent

Ascospores: Dark-olive-brown, lemon shaped, apiculate at both ends.

Cytochalasins: Chaetoglobosin V

Diagnostic features: Colonies dark –greenish- grey to greenish black perithecia terminal hairs dark-olive-brown, branched at right angles and reflexed, ascus club shaped ascospores dark- olive- brown lemon shaped, apiculate at both ends.

15. Chaetomium globosum Kunze

Synonymy= *C. coarctatum*
C. olivaceum
C. rectum
C. subglobosum

Colonies: Growing well on PSA.

Mycelium: Pale brown, producing perithecia after five days.

Perithecia: Scattered or gregarious, globose to subglobose to broadly oval, olive green to grayish green, 160-390 x 230-300 μm, thickly clothed with hairs,

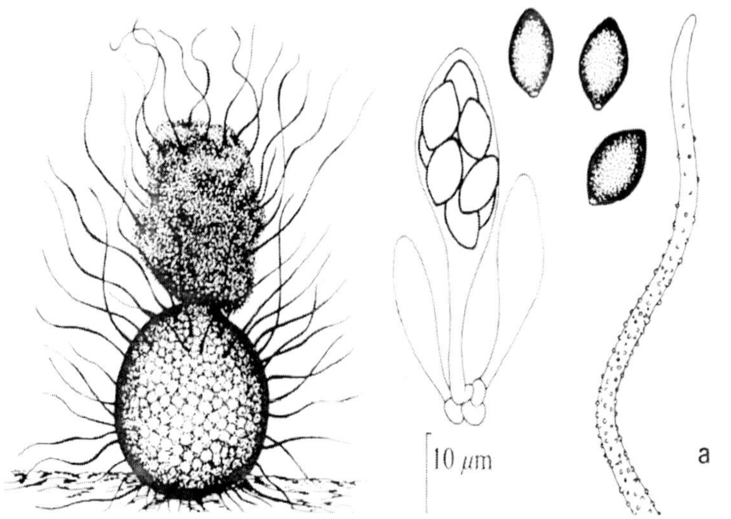

Perithecium with hairs, asci and ascospores

Terminal hairs: Abundant, light coloured, finely roughened, obscurely septate, slender, about 3-4 μm in diam, up to 77 μm long, tip blunt, undulate throughout, forming a dense inter-woven bushy head;

Lateral hairs: Light colored, finely roughened, septate, 2.5-3.5 μm in diam, long, slender, straight to slightly flexed or undulate.

Asci: Oblong, clavate, 8-spored, 28-35 x 53-55.5 μm,

Ascospores: Dark, lemon shaped, broadly ovoid, apiculate at both ends, 8.5-10.5 x 8-9 μm.

Cytochalasins: Chaetoglobosin A, B, C, D, E, F, G, Q, R, T, U, V, W and Y

Diagnostic features: Colonies varingly colored, olive- grey, olive-green, olive-black, olive- brown, flexuous becoming sinulate towards the ends minutely rounded terminal hairs, asci club shaped, ascospores olive -brown, lemon shaped, sub-apiculate at both ends.

16. Chaetomium mollicellum Ames

= *Ovatospora mollicella*

= *Chaetomium apiculatum*

Colonies: Spreading light-brown, producing perithecia after 5-6 days.

Perithecia: Dark gray, gregarious, translucent, globose or ovate, 64-160 x 54-130 µm, fixed to the substratum by delicate rhizoids, rounded at base, densely covered with hairs,

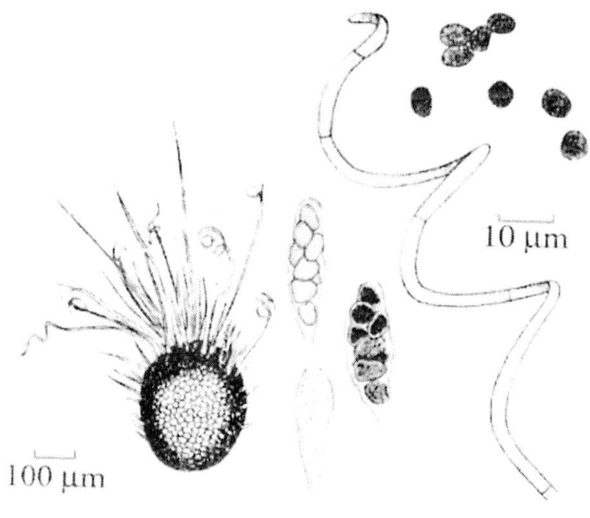

Perithecium with hairs, asci, coiled hair and ascospores

Terminal hairs: of 2 types, one type straight at the base forming loose spirals at the apex like corkscrew, 4-5 µm wide at the base, occasionally branched, dark brown, septate, with dark thorny granules, second type long, straight to undulate, brown, septate, finely roughened, 3-4 µm at the base,

Lateral hairs: Few, long, pale-brown, septate, tip finely pointed, hyaline, straight or undulating,

Asci: Long, cylindrical, 8-spored, pedicellate,

Ascospores: Brown when mature, ellipsoid to almond shaped, apiculate at one and rounded at the other end, 5-9-8.5 x 4.2-8.5 µm.

Cytochalasins: Chaetoglobosin A, B, C, D, E, F and G

Diagnostic features: Perithecia globose or ovate with crowded hair at the apex, hairs of 2-types straight hairs, terminal hairs long pale-brown finely pointed straight or undulating, asci cylindrical, ascospores ellipsoid to almond shaped, apiculate brown at the maturity.

17. Chaetomium subaffine Sergeeva

Colonies: Heterothallic, with a daily growth rate of 7.5-9 mm, without coloured exudates, with a grey or pale-brown aerial mycelium. Conidiophores comprising individual conidiogenous cells.

Conidiogenous cells: Lateral on aerial hyphae, discrete, ampulliform.

Conidia: Formed in basipetal succession, obovate or spherical, aseptate, colourless or pale, 2-3 µm.

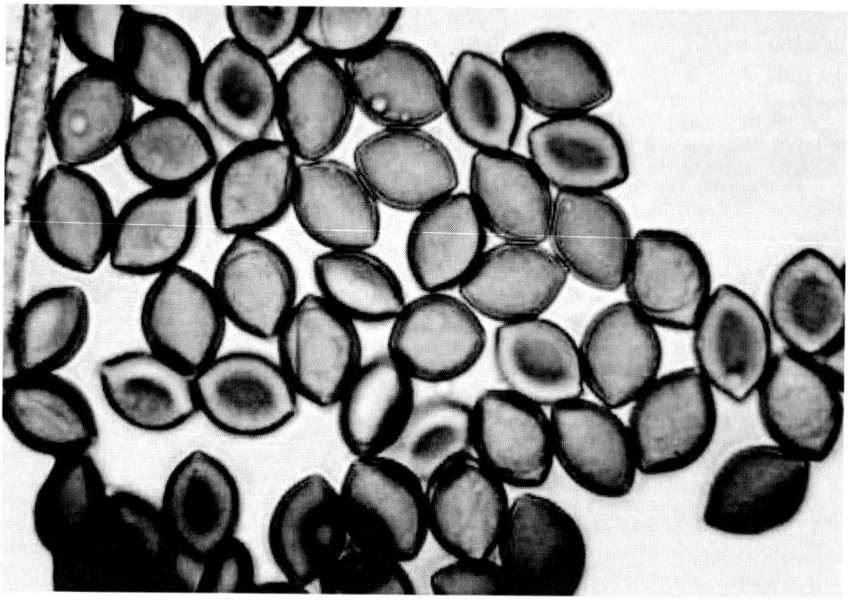

Ascospores

Ascomata: Maturing within 14 days, dispersed, pale-olivaceous in reflected light, superficial, obovate or ovate, ostiolate, 220-380 µm, with a brown wall of textura intricata or epidermoidea; ascomatal hairs numerous, flexuous or undulate, unbranched, long, verrucose or nearly smooth, brown; indistinctly septate, 3-4 µm broad near the base.

Asci: Fasciculate, clavate or fusiform, stalked, 8-spored, evanescent, 36-50 × 9-13 µm.

Ascospores: Limoniform, biapiculate, bilaterally flattened, brown and rather thick walled when mature, containing numerous droplets, 11-15 × 8-11 × 7-8.5 μm, with an apical germ pore.

Paraphyses: Not observed.

Cytochalasins: Chaetoglobosin A, B, C, D, E, F and G

Diagnostic features: Colonies spreading grey or pale-brown, perithecia superficial, ovate or obovate. Asci clavate or fusiform with 8-ascospores evanescent ascospores limoniform.

18. *Chalara* (Corda) Robenhorst

Colony: White, spreading

Conidiophores: Short, micronematous, manonematous cylindrical cell with an uppermost cell giving rise to an elongate conidiogenous cell.

Conidiogenous cells: Fragmenting to form arthroconidia sometimes phialides

Conidia: Terminal or intercalary, determinate, catenate, cylindrical rarely with a hyaline longitudinal slit, sometimes slightly inflated in the lower half, long and tubular in the upper half or semi endogenous, cylindrical, truncate at both ends, hyaline or pigmented, variable in width and length, produced in long fragile chains.

Teleomorph: *Ceratocystis fimbriata* Ellis and Halst

Colony: White turning to black with the production of ascomata

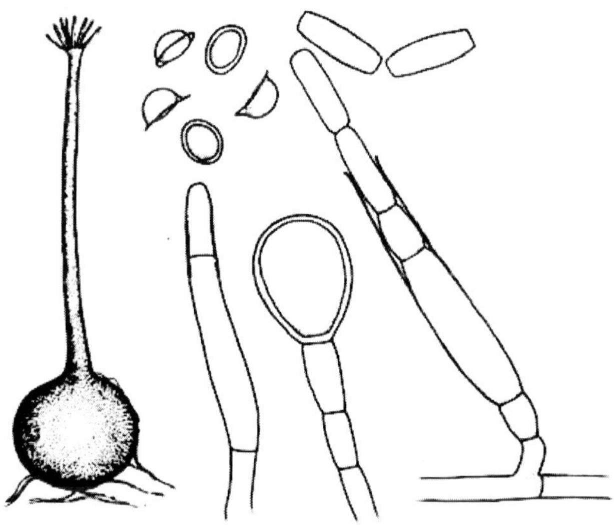

(a) Mature ascoma with ostiolar neck, (b) Scattered globose asci, (c) Ascospores, (d) Conidia and conidiophores

Ascomata: Single to aggregated, superficial or immersed in substratum, ostiolate or non-ostiolate, globose to sub-globoase, hyaline to dark brown, ostiolar neck is long and cylindrical

Ascus: Globose to sub-globose, scalled 8-spored and evanescent

Ascospores: Hyaline, 1-celled with or without gelatinous sheath, variable in shape, curved, lunate, remiform or hat shaped

Cytochalasins: Cytochalasin K, L and M.

Diagnostic features: Globose to sub-globose perithecium with long neck ascus with single celled ascospore, remiform or hat shaped

19. ***Cochliobolus tuberculatus*** Sivan

Colonies: Blackish-brown

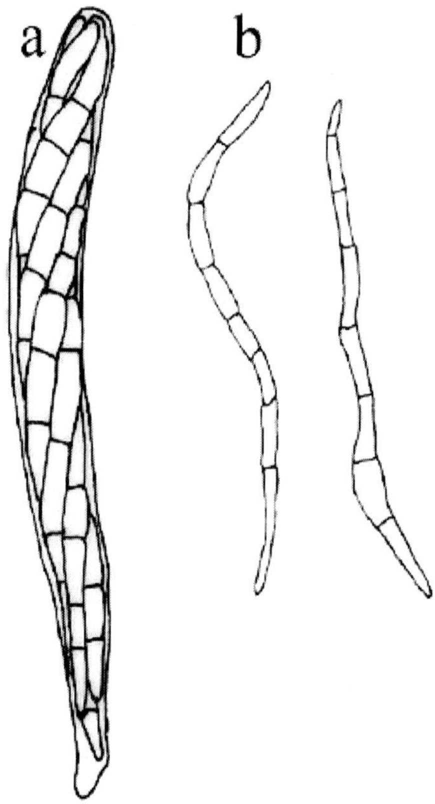

(a) Ascus with spirally coiled ascospores (b) ascospores

Stromata: Simple or branched,

Psudothecia : Black, globose, usually forming on a columnar basal stroma or a flattened crust, 500-720μm high, 400-490 μm wide, beaked with a conical truncate beak up to 300 µm high, 115-140 µm wide at the base, often hairy in the globose part, hairs septate, simple, brown;

Asci : Cylindrical, short-stalked, bitunicate. 2-8 spored.

Ascospores: Filiform, hyaline, helically coiled in the ascus and straightening at one or both ends, more at the base, sometimes with truncate base, mucilaginous, sheath up to 4 µm thick, 13-23 septate, 160-460 x 3-4.5 µm.

Pseudoparhysis: Cellular, hyaline, filiform, branched above.

Anamorph = *Curvularia tuberculata* Jain

Conidiophores: Macro-or mononematous, unbranched, terminal, often geniculate above, sympodial, cylindrical, cicatrized.

Conidia: Acropleurogenous, straight, ovoid, obclavate, 3-5(-8) mostly 3-septate, middle cells darker, end cells sub hyaline to pale or dark brown, mature conidia tuberculate, 23-52 x 13-20 µm, young conidia sub hyaline and smooth walled.

Cytochalasins: Cytochalasin E

Diagnostic features: Pseudothecia black, globose, asci short stalked, bitunicate, Ascospores filiform, septate and compacting coiled in ascus Curvularia is the anamorph. Conidiophores unbranched, geniculate, conidia acropleurogenous, straight, ovoid to obclavate, 3-septate and tuberculate.

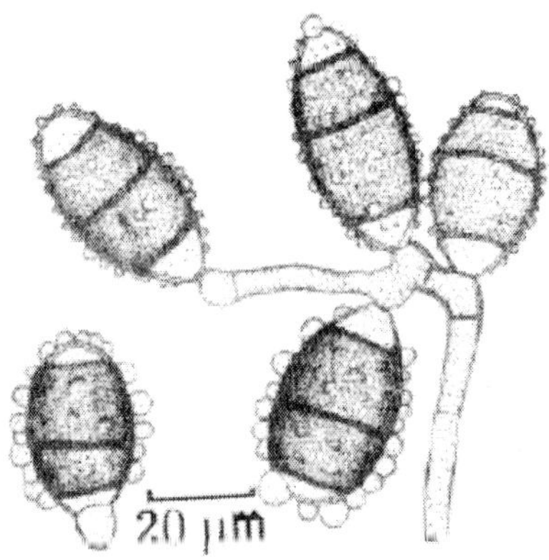

Conidiophores with tuberculata conidia

20. *Cordyceps taii* Liang and Liu

Colonies: Flat with little aerial mycelium, dull grey-green reverse centre deep-green.

Ascomata: Stromatic emerge free the head of insect host, cylindric, with yellow fertile head.

Perithecia: Flask shaped with curved neck, obliquely immersed.

Ascus: Cylindrical with a ascus cap, ascospores-8 after fragmented into 1-celled structure while in ascus itself.

Ascospores: Cylindrical, 20.0-(28.6)-34.6x1.4-(2.5)-3.1μm

Anamorph: *Metarrhizium*

Cytochalasins: Deacetylcytochalasin C and Zygosporin D

Diagnostic features: Ascomata stromatic, cylindric, perithecia flask shaped, ascus cylindrical with a cap, ascospores get fragmented to 8 ascospores, ascospores cylindrical. *Metarrhizium* is an Anamorph.

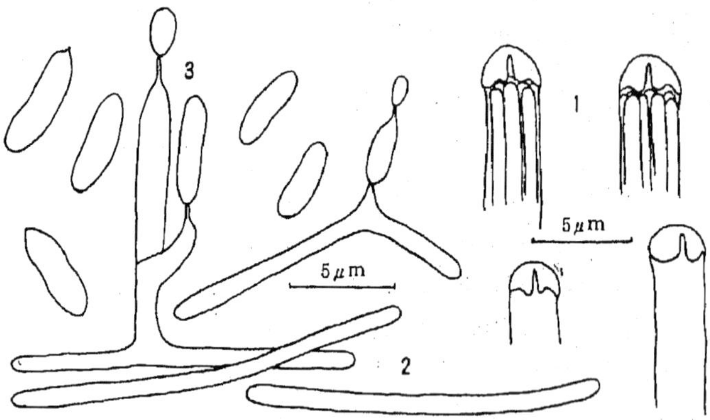

Conidiophore and conidia, asci with apical apparatus

21. ***Coriolus vernicipes*** Berk

= *Microporus vernicipes* Berk

Basidiomata: Annual, solitary or in smaller groups, often several pilei grow together, laterally stipitate, pileus.

Pileus: Light-brown to chestnut or bay, glabrous, smooth to strongly veined radially or even slightly undulated in old specimens, narrowly concentrically zoned, rather indistinct in some specimens, usually lighter towards the margin which is thin, even to lobed or incised, from the base there is a distinct light adpressed tomentose, ochraceous layer spreading towards the margin, rarely extending more than 1 cm from the base in irregular tongues, this layer extends further down the upper side of the stipe

Stipe: Short and lateral, rarely above 3 cm long, 2-8 mm in diameter, often almost lacking, appearing more like a contracted base between the pileus and the substrate, on the upper side covered with a white pad of mycelium spreading from the pileus, cream to ochraceous in colour, on the lower side either covered with decurrent pores or with a light-yellowish smooth and glabrous cuticle, distinctly delimited towards the upper pad of mycelium, at the base expanded into a broad, smooth disc, up to 2 m in diameter.

Pore Surface: Cream to light-brownish, often discoloured by dark spots, lighter towards the margin, sterile margin if present, very narrow and white, pore round and entire, 6-7 per mm, tubes concolorous with the pore surface or lighter, up to 1 mm deep.

Cortext: Both in stipe and pileus pure white, up to 1 mm thick in the pileus, with a distinct dark cuticle which also is present under' the pad of adpressed mycelium.

Hyphal System: Trimitic, generative hyphae with clamps, 2-3 μm in diameter, skeletal hyphae dominating in the fruit body, thick-walled to solid, 3-7 μm in diameter, binding hyphae few, tortuous and tapering towards the outer parts, 1-4 μm in diameter, Coralloid dichophytic elements present along the dissepiments.

Spores: Cylindrical, hyaline, smooth and non-amyloid and non-cyanophilous.

Cytochalasins: Cytochalasin D

Diagnostic features: Basidiomata laterally stipitate in smaller groups, pileus light-brown to chestnut-brown with pore surface cream to light-brown. Pore round and entire, hyphal trimitic, spores cylindrical hyaline, smooth and non-amyloid and non-cyanophilous.

Fruit bodies on dead wood logs

22. ***Curvularia lunata*** (Wakker) Boedin

Synonyms = *Curvularia aeria* (Bat. Lime & Vascone) Tsuda

= *Mallestela aeria* Bal-lima & Vasume

Teleomorph: *Cochliobolus lunatus* Nelson & Haasis

Colonies: Dar-grey, usually-zonate

Stromata: Regularly and abundantly formed in culture

Mycelium: Branched, septate

Conidiophores: Macronematous, mononematous, straight geniculate, smooth, brown, septate.

Conidia: Solitary, acropleurogenous, 4 celled, middle cells darker, curved pseudoseptate.

Cytochalasins: Cytochalasin B

Diagnostic features: Acrogenous, porospore, 4 celled, 2 middle cells darker, curved, with protuberant hilum. *Cochliobolus* is the teleomorph.

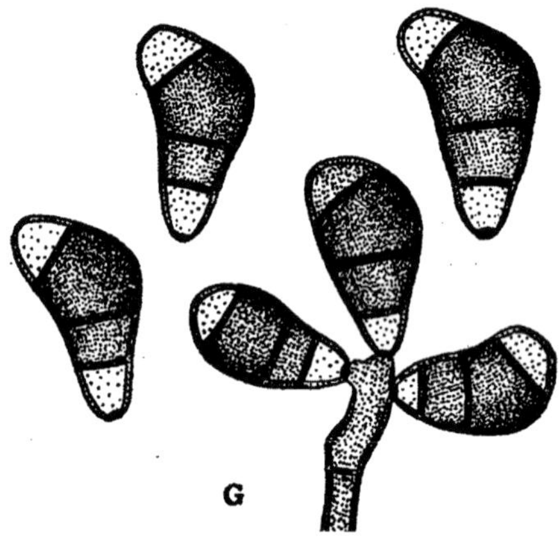

G

Conidiophore with conidia

23. ***Cylindrocladium floridanum*** Sobers

Colonies: Irregular, margins, thc aerial mycelium white and effuse do not changed with age and the microsclerotia were formed in large, non-zonate masses.

Conidiophores: With lateral stipes arising from secondary and tertiary conidiophores, globose to subglobose, obovate, apex

Phialides: Doliform 7-18 x 3.1-6.0 μ.

Conidia: Cylindrical, erect, small size, mostly 1- septate

Lateral stipes: Commonly present

Cytochalasins: Chaetoglobosin O

Diagnostic features: Colonies white and effuse large number of microsclerotia, conidiophores branches with obovate or globose apex, phialides doliform, conidia cylindrical small size with one septa.

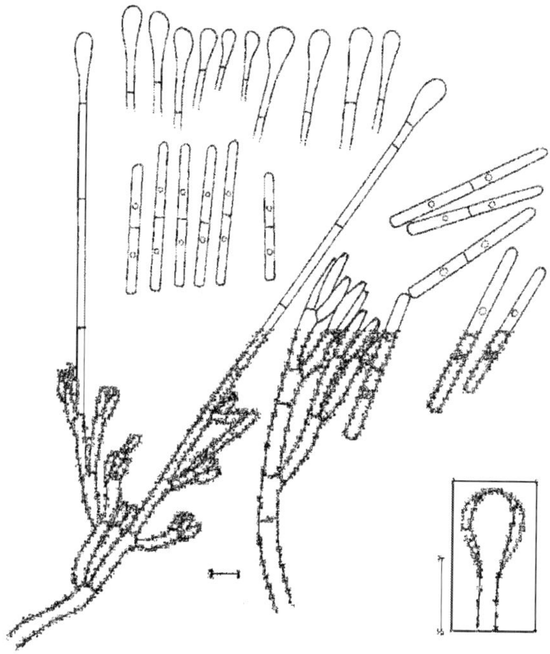

(a) Conidiophores with elongated stipe (b) septate cylindrical conidia apex of style

24. ***Daldinia vernicosa*** (Schw.) Ces. & De Not.

= *Sphaeria vernicosa* Schwein

= *Hypoxylon vernicosum* Ces. & De Not.

Stromata: Pulvinate, subglobose, subturbinate to turbinate, frequently with a distinct stipe, stromata usually single, occasionally confluent, 6'0-30 x 8-20 x 4-12 mm, erumpent from bark or superficial on decorticated wood; ectostroma thin and carbonaceous, at first reddish-brown becoming dark-brown to black and highly polished; entostroma fibrous, becoming brittle, conspicuously zonate with lighter grey and white zones showing collapsing with subsequent formation of cavities.

Ascomata: Perithecia small, monostichous or polystichous, subglobose or oblong-ovate when compressed, 0.5-1.0 x 0.2-0.7 mm; ostiole distinct, often punctuate.

Asci : Cylindric, with long stipe, 8-spored.

Ascospores: Uniseriate, inequilaterally elliptic, dark-brown, 7-12 (15.5) x 4.5-7 μm.

Anamorph: ***Nodulisporium*** spp. Preuss

Colonies: Effuse, grey, brown, olivaceous or blackish brown, often velvety. Mycelium partly immersed, partly superficial.

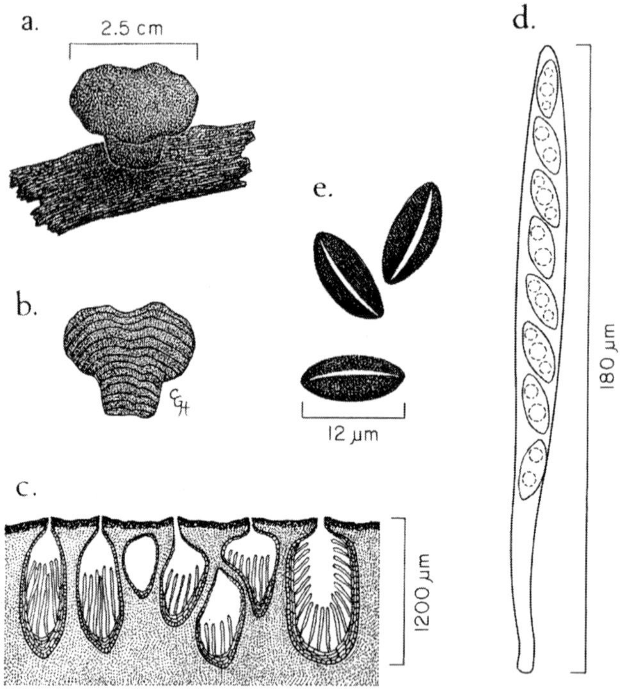

(a) Perithecium in natural habitat (b) striate perithecium (c) cross section of perithecia (d) Ascospores (e) ascus with ascospores

Stroma: None seen.

Setae and hyphopodia: Absent.

Conidiophores: Macronematous, mononematous or synnematous, individual threads often much branched especially towards the apex, flexuous, pale to dark brown or olivaceous brown, smooth or verrucose.

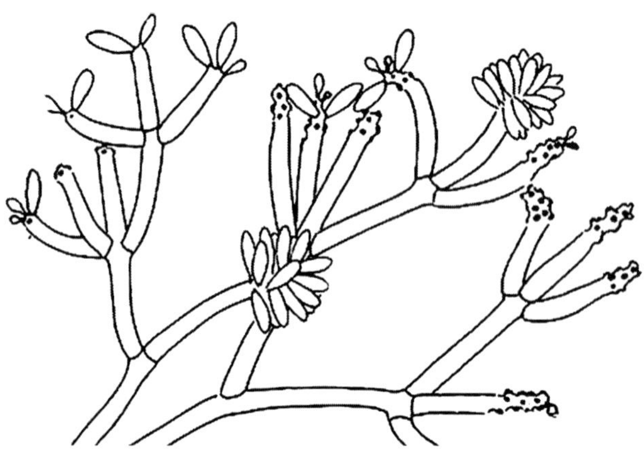

Conidiophores and conidia

Conidiogenous cells: Polyblastic, integrated and terminal becoming intercalary, or discrete, solitary or arranged penicillately, sympodial, cylindrical to clavate, denticulate; denticles generally very short, fragile.

Conidia : Solitary, dry, acropleurogenous, simple, ellipsoidal or obovoid, truncate often with a small frill at the base when detached, hyaline to brown or olivaceous brown, smooth or verrucose, 0-septate

Cytochalasin: Cytochalasin I

Diagnostic features: Ascomata stromatic, dark-brown to black, perithecia small, monostichous or polystichous, asci cylindric, with long stipe, inequilaterally elliptic, dark-brown, one celled. *Nodulisporium is the Anamorph.*

25. *Diplodia macrospora* Sacc

Colony: Thick reverse, cottony, aerial mycelium. Reverse of the colony becomes yellow brown

Pycnidia: Dark-brown to black, globose produced on surface later gets immersed

Conidia: 1-3 septate, cylindrical to elongate

Cytochalasins: Chaetoglobosin K, L and M

Diagnostic features: Produce cylindrical conidia with 1-3 septa.

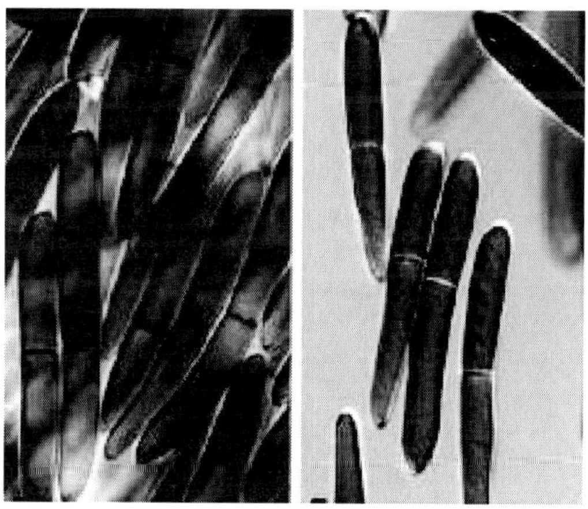

2-3 celled conidia

26. *Discosia* Lib

Mycelium: Immersed, branched, septate, pale-brown.

Conidiomata: Eustromatic, amphigenous, separate or aggregated and confluent, flattened, black, epidermal, unilocular or multilocular, locules separated by textura porrecta; wall very thick at the base, pale-brown, thin-walled textura angularis, upper wall thinner, dark-brown, thick-walled, textura angularis in which lumina are occluded. Ostioles l-many to each conidioma, circular, papillate.

Conidiophores: Absent.

Conidiogenous cells: Holoblastic, each forming a single conidium, discrete, determinate, conical, hyaline to pale-brown, smooth, restricted to the basal conidiomatal wall.

Conidia: Straight or slightly curved, dorsiventral, 3-4 euseptate, smooth, tapered to a truncate base, apex obtuse, hyaline to pale-brown, cells of varying length; apical and basal cells each with a subapical and suprabasal unbranched, filiform setula formed on the concave side of the conidium.

Cytochalasins: Chaetoglobosin P

Diagnostic features: Conidiomata: eustromatic, black, unilocular or multilocular, l-many Ostioles, Conidiogenous cells holoblastic, forming a single conidium, restricted to the basal conidiomatal walls, Conidia3-4 euseptate, , hyaline to pale-brown, apical and basal cells each with a subapical and suprabasal unbranched, filiform setula.

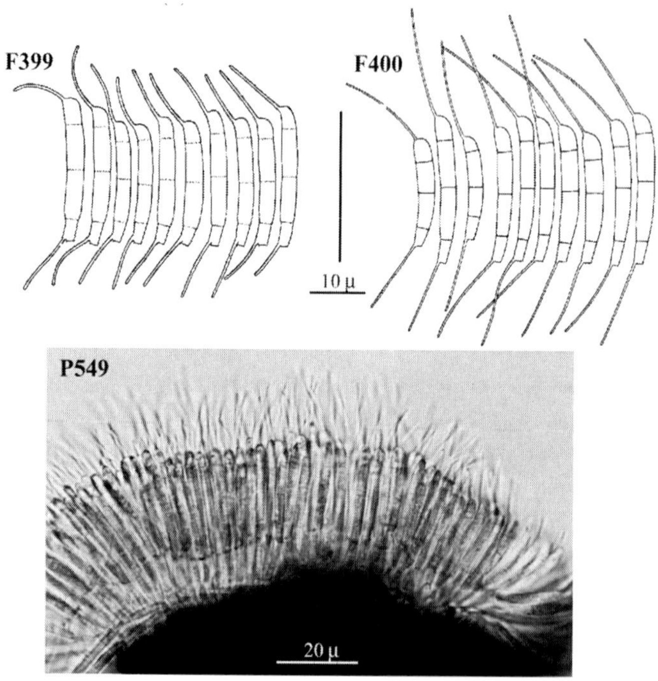

Stromatic pycnidium with conidia and conidia with filiform setula

27. ***Drechslera biseptata*** Sacc. & Roum

= *Helminthosporium biseptatum*

= *H. biforme*

Colonies: Colonies growing rapidly, forming uniform dark grey to black, low, dense mycelia; margin light grey, forming numerous conidiophores and conidia.

Conidiophores: Arising either solitarily from hyphae or in large fascicles from very dark-brown pulvinate stromata. Conidiophores are of two kinds, flexuous or geniculate, pale to mid brown, thin-walled ones to about 80μ long and 3-8μ thick, and subulate, straight or flexuous, dark-brown, thick-walled ones up to 800μ long, 10-14μ thick at the base tapering to 6-8μ at the rather paler apex.

Conidia: Straight, typically obovoid or broadly clavate, occasionally ellipsoidal, pale to mid brown, smooth or verrucose, with almost always 2-3 pseudosepta, 20-30μ long, 11-19μ thick in the broadest part, 23-33 x 14-17μ.

Cytochalasins: Cytochalasins A

Diagnostic features: Conidiophores solitary or originating from very dark-brown pulvinate stroma, two types, obovoid or broadly clavate, pale to mid-brown, 2-3 septate conidia.

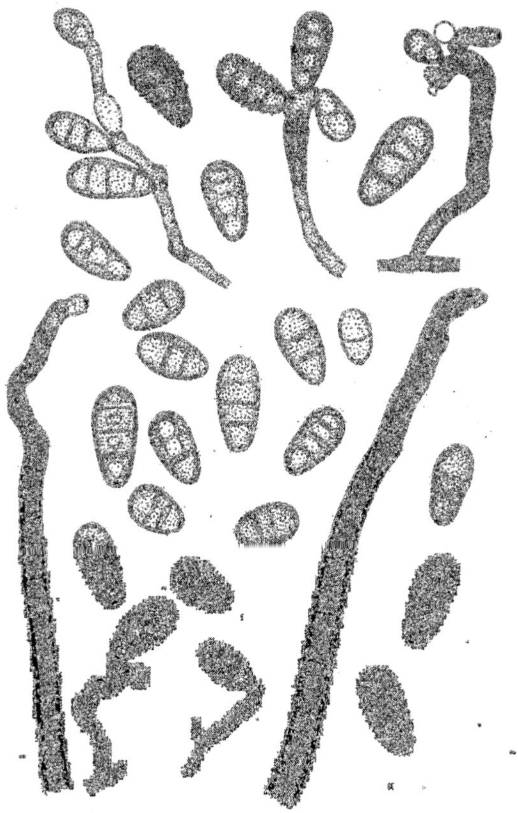

Conidiophores with conidia

28. ***Drechslera dematioideum*** (Bubak & Wrobl.) Subram & Jain (1966)

Synonyms: *Helminthosporium dematioideum* Bubak & Wroblewski.

Colonies: Olive-brown to black

Conidiophores: Macronematous, mononematous, arising single or in small groups, erect or flexuous, geniculate with several conidial scars, brown.

Conidia: Straight, cylindrical to clavate, rounded at the ends, golden-brown to dark-brown, smooth, thick walled 2-7 pseudosepta, distinct hilum.

Cytochalasins: **C**ytochalasins A, B, D & F

Diagnostic features: Conidia straight, cylindrical to elevate rounded at the edges, golden–brown to dark-brown, smooth, thick walled, 2-7 pseudoseptate.

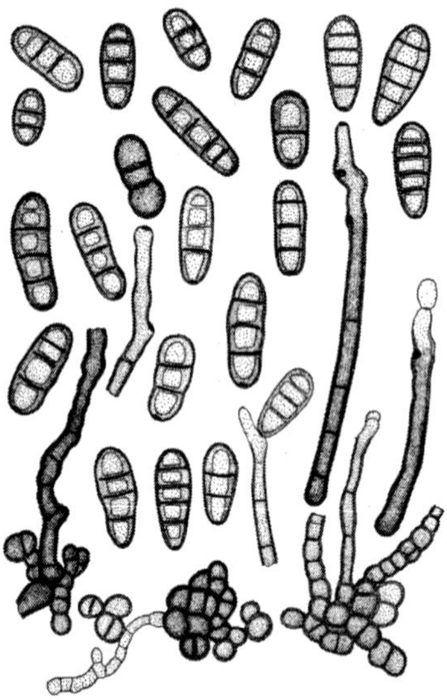

Conidiophores with conidia

29. ***Drechslera erythrospila*** Shoemaker

= *Helminthosporium erythrospilum* Drechsler

Conidiophores: Conidiophores emerging usually singly or in pairs, cylindrical, sometimes repeatedly geniculate near the apex, mid to dark-brown, up to 340μ long, usually 6-8μ thick, swollen at the base to 10-15μ.

Conidia: Conidia typically straight, occasionally curved, cylindrical or almost cylindrical rounded at the ends, yellowish to mid-pale olivaceous brown, with 2-10, most frequently 4-6 pseudosepta, occasionally up to 100 x 16μ but most commonly 40-70 x 11-13μ, the basal ell is often slightly longer than the others and may be cut off by a dark transverse septum.

Cytochalasins: Cytochalasin A and B

Diagnostic features: Conidiophore straight branched at the apex and becomes geniculate conidia cylindrical, 2-10 septate, basal cells is longer than other cells.

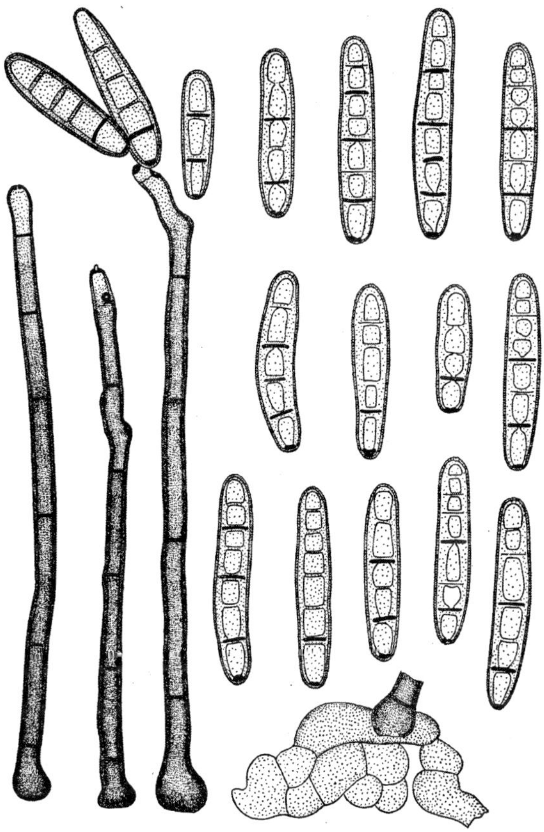

Conidiophores with conidia

30. *Endothia gyrosa* Fr.

Colonies: Asoma an ostiolate perithecium, immersed in a diatrypoid, pseudoparenchymatous stroma.

Stroma: Immersed in host tissue, brightly colored, erumpent through host periderm at maturity.

Ascomata: Perithecia in one or several layers, subglobose, upright, with ostiolar necks erumpent separately. Ascomal wall dark brown, outer cells flattened and thick-walled, inner cells hyaline and thin-walled.

Asci: Unitunicate, clavate to ellipsoid, with nonamyloid refractive ring in apex, often with deliquescent base, and lying free in perithecium at maturity, 8-spored.

Ascospores: Hyaline, 1-celled, allantoids to cylindrical and usually with a slight curve, nonseptate, 7.5-11.5 x 1.0-1.8µ m.

Anamorph = *Endothiella* Sacc

Conidiophores: Cylindrical or slightly tapering towards the apex, give rise to abundant rod-shaped, hyaline, thin-walled.

Conidia: 4-4.5 x 1.5-2µm.

Cytochalasins: Cytochalasin H, J and Z_{10}- Z_{11}

Diagnostic features: Stroma immersed in substratum, ascomata is an ostiolate perithecium, asci unitunicate, clavate to ellipsoid, deliquescent base, ascospores hyaline, 1-celled, allantoids to cylindrical. *Endothiella* is anamorph.

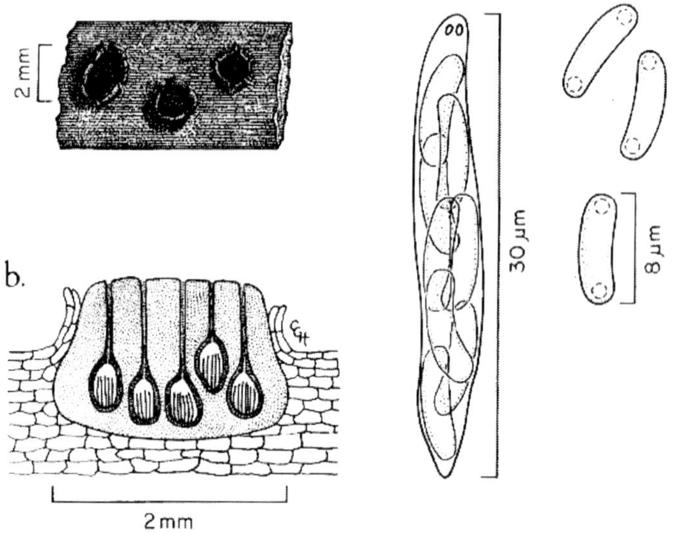

(a) Perithecia in natural habitat (b) cross section of perithecia (c) ascus with Ascospores

31. *Eutypella* (Nitschke) Sacc.

Ascomata: an ostiolate perithecium, immersed in a stroma composed of both host and fungal tissue; stroma valsoid, circular to elliptical, delimited in host by a blackened zone, with a black, erumpent ostiolar disc.

Perithecia: polystichous in stroma, with elongate ostiolar necks that converge; ostioles indistinctly 3-5-sulcate.

Asci : unitunicate, clavate to spindle-shaped, long-stipitate, with nonamyloid apical rings, 8-spored.

Ascospores: 1-celled, allantoid, subhyaline to subolivaceous.

Cytochalasins: Scoparasin A, B and C.

Diagnostic futures: Ostiolate perithecium, immersed in a stroma composed of both host and fungal tissue; stroma valsoid, circular to elliptical, delimited in host by a blackened zonate, perithecia polystichous in stroma neck, ostioles indistinctly 3-5-sulcate, asci unitunicate, clavate to spindle-shaped, long-stipitate, ascospores 1-celled, allantoid, subhyaline to subolivaceous

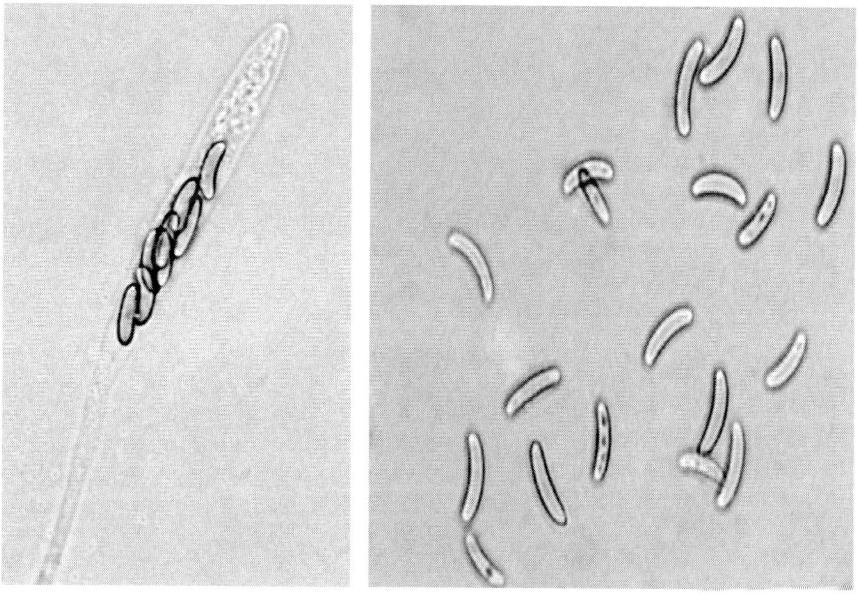

(a) Ascus **(b) Ascospores**

32. ***Geniculosporium*** Chesters & Greenhalgh

Colonies: Effuse, grey, brown or dark-blackish-brown, hairy or velvety.

Mycelium: Partly superficial, partly immersed.

Stroma: None.

Setae: Hypopodia absent.

Conidiophores: Macronematous, mononematous, flexuous, irregularly or sometimes dichotomously or trichotomously branched, very pale to mid brown, smooth.

Conidiogenous cells: Polyblastic, integrated and terminal on branches or discrete, sympodial, geniculate, denticulate with very short, thin walled separating cells which break across the middle leaving a minute frill or collar at each geniculation which corresponds to a frill at the base of each conidium.

Conidia: Solitary, dry, acropleurogenous, simple, ellipsoidal or obovoid, colourless to pale-brown or olivaceous brown, smooth, 0-septate.

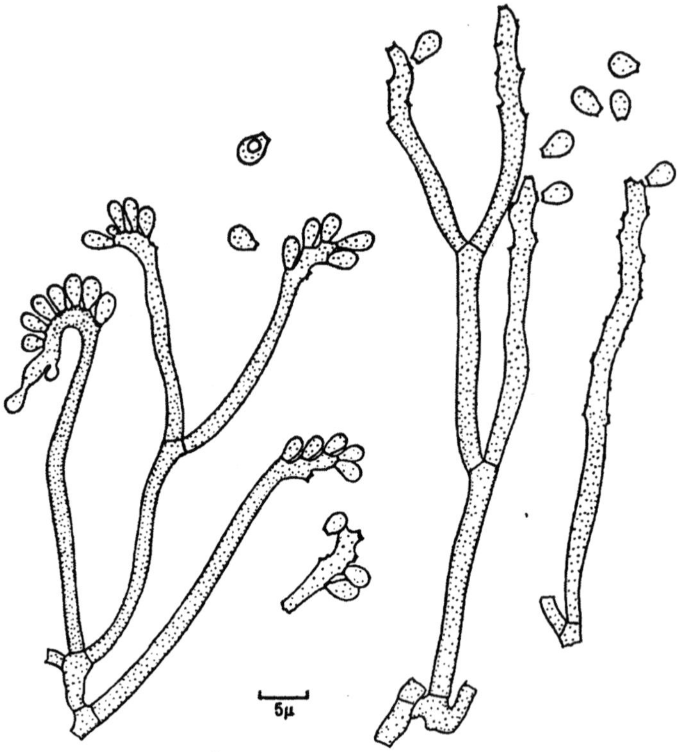

Conidiophores with conidia

Teleomorph = *Hypoxylon serpens* Pers

Stromata: Superficial or erumpent through bark, discrete to effuse, often forming extensive patches 1-1,5 mm thick, surface whitish-grey at first, becoming bronze and finally dark purplish-brown to black when mature, smooth but with conspicuous projecting perithecia, matt to slightly nitid, carbonaceous.

Ascomata: Perithecia projecting from the stroma, usually numerous, globose to ovate, 0,5-1 mm diam., rarely to 1,5 mm high, ostioles papillate.

Asci : Sub-cylindrical to cylindrical, stalked, 8-spored, Paraphyses numerous but gelatinizing at maturity, simple, filiform, sparsely septate, 2-4 μm wide.

Ascospores : 8-spores, uniseriate, inequilaterally ellipsoid, brown to dark-brown with a longitudinal furrow which may be indistinct.

Cytochalasins: Cytochalasin H and U

Diagnostic features: Colonies hairy or velvety, brown or dark-blackish-brown, Conidiophores: macronematous, mononematous, irregularly branched, conidiogenous cell polyblastic, denticulate, conidia acropleurogenous, ellipsoidal, to pale-brown, 0-spetate. Teleomorph is *Hypoxylon.*

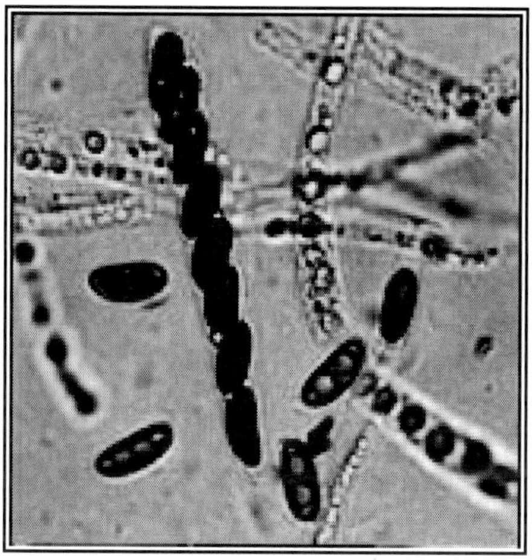

Ascus with ascospores

33.*Gnomonia* sp. Ces & De Not

Ascomata: Ascostromata ostiolate perithecium, non-stromatic, brown, immersed in host tissue, globose to subglobose, usually with long ostiolar neck, ascomal wall composed of several layers of dark, compressed cells.

Asci: Numerous, unitunicate, ellipsoid, with refractive, nonamyloid, apical ring, with deliquescent bases, lying free in perithecium at maturity, 2-many-spored.

Ascospores: Hyaline, occasionally brownish in age, 2-celled, ellipsoid, fusoid, or cylindrical, septum median, or sub or supra-median, sometimes developing additional septa with age, biguttulate, smooth, sometimes with appendages.

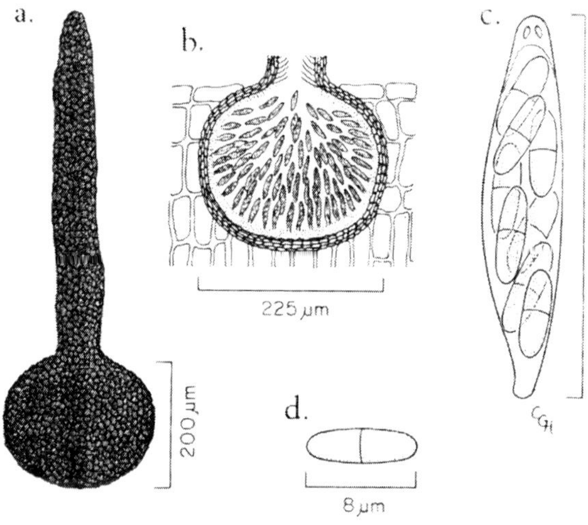

(a) Ascomata with long neck (b) cross section of perithecium, (c) ascus with Ascospores (d) 1-septate ascospores

Anamorphs = *Cylindrosporella* Hohn

Colonies: Acervuli produced in grayish-brown to brown, circular, or irregular shaped lesions on both leaf surfaces, particularly on upper side, circularly gregarious, subcuticular, flat, dark brown, 55–200 μm in diameter, composed of pale yellowish-brown textura angularis.

Conidiogenous cells: Densely crowded at the base of acervuli, phialidic, ampulliform, hyaline, 4–6 x 2–4 μm.

Conidia: Hyaline, thin-walled, filiform, or fusiform, often curved, 7–12 x 1–1.5μm.

Cytochalasins: Cytochalasins A and B

Diagnostic features: Non-stromatic, ostiolate perithecium, many layered wall, asci lying free in perithecium at maturity, with many ascospores, ascospores ellipsoid to fusoid with single septa, biguttulate, smooth, sometimes with appendages. *Cylindrosporella* is the anamorphs.

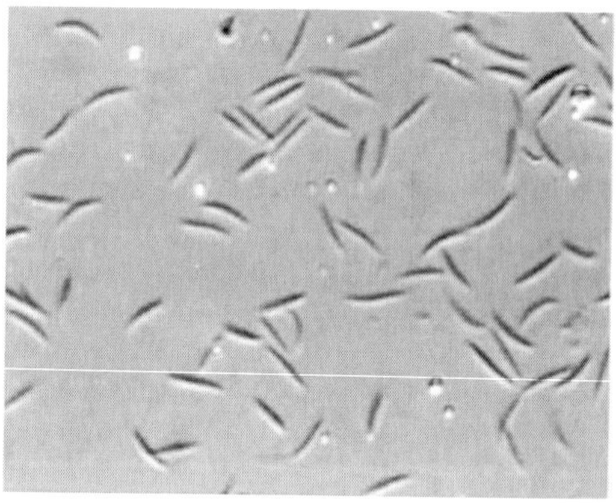

Conidia

34. *Halorosellinia oceanica* Schatz.

Colonies: On oat meal agar covering Petri dish in 4 weak, at first white, velvety, appressed, irregularly zonate, with dense, entire margins, then darkening on the zonation lines. Reverse uncoloured. Stroma-like pustules produced on the zonation lines, pulvinate, up to 1 mm diam. 0±8 mm high, pinkish at first, blackened from bases upwards.

Ascomata: Immersed in pseudostroma, subglobose to hemispherical, soft to leathery, black, ostioles papillate Peridium 25–35 μm wide, comprising tissue of *textura porrecta*, fusing at the outside with the pseudostromata. Paraphyses 2–2.5 μm wide at the base, abundant, persistent, remotely septate.

Asci: Eight-spored, 177–219 μm, spore-bearing part 135–140 μm, stipe 37–79 μm, cylindrical, unitunicate, subapical apparatus dark blue in Melzer's reagent, tapering with a distinct apical rim, (4.7–)5.6–6.6 x 4.2–4.7 μm.

Ascospores: Uniseriate to obliquely uniseriate or partially biseriate at the upper end of the ascus, dark greyolive to opaque brown, more or less inequilaterally ellipsoid, ventral side varying in degree of convex curvature, upper end broadly rounded, lower end slightly pointed, one-celled throughout ascospore development, (17.9–)18.7–26(–28) x 7.5–13(–13.5) µm biguttulate, wall smooth and relatively thick, without appendages or loosening perispore, germination slit usually clearly seen on the ventral side, straight, conspicuous, 1/2 – 3/4 total length of spore.

Cytochalasins: Cytochalasin Q

Diagnostic features: Perithecia immersed in pseudostroma unitunicate, black, osliolate, asci cylindrical, unitunicate, ascospores hyaline, 1-celled, biguttulate, *Geniculosporium* is the Anamorph.

Perithecia

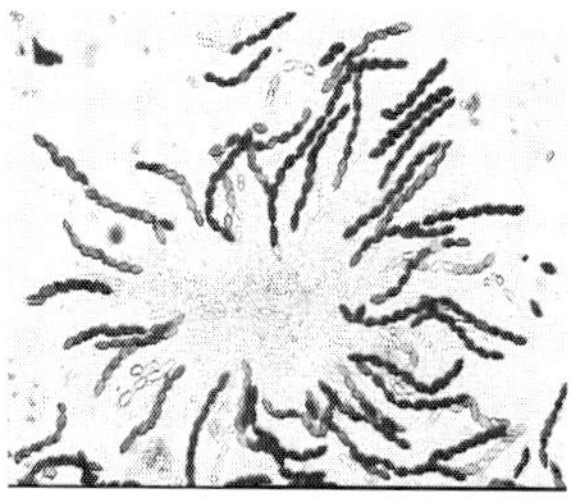

Asci with ascospores

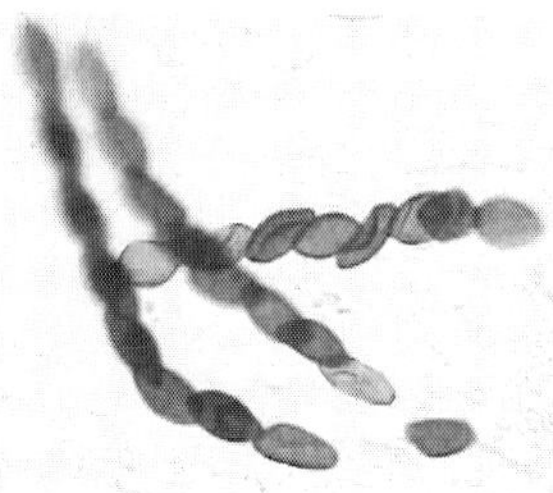

Liberated ascospores

35. ***Hirsutella sp***

Mycelium: Margin spreading, cottony, white. Hyphae hyaline, septate, smooth-walled, sparingly branched, 3.0-5.6µm wide. Synnemata sparse, arising from the dorsal side of host cephalothorax, erect, 1.5 mm in length, subulate, white. Infects insects.

Conidiophores: Arising laterally from hyphae or synnemata, integrated, mostly monophialidic, sometimes polyphialidic, with 2-4 necks.

Conidiogenous cells : Erect, septate, hyaline, smooth, solitary, ampulliform to subcylindrical, 20.6-46.0 µm long, base somewhat inflated, (12.7-)17.5-36.5 × 2.8-4.0 µm, mostly apex abruptly tapering, forming a thin, long, smooth or rough neck, 9.5-17.5 × 0.6-1.0 µm.

Conidia: Hyaline, smooth, one-celled, fusiform to obovoid or obpyriform, sometimes fusiform, apiculate, produced singly or rarely in groups at the neck apex, sometimes embedded in mucous sheath, 5.6-6.4 × 2.0-3.2 µm.

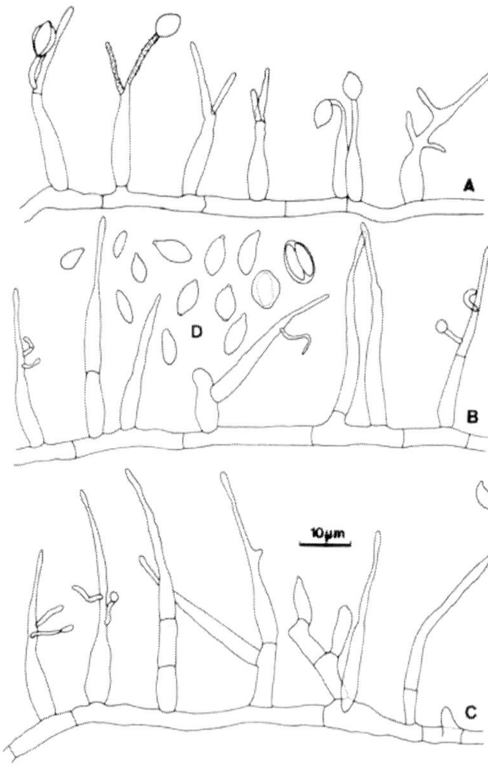

Conidiophores and conidia

Teleomorph: ***Cordyceps*** (Fr.) Link

Colonies : Ascomata formed in or on a stroma formed from mycelia masses in insects, spiders or ascomata of *Elaphomyces,* cylindric, clavate or capitates-stipitate, simple or sometimes branched, variously colored, white, yellow, orange, red to brown or black. Ascoma an ostiolate perithecium, superficial and free or partly or completely immersed in a fleshly stroma composed entirely of fungal tissue, obpyriform, ovoid, conoid or subglobose. Centrum containing lateral paraphyses.

Asci: Unitunicate, cylindrical, subfusoid, or narrowly clavate, usually with a thickened, nonamyloid, apical cap, 8-spored.

Ascospores: Filiform, hyaline, multiseptate, often separating into 1-celled part-spores.

Cytochalasins: Cytochalasin Q

Diagnostic features: Ascomata formed from fungal hyphae produce ostiolate perithecium, asci unitunicate, cylindrical, subfusoid, ascospores filiform, hyaline, multiseptate, often separating into 1-celled part-spores, conidiophores arising laterally from hyphae or synnemata, monophialidic sometimes polyphialidic phiatide tapering to long neck producing conidia singly or in groups, conidia one celled, apiculate embedded in mucous sheath, insectcolous.

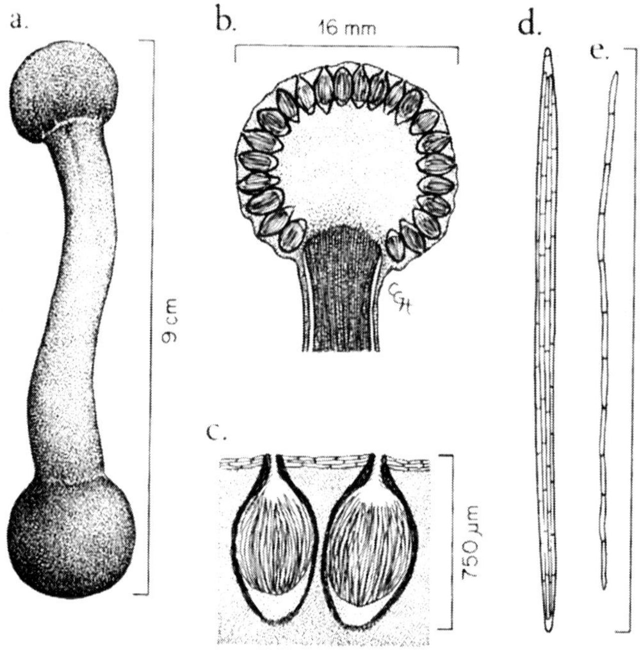

(a) Stromatic ascomata (b & c) Vertical section of stroma showing perithecia (d) slender ascus (e) filiform ascospores

36. *Hormiscium* Kunze

Colonies: Usually effuse but sometimes small and discrete, olive, brown, dark blackish brown or black, often velvety.

Mycelium: Superficial and immersed.

Stroma: None.

Setae and hyphopodia: Absent.

Conidiophores: Macronematous or semi-macronematous, unbranched or irregularly branched, straight or flexuous, subhyline to mid brown, smooth or verruculose.

Conidiogenous: Cells polyblastic or sometimes monoblastic, integrated and terminal, or more commonly discrete, determinate, usually spherical, sometimes becoming cupulate, smooth, verruculose or echinulate, distal fertile part thin-walled, sometimes collapsing, proximal sterile part dark brown or reddish brown, thick-walled.

Conidia: Dry, in simple or branched chains arising from the surface of the upper half of the very characteristic conidiogenous brow, smooth, verruculose or echinulate, with 0, 1 or several transverse septa, usually strongly constricted at the septa. The terminal cell of a multiseptate conidium is frequently a conidiogenous cell.

Cytochalasins: Cytochalasin B

Diagnostic features: Entire mycelium transforming into chains of conidia. This has been merged with *Torula* and it is no more valid genus.

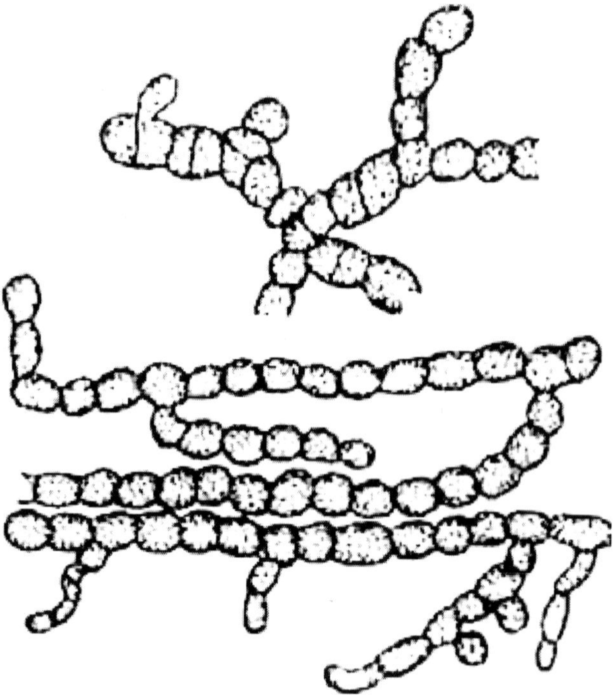

Catemlated hyphae

37. ***Hypocrella*** Sacc.

Colonies: Colonies on PDA at 25°C after colonies are initially white and then may turn greyish brown or light brown.

Stromata: Generally pulvinate, with or without hypothallus, in various shades of whitish, yellow, orange, and green; stroma surface generally minutely pulverulent or pruinose; stromatic tissue generally changing colour in 3% KOH, to reddish or brown; perithecia generally embedded in stroma or half-embedded.

Perithecia: Obpyriform to subglobose, paraphyses present

Asci : Mostly cylindrical or clavate, generally capitate.

Ascospores: Hyaline, multiseptate, smooth, filiform to long fusiform, non-disarticulating.

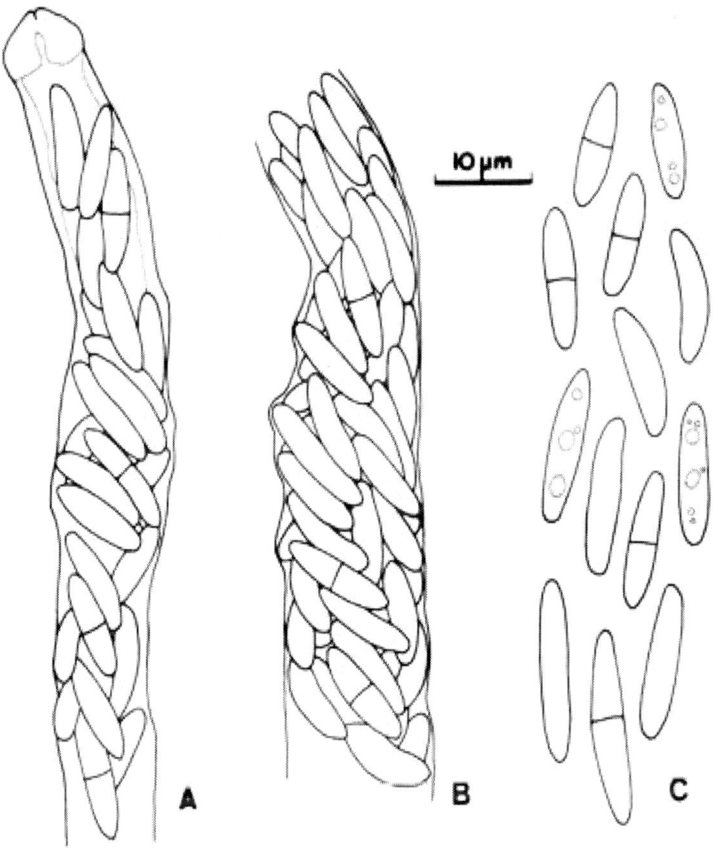

(A) Ascus showing apex with large number of ascospores (B)Ascus liberating ascospores (C) ascospores single celled and bicelled.

Anamorph: ***Aschersonia*** Mont.

Conidiomata: Pycnidia flask-shaped

Conidiophores: Sometimes indistinct, branching once monochasial, twice monochasial, monoverticillate, or two level monochasial; phialides cylindrical, slightly tapering towards tip, conidial masses formed in mucilaginous.

Conidia: Hyaline, smooth, unicellular, fusoid, generally with acute ends.

Cytochalasins: Cytochalasin D

Diagnostic features: Stroma pulvinate in various shades of colour, perithecia embedded or half immersed, perithecia obpyriform to subglobose, asci cylindrical to clavate with a cap, ascospores hyaline, multi septate, filiform to fusi form, generally breaks into single celled structures. *Aschersonia* is the Anamorph.

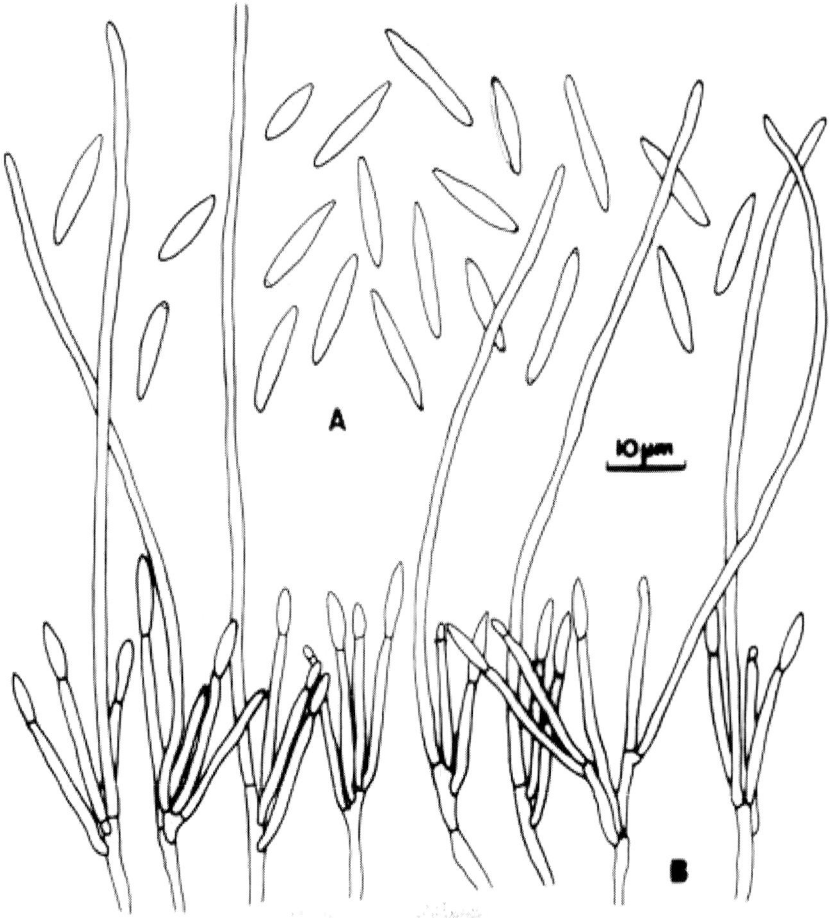

(A) Conidia (B) conidiophore with elongated stipe

38. ***Hypomyces*** (Fr.) Tul.

Mycelium: Forming an effuse, often cottony subiculum on surface of substrate, brightly colored, somewhat fleshy.

Ascomata: Ostiolate, perithecium, immersed in subiculum, with protruding ostiolar necks.

Ascomal wall: Thin, composed of elongated cells, slightly darker on exterior. Centrum containing apical paraphyses.

Asci : Unitunicate, cylindrical, with thick, nonamyloid apical cap, 8-spored.

Ascospores: Hyaline, 1-2 celled, fusoid to fusiform, with an apiculus at each end, usually roughened.

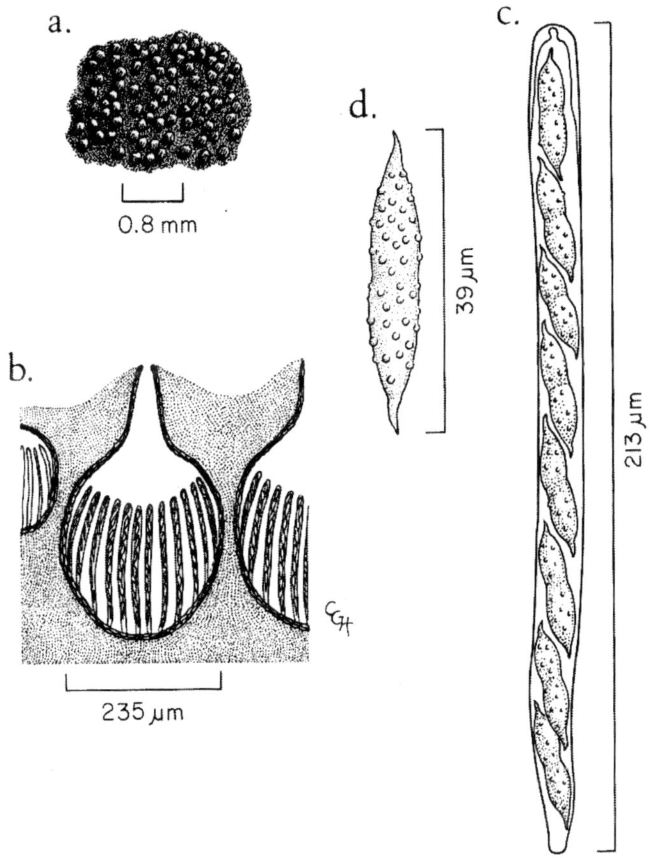

(a) Ascomata in stroma (b) cross sectional view of perithecia (c) ascus with ascospores (d) ascospores

Anamorph = *Trichothecium roseum* Pers

Colonies: Fast growing, at first white later pinkish, pulvinate, velvety, powdery from conidial formation, zonate in diurnal rhythm

Conidiophores: Erect or sub erect, single or in groups, unbranched or sparingly branched, hyaline, the tip near basipetal, imbricate zigzag chains of conidia held together by mucilage.

Conidia: Ellipsoidal, 2-celled, thick walled, hyaline, with well marked truncate attachment point.

Cytochalasins: Cytochalasin-A and B

Diagnostic features: Mycelium effuse, often cottony subiculum, brightly colored, somewhat fleshy; Ostiolate, perithecium with protruding ostiolar necks, unitunicate, cylindrical, 8-spored, ascospores hyaline, 1-2 celled, fusiform, with an apiculus at each end. *Trichothecium* is the Anamorph.

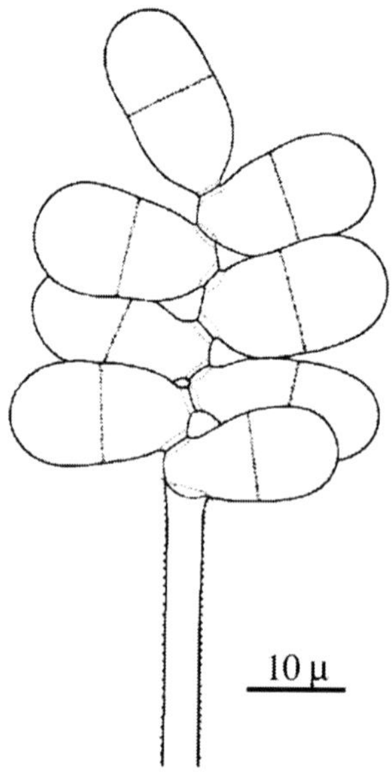

Conidiophore with 2-celled conidia

39. ***Hypoxylon terricola*** Miller

Colonies: Dark grey, sometimes when old with pink to grey sectors, surface velvety with grey aerial mycelium.

Stromata: Superficial on coniferous needles litter, effused-pulvinate, 2-15 mm broad x 0.8-1.3 mm thick, soft leathery; surface silver grey when immature, greyish brown to blackish brown when mature, with conspicuous perithecial mounds; internal tissue soft, whitish, delimited by a black outer and basal layer 30-40 µm thick; margin abrupt.

Perithecia: Subglobose, 0.4-0.5 mm diam.

Ostiole: Papillate, obtuse, black.

Asci: Cylindrical, short-stipitate, the spore-bearing parts 100-110 µm long x 8-10 µm broad, the stipes 35-55 µm long, with apical apparatus amyloid, cuboid to slightly broader than high, 2.5-3 µm high x 3-3.5 µm broad.

Ascospores: 11.5-16.5 x 5.5-7 µm, dark brown, ellipsoid-inequilateral with narrowly to broadly rounded ends, with a conspicuous germ slit spore-length on the less convex side and a gelatinous sheath up to 1.3 µm thick visible on the less convex side of young ascospores. *Geniculosporium* is the Anamorph.

Cytochalasins: Cytochalasin C, D, N,O,P,Q and R

Diagnostic features: Colonies subiculum on the surface of the substratum, brightly colored, fleshy, Ostiolate perithecium, immersed in subiculum with producing neck, asci unitunicate, cylindrical, non-amyloid apical cap, ascospores hyaline, 1-2 celled, fusoid to fusiform

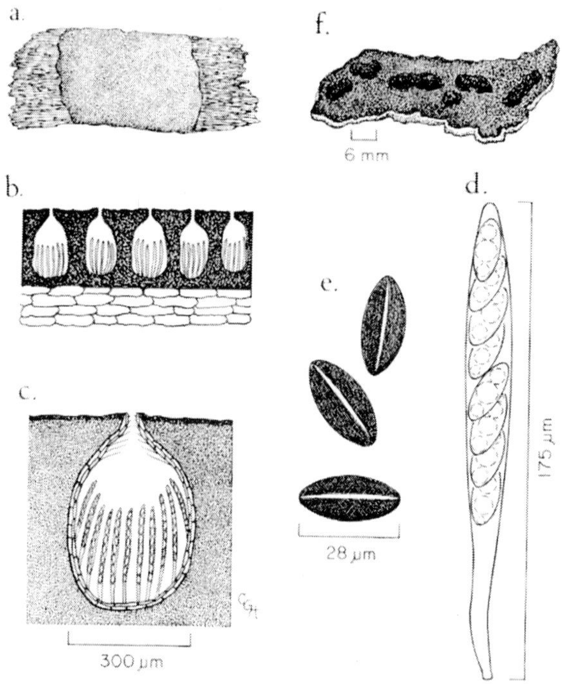

(a) Stroma with perithecia (b) and (c) cross sectional view of perithecium (d) ascospores (e) ascus with ascospores (f) nutrional habitat

40. ***Metarhzium anisopliae*** (Metschn) Sorok

Synonymys = *Metarhzium velutinum* Borowrka

= *Entomorphthora anisopliae* Metsch

= *Isaria destructor* Mtschn

= *M. album* Pelch

= *M. brunneum* Pelch

= *Mepothecium communae* Pidopl. Kiril

Parasitic on insects belonging Coleoptera (of families Elateridae and curculionidae)

Colony: Slow growing

Conidiophores: Penicillium type conidiophores are aggregated to form synnemata, aporogenous cells, flask shaped with thickened neck blows out a conidia are delimited at the base by completely closed septum

Conidia: Conidia single celled, hyaline, smooth walled

Cytochalasins: Cytochalasin C and D (Zygosporin A)

Diagnostic features: Parasitic on coleopteran group of insects, synnematous, phialides, single celled, hyaline, smooth walled.

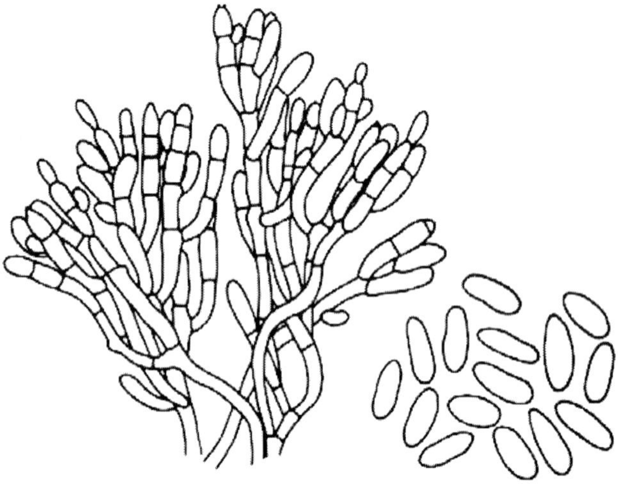

Conidiophores and conidia

41. ***Microporus affinis*** Blume & Nees ex Fr.

Basidiomata: Annual, solitary or in groups, laterally stipitate or with a distinct stipe, flabelliform, spathulate, semicircular dimidiate, flat or depressed in the area around the stipe, pileus up to 10 cm long and 8 cm wide, sometimes imbricate with several pilei arising from the same point of attachment.

Pileus: Glabrous or tomentose, strongly zoned or banded, either as narrow bands or as slightly raised sulcate zones, colour variable from light-yellowish via brown and chestnut and bay to almost black, usually darker at the center than along the margin, the tomentum is very variable, but when present, it is light-greyish and adpressed, more or less velutinate, tomentum is most persistent and thickest near the stipe.

Stipe: Lateral, up to 4 cm long and prominent to almost lacking, 2-8mm in diameter, usually expanded both towards the base and the pore surface, round to slightly flattened, first greyish and finely adpressed tomentose, later glabrous in parts and then almost black with a distinct crust over a white context, at the base usually expanded into a circular disc, more or less persistently covered with a tomentum, cuticle up to 150 μm thick and black in section, below the cuticle there is a looser layer of hyphae around a very dense and hard core, the latter is most prominent in old and thick specimens.

Pore Surface: Light cream, later pale-ochraceous and even grayish in some specimens with the grey colour spreading form the base, margin 1-3 mm wide and pure white, pores round and entire, very minute 7-10 per mm, tubes light cream, up to 1 mm deep.

Context: Pure white and dense, up to 5 mm thick near the base in old specimens covered with a distinct dark cortex from which the tomentum develops as single projecting skeletal hyphae

Hyphal System: Trimitic, generative hyphae hyaline and with clamps, 1.5-3 μm in diameter, skeletal hyphae in the context straight to slightly flexuous, 3.5-6 μm in diameter, thick-walled to almost solid, with a more distinct lumen in the tomentum and up to 10 μm in diameter, binding hyphae common in the context and trama, tortuous and much branched, mostly 1.5-3 μm in diameter, but near the base thicker and up to 6 μm in diameter and here with fewer side branches, very finely coralloid elements present along the dissepiments partly occluding the pore mouths.

Spores: Small cylindrical to oblong ellipsoid, some slightly concave, hyaline and thin-walled, non-amyloid and non-cyanophilous.

Cytochalasins: Cytochalasin D

Diagnostic features: Basidiomata solitary or in groups, laterally stipitate or with a distinct stipe, pileus glabrous or tomentose, pore surface cream, pores minute, small cylindrical spores, hyaline, non-amyloid and non-cyanophilous.

Basidiocarp on wood logs

42. ***Nigrosabulum*** Malloch & Cain

Colonies: Grwoing restrictedly, granular, grey to dark green, closely zonate; reverse dark green.

Ascomata: Cleistothecial, non-stromatic, superficial, sub-spherical to spherical, non-ostiolate, black, glabrous, 100-250μm diam. Asci borne in irregularly disposed culsters, non-stipitate, 6-9 μm diam.

Ascospores: spherical, 3-4 μm diam, often with small particles adhering to the wall.

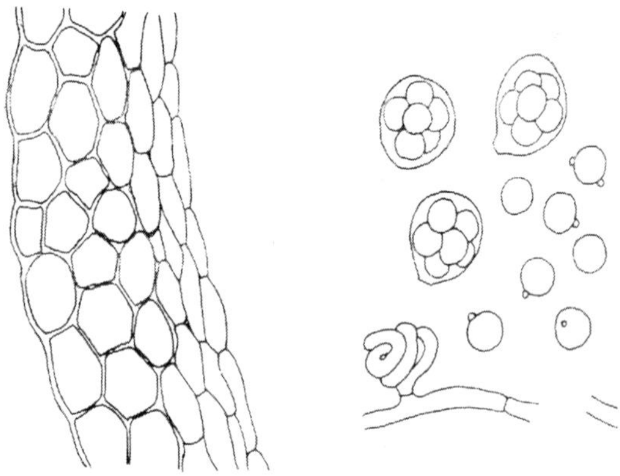

Cleistothecium wall, crozier, asci with 8-ascospores

Anamorph: *Acremonium* spp

Conidiophores: Conidiophores arising from the aerial or substrate mycelium, unbranched, septate, tapering from the base to the apex. Phialides 20-32 x 2-3 μm.

Conidia: Conidia forming slimy heads at the tips of the conidiophores, hyaline, 1-celled, ellipsoidal, fusiform, ovoid or allantoids, 3.5-8.5 x 1-3 μm, smooth-walled.

Cytochalasins: Cytochalasin F and G

Diagnostic features: Ascocarp Cleistothecium, non-stromatic, 3-4 layered, asci globose, scallered, ascus with 8-ascospores, ascospores with small particle, *Acremonium* is the Anamorph.

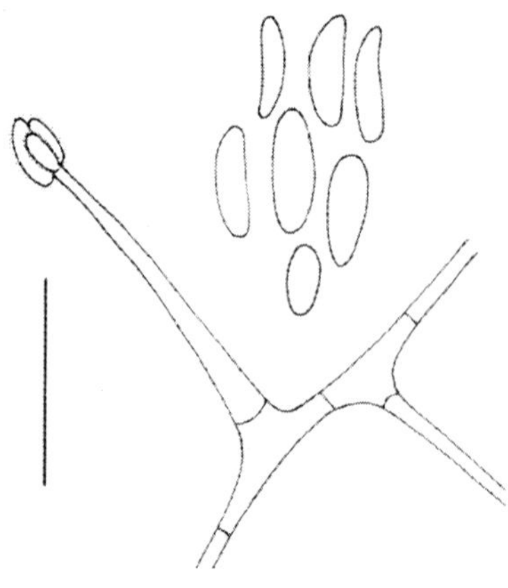

Conidiophore and 1-celled conidia

43. *Penicillium aurantiovirens* Biourge

Synonymy = *P. aurantiocandidum*

P. aurantiogriseum

P. brunneoviolaceum

P. cyclopium var. *aurantiovirens*

P. johanniolii

P. lanosocoeruleum

P. martensii

P. puberulum

P. verrucosum

Monilia formosa

Colonies: Growing 30-35mm on CYA in 7 days at 25ºC, sometimes smaller, radially sulacte, moderately deep, surface texture smooth to granular, fascicles seen in young growth at low magnifications, margins usually entire, deep, white, under low magnifications seem to consist predominantly of fascicles of immature conidiophores.

Mycelium: White, usually inconspicuous, occasionally dominating the colony appearance in floccose isolates;

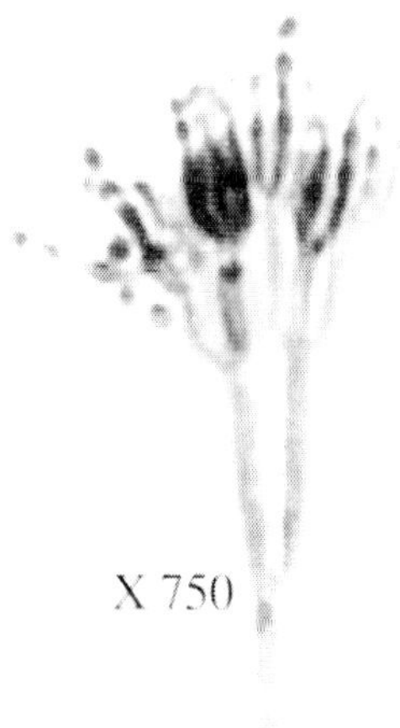

Conidiophores with chain of conidia

Conidiophores: Borne singly or in fascicles, mostly from substrate hyphae, with stipes commonly 200-400 x 3-4 μm, sometimes longer or of indeterminate length because terverticillate or less commonly biverticillate penicillin; rami usually brown singly. 15-25 x 3-3.5 μm, metulae in verticils of 3-4, 10-12 (-15) x 2.8-3.5 μm, phialides 5-8 per metula, slender, ampulliform, typically 9-10 x 2.5-2.8 μm, with indistinguished collula.

Conidia: Grayish-blue; exudates typically conspicuous near the margins, rarely completely absent, clear or pale-brown; soluble pigment produced by some isolates, brown to reddish brown, reverse varying from isolate to isolate, pale, light to brilliant orange, or reddish to violet brown. Conidia subspheroidal to ellipsoidal, commonly 3.5-4 x 2.5-3.2 μm, smooth, usually borne in long well defined columns.

Cytochalasins: Chaetoglobosin C

Diagnostic features: Conidiophores arise in fascicles, conidia 1-celled, subspheroidal to ellipsoidal produced in long chain.

44. *Penicillium chrysogenum* Thom

Subgenus: *Penicillium*

Section: Chryscgenea

Series: Chrysogena

Teleomorph: None

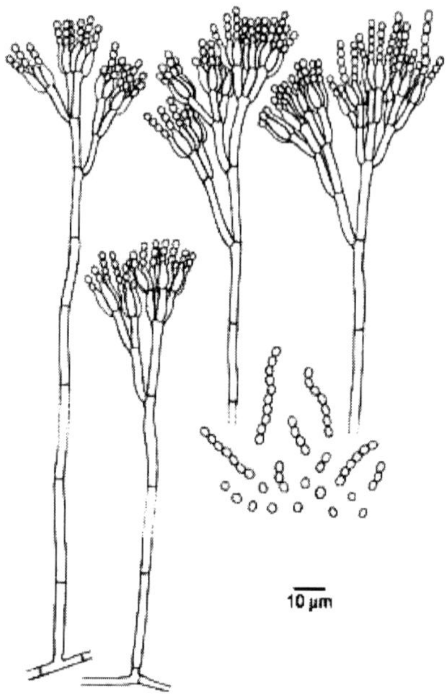

Conidiophores and conidial chains

Colony texture: Floccose to velutinous

Colony colour on CYA: Blue-green to green

Exudates droplets on CYA: Often present, copious, yellow

Reverse of colony on CYA: Cream, yellow rarely brown

Reverse of colony on YES: Citrine yellow

Diffusible colour: Yellow pigment often produced

Ehrlich reaction: No reaction or yellow reaction

Conidiophores: Bi-, ter- and quarter verticellate both appressed and divergent ram borne from aerial and sub-surface hyphae.

Conidia: Smooth-walled, globose to sub-globose to broadly ellipsoidal.

Phialides: Cylindrical with short broad collula

Metulae: Cylindrical

Rami: Cylindrical

Stipes: 200-300 μm X 3-4 μm

Synnemata or fasciculation: None

Sclerotia: None

Cytochalasin: Penochalasin K

Diagnostic features: Globose to sub globose to broadly ellipsoidal, smooth walled conidia, relatively short phialides with short broad collula, high growth rate on YES with a yellow reverse and strong sporulation.

45. *P. discolor* Frisvad & Samson

Colony texture on CYA: Velulinous to fasciculate.

Colony colour on CYA: Dark-green

Exudate droplets on CYA: Clear

Reverse colour on CYA: Cream yellow

Reverse colour on YES: Orange turning to deep red with age.

Diffusible colour: Brillant-red on YES

Ehrlich reaction: Weak, violet

Conidiophores: Terverticillate, appressed elements borne from subsurface hyphae

Phialides: Cylindrical, tapering to a distinct collulum

Metulae: Cylindrical

Stipes: Rough walled

Synnemata or fasciculation: Fasciculate on MEA and oat.

Sclerotia: None

Conidia: Rough walled, globose to subglobose

Cytochalasins: Chaetoglobosin A, B, C, D, E, F and G

Diagnostic features: Colonies velulinous to fasciculate, dark-green reverse of the colony orange changing to deep red with age, conidiophores terverticillate, conidia rough walled, globose to subglobose

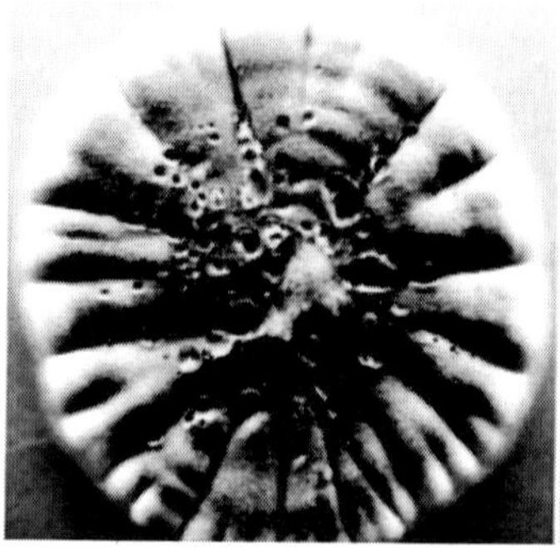

Colony surface

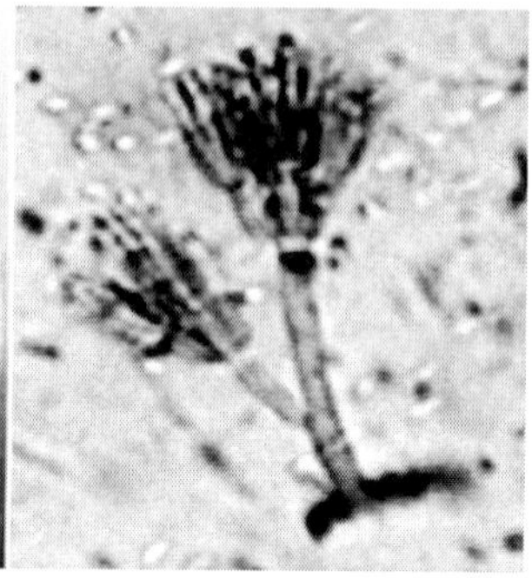

Conidiophore with conidia

46. ***Penicillium expansum*** Link

Subgenus: *Penicillium*

Sections: *Penicillium*

Series: Expansa

Teleomorph: Absent

Colony texture: Floccose to weakly fasciculate

Colony colour on CYA: Blue-green to green

Exudate droplets on CYA: Copious, clear

Reverse colour on CYA: Cream to yellow with brown centre, orange – brown or dark-brown

Reverse colour on YES: Cream yellow or orange

Diffusible colour on CYA: Common, rosy-brown

Ehrlich reaction: Strong, violet

Conidiophores: Terverticillate, appressed elements borne sub-surface and aerial hyphae

Conidia: Smooth walled, ellipsoidal, rarely exceeding 3.5 μm in diameter

Phialides: Cylindrical tapering to a distinct collulum

Metulae: Cylindrical

Rami: Cylindrical

Stipes: Smooth walled, occasionally rough walled at lower part.

Synnemata or fasciculation: Capitulate synnemata produced in fresh strains or after cold storage on MEA or OAT

Sclerotia: None

Cytochalasins: Chaetoglobosin A, B, C, D, E, F, G and O

Diagnostic features: Ellipsoid smooth-walled conidia, expan solids, patulin, Chaetoglobosin, commonsense A and B.

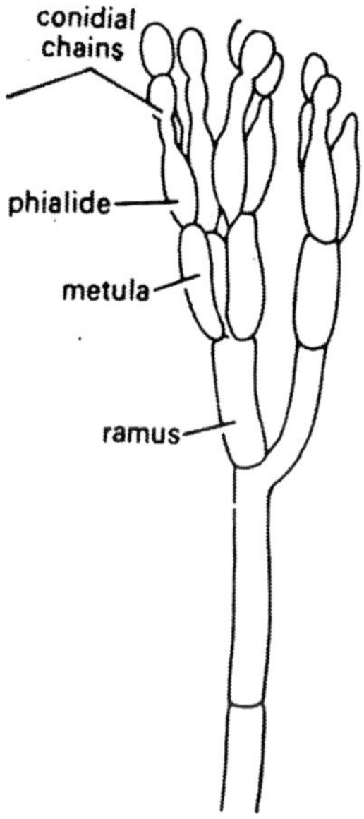

Conidiophore with budding conidia

47. ***Periconia*** spp. Tode

Colony: Effuse, small and compact, grey-brown, olivaceous-brown or black, hairy

Hyphae: Immersed, partly superficial

Stroma: Pseudoparenchymatous, mid to dark -brown

Setae: Absent

Hyphopodia: Absent

Conidiophores: Macronematous or semi macronematous, mononematous, brown mostly with stipe and sphaeroid head, branches present or absent

Stipe: Straight of flexous, pale to dark-brown, smooth, rarely verrucose, sometimes apex sterile or setiform

Conidiogenous cells: Monoblastic and polyblastic discrete on stipe and branches, determinate ellipsoidal, subsphaerical to spherical.

Conidia: Catenate, chains often branched arising at one or more points, subsphaerical to spherical, ellipsoidal, oblong or broadly cylindrical, pale to dark-brown, verrucose or echinulate, non-septate.

Cytochalasins: Periconiasins A-C

Diagnostic features: Brown, macronematous, conidiophore with stipe and sphaeroid head, conidiogenous cells, monoblastic or poly blastic, spherical conidiogenous cells which bear conidia in chain, conidia nonseptate, verruculose or echinulate.

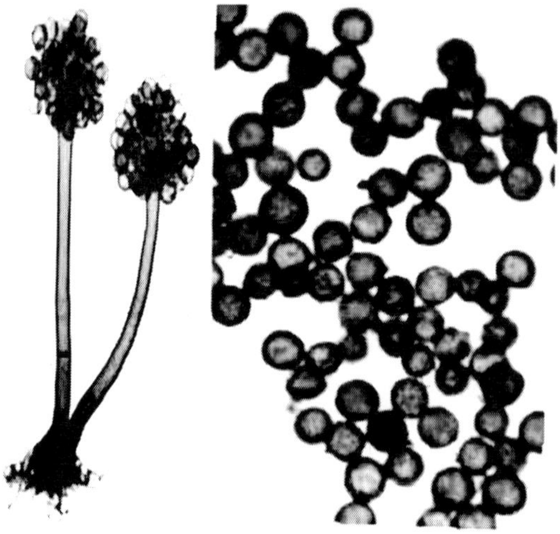

Conidiophore with conidial head, globose conidia in chain

48. ***Pestalotiopsis*** De Notaris

Conidiomata: Cupulate, immersed at first, then erumpent, unilocular, glabrous, dark brown to black, dehiscing by irregular splits in the apical wall; wall tissue of textura angularis, cells thick-walled and dark brown in the outer layers, thin-walled and paler in the inner layers, merging with a thin textura intricata of thin-walled and paler elements toward the margins of the conidiomata.

Conidiophores: Arising in the concavity of the conidioma, septate, irregularly branched, pale brown to brown, thick-walled and verrucose in the lower part, thin-walled and colourless above, invested in mucus. Conidiogenous cells discrete or integrated, cylindrical, colourless, smooth.

Conidiogenesis: Ontogeny holoblastic by apical wall-building in the first conidium and by replacement wall-building in subsequent conidia; maturation by diffuse wall-building synchronous with ontogeny; delimitation by a double septum; secession schizolytic; proliferation enteroblastic-percurrent to produce additional conidia higher levels; annellations present on conidiogenous cells; regeneration of conidiogenous cells absent.

Conidia: fusiform, distoseptate, smooth-walled, bearing apical and basal appendages of type A; basal cell obconic with a truncate base, thin-walled, smooth, colourless; median cells with reduced lumina, thick-walled, smooth, mid-brown; apical cell conic, thin-walled, smooth, colourless; appendages cellular and maintaining protoplasmic continuity with the conidium body, dichotomously branched and often recurved, rarely unbranched, filiform; apical appendages 3-9; basal appendage single, centric, often branched.

Cytochalasins: Cytochalasin U

Diagnostic features: Disc shaped acermulin, conidiophore is annellate, phialides producing 5-celled conidia, middle 3 cells bigger darker in colour, apical and basal cell hyaline, basal cell with start-pedicel, apical cell with setulae.

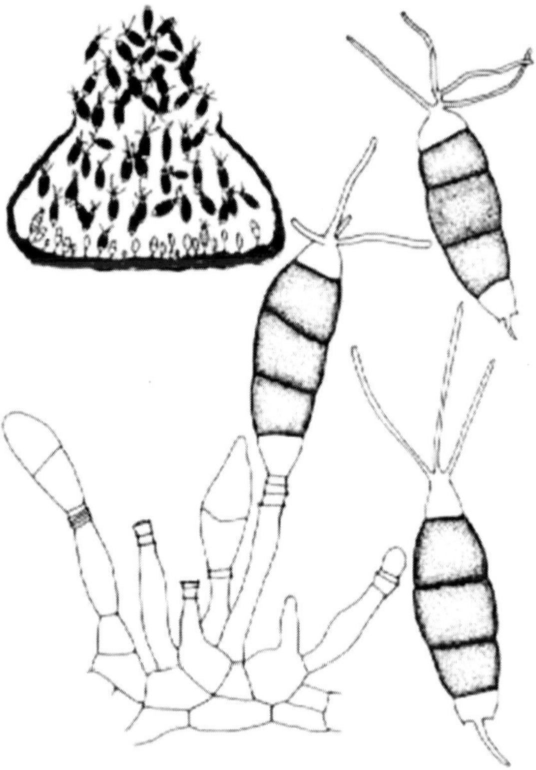

Disc shaped acervuli and setulate conidia

49. Phoma exigua Desm

= *Boeremia exigua*

Colonies: Black, dense with irregularly scalloped margin, attaining a diameter of 6.6-7.0 cm in days on oat agar,

Pycnidia: Black, globose to sub-globose, some colorless to form irregular fruitfriction, parenchymatous ostiolate, 147-294μm,

Pycnidiospores: Hyaline, 1-celled, mostly ellipsoid, few ovoid, 4-9 x 2.0-5.0 μm, red pigment produced which on application of NaOH change to dull-green or finally red.

Cytochalasins: Cytochalasin F, W, Z_4, Z_5 and Z_6.

Diagnostic features: Pycnidia black, pycnidiospores 1-celled, hyaline, ellipsoid, red pigment produced.

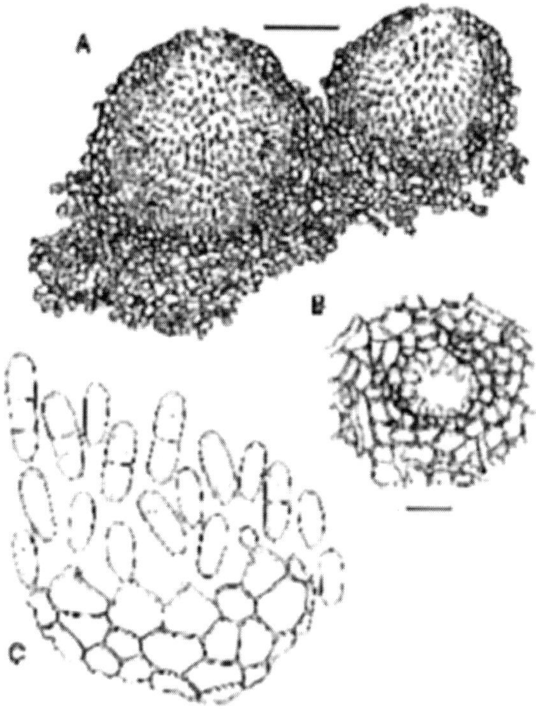

(A) Pycnidia with conidia (B) Pycnidial cells with sense of phialides (C) Pycnidiospores

50. *Phoma herbarum* var. *medicaginis* Westend

Colony: Grow rapidly produce dark mycelium forming pycnidia

Pycnidia: Pycnidia globose, or ovoid with ostioles at the tips of small papillae

Conidiophores: In conspicuous produce numerous pycnidiospores.

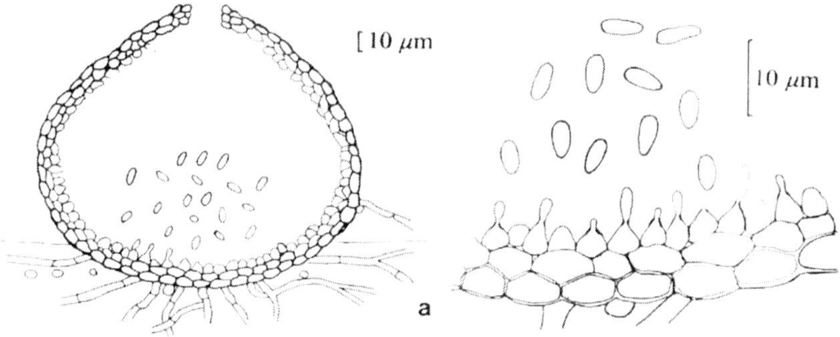

(a) Pycnidia with conidia (b) Pycnidial cells with sense of phialides

Conidia: Ovoid, colourless, guttulate, either, unicellular, or bicelled, presence of both one celled and two celled conidia in the same pycnidium it is assigned to both ascochyta.

Cytochalasins: Cytochalasins C and D

Diagnostic features: Pycnidia globose, ostioles at the tips of small papillae, conidia ovoid, hyaline biguttulate, 0 or 1-septate

51. ***Phomopsis leptostromiformis*** J.G. Kuhn

Colonies: forms a thick white aerial felt.

Mycelium: hyaline, branched, stromata or protopycnidia are white at first and darken eventually to black at maturity. They are irregular in shape, varying from spherical to elongate and can even be bifurcate.

Pycnidia: One or more ostioles which are either schizogenous or conical

Conidiophores: hyaline, slender, simple or branched, tapering and 8-14 x 1.4-2.5.

Conidia: Commonly extruded through the ostioles. They are salmon pink to orange in mass, straight or very slightly curved, cylindric with tapered ends and 8.0-8.2 x 1.5-1.7μ

Cytochalasins: Chaetoglobosin M and Phomopsin A, B

Diagnostic features: Pycnidium pseudoparenchymatous, surrounded by black sclerotia, locula a dense palisade of slender, simple or branched tapering phialides produced.

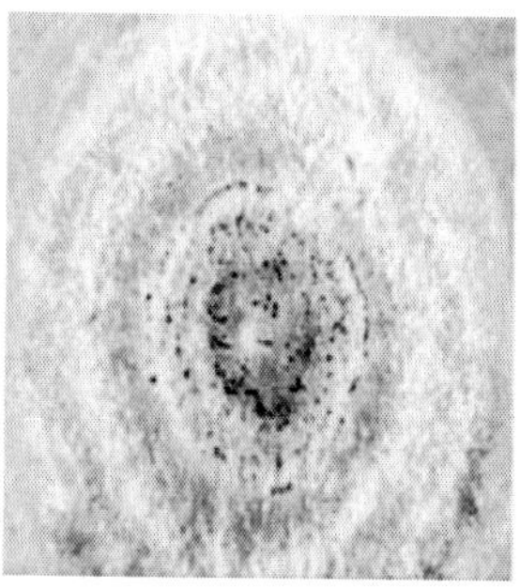

Zonate pycnidia

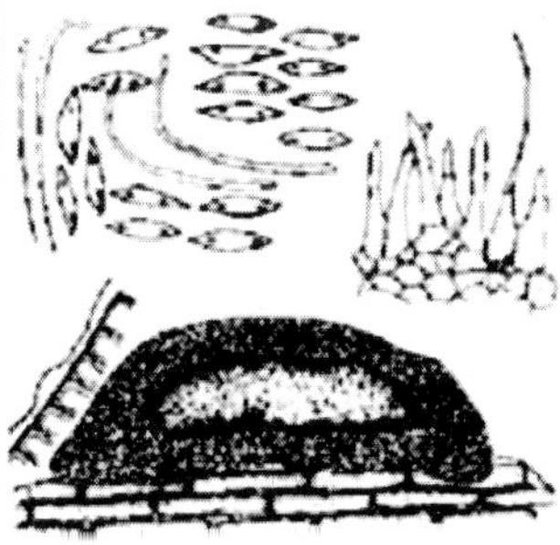

cross section of pycnidium

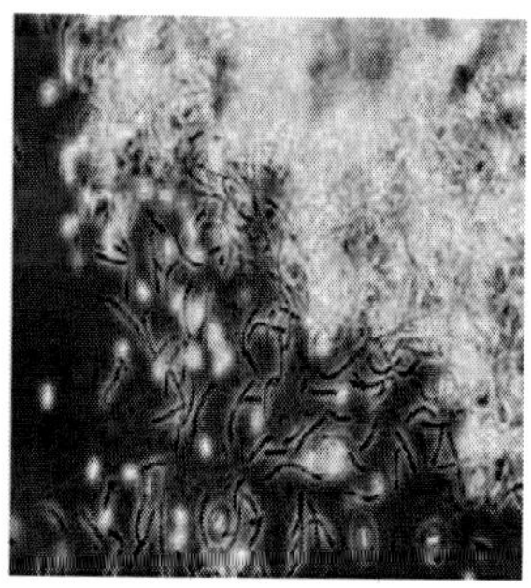

α and β conidia

52. ***Phoma multirostrata*** (P.N. Mathur, S.K. Menon & Thirum.)

Colonies: Smooth with entire sharp margin. Aerial mycelium sparse, floccose or tufted, white to gray or absent. Agar surface olivaceous to chestnut with colorless sectors. Reverse concolorous.

Pycnidia: solitary or confluent, globose to subglobose or irregular, glabrous, brown to black, superficial or immersed, Ostioles conspicuous, non-papillate or on elongated necks. Pycnidial wall pseudoparenchymatous, composed of oblong to cylindrical or elongated cells, 4-5 layers, thick.

Conidiogenous cells: phialidic, hyaline, simple, smooth, flask-shaped.

Conidia: oblong to ellipsoidal, thin-walled, smooth, hyaline, aseptate, highly variable in size, with 0-3(-4) polar guttules. Conidial matrix white to buff or rosy-buff.

Chlamydospores: mostly unicellular, 5-15 μm diam, ellipsoidal to oblong to somewhat pyriform, olivaceous or pale- brown with greenish guttules, solitary or in chains, intercalary but incidentally also terminal. A bunch of clustered unicellular chlamydospores can be observed regularly, especially in older cultures.

Cytochalasins: Multirostrain A

Diagnostic features: Colonies white to gray, pycnidia solitary or confluent, brown to black, glabrous pycnidia, pycnidial wall pseudoparanchymatous, 2-4 layered, conidiogenous cells phialidic, flask-shaped, conidia oblong to ellipsoid, 3-4 polar guttulate, chlamydospores: mostly unicellular, solitary or in chains.

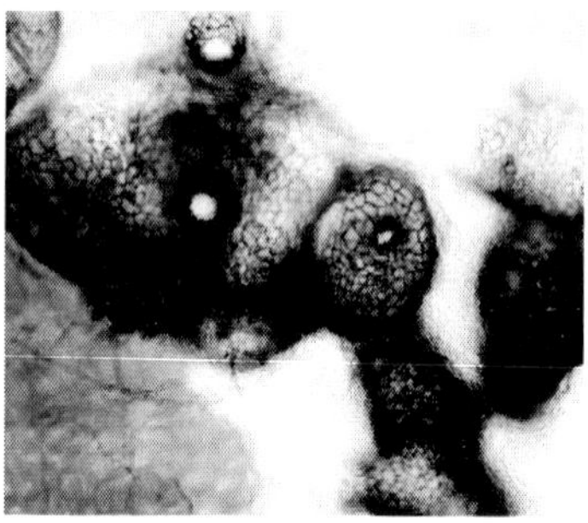

Pycnidia

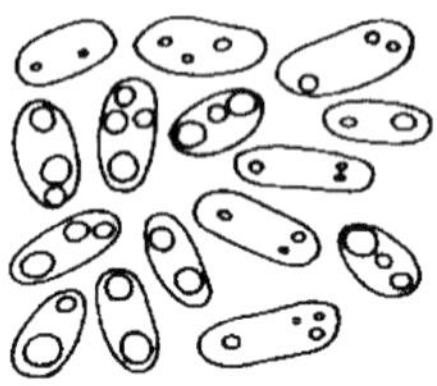

Conidia

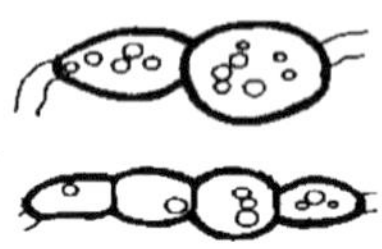

Chlamydospores

53. ***Phomapsis paspali*** Johnst

Colonies: On oatmeal agar and malt extract agar more than 90 mm diam. in 7 days at 20°C

Mycelium: Cottony, grey; hyphae erect, long, reaching the lid of the petri dish. Reverse; dark grey or black

Pycnidia: Light to dark-brown, immersed to superficial, solitary to gregarious, 120-250 μm diam. Solitary, dark walled with a single ostiole.

Conidia: hyaline, unicellular, oblong-elliptic, egguttulate, 6.5-11 x 3-4 μm.

Alpha conidia: single celled, clear, and oval to fusoid (football-shaped).

Beta conidia: are single celled, clear, long and thin with a characteristic curve or bend

Chlamydospores: None

Cytochalasins: Cytochalasins G, H, J

Diagnostic features: Non stromatic single or group of pycnidia, dark-walled ostiolate, conidia hyaline, unicellular, oblong to elliptic egguttulate.

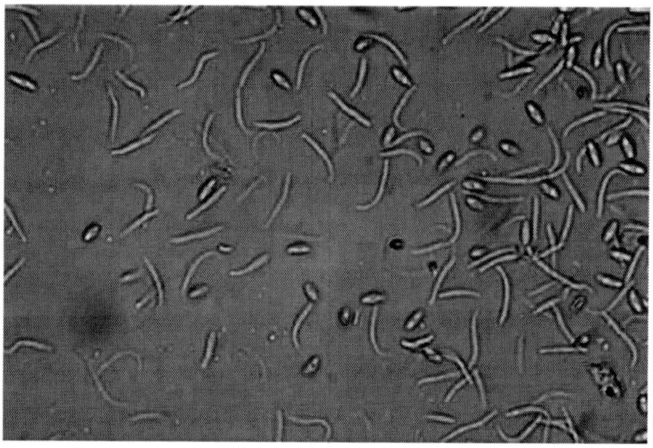

Alpha and beta-conidia

54. ***Phomopsis sojae*** Lehman

Colonies: growing mycelium at first white turning orange-yellow

Pycnidia: On pods, stems and petioles of soybean, mostly simple, 1-chambered, often boat-shaped, 82-225 x 82-375μm on pods and stems and 120-180 x 135-240 μm in leaf tissue, short necks, stroma diffuse or absent.

Alpha pycnidia: Ovoid, 7.5 x 3 μm, hyaline.

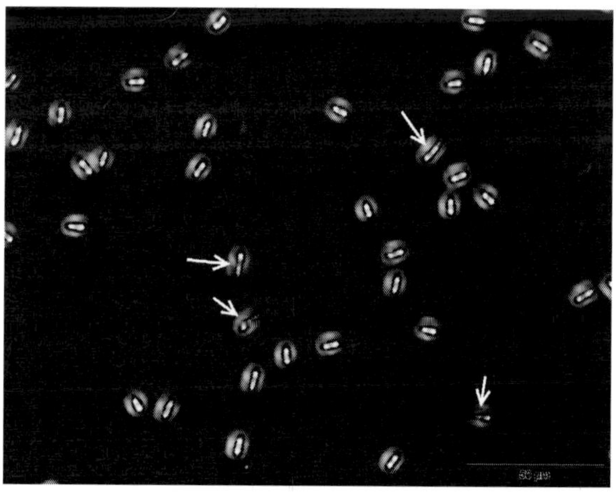

Alpha and beta-conidia

Alpha pycnidiospores: 1-celled, straight, fusiform, commonly rounded at one end and noticeably pointed at the other, 4.5-10 x 1.7-2.6 μm (pods), 5.3-8.5 x 1.8-2.8 μm (leaves), 6-9 x 1.8-2.7 μm (stems), 2-guttulate, hyaline.

Beta pycnidia: Sickle-shaped, 16.6 x 1.6 μm

Beta pycnidiospores: Long, slender, curved or hooked, hyaline.

Cytochalasins: Cytochalasin H and J

Diagnostic features: Pycnidia with short neck, with two distinct pycnidia, pycnidia producing α-conidia are ovoid while β-conidia producing pycnidia are sickle shaped, α-conidia, hyaline, 1-celled, fusiform egguttulate, β-conidia long slender curved or hooked, hyaline.

55. **Physalospora** Niessl

Colonies: In the dark, colonies of all isolates on PDA were initially white, turning grey and then black.

Ascomata: An ostilate perirhecium, immersed in host tissue, with erumpent ostiolar neck, perithecia single, globose, obpyriform or vertically ellipsoidal, brown, glabrous or with hairs, often with short setae around ostiole. Ostiolar neck papillate or conical, often darker than rest of ascoma, lined with periphyses.

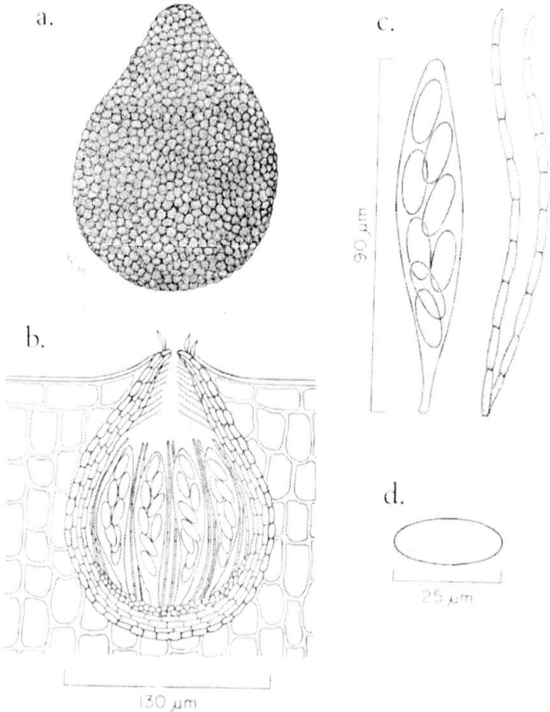

(a) Perithecium (b) cross section of perithecium showing arrangement of asci, ostiole paraphyses (c) ascus with two showing ascospores (d) ascospores

Perithecial: Wall of flattened cells of flattened cells, outer cells subhyaline or olive-brown, inner cells hyaline. Paraphyses filiform, often with slightly calvate tips, sometimes with mucilaginous, coalesced walls.

Asci: Unitunicate, clavate-cylindrical, truncate at apex, broadest in middle, tapering to a short stipe, some species with a simple, amyloid or nonamyloid apical ring, 8-spored.

Ascospores: 1-celled, hyaline or lightly colored, obovoid to elliptic, smooth or finely roughened, with granular contents, sometimes with a thin, gelatinous sheath.

Cytochalasins: Cytochalasin J

Diagnostic features: Ascomata perirhecium with glabrous wall, erumpent ostiolate, papillate or conical neck, asci unitunicate, clavate-cylindrical with short stipe, 8-spored, ascospores hyaline, one celled, ovate elliptic.

56. *Pseudeurotium zonatum* van Beyma

Ascomata: Zonata, superficial immersed, globose, 30-150 µm diam.,

Ascocarp: Cleistothecium, spherical, brown, 100-230 µm diam.

Asci: spherical, pyriform to ellipsoid, 7-9 x 6-6.5 µm, with 6-8 ascospores.

Ascospores: globose, smooth, hyaline becoming brown, 3-3.3 µm in diam.

Anamorph: *Teberdinia*

Conidial state: Sporotrichum-like with hyaline, ellipsoid to ovoid.

Conidia: Glabrous, broadly ellipsoid, ellipsoid or obovoid.

Cytochalasins: Cytochalasins G, X, Y and Z

Diagnostic features: Ascomata cleistothecium, spherical, brown, asci spherical, pyriform to ellipsoid, 8-spored, ascospores smooth, globose, 1-celled.

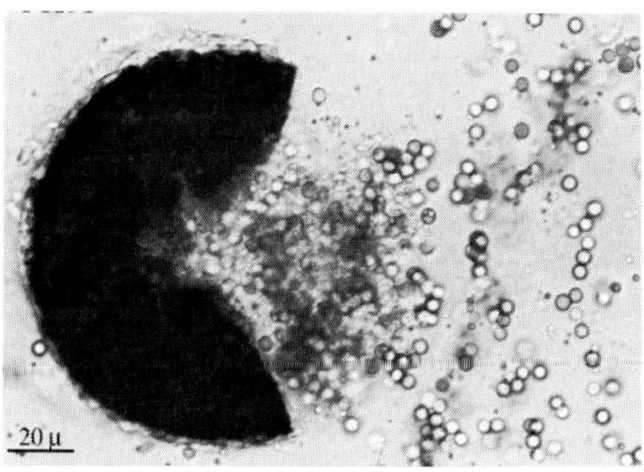

Cleistothecium and ascospores

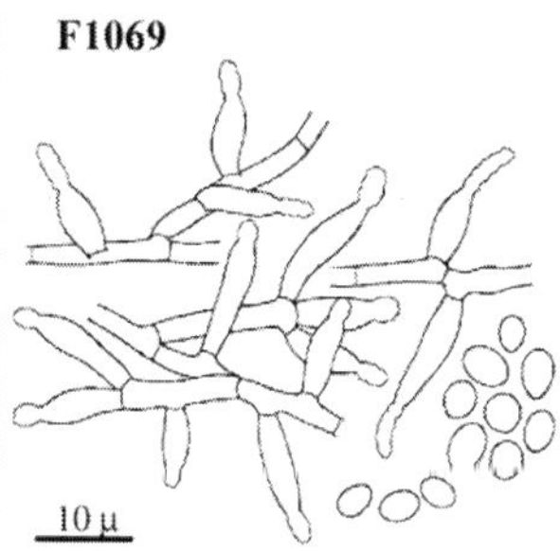

Conidiophore and conidia

57. *Pyrenophora semeniperda* Brittlebank & Adam

Colonies: Colonies grow slowly and cover half of the slant with white loose mycelium, no synnemata, a few conidia are formed, reverse dark-green with a few stands penetrating the agar medium.

Ascomata: Perithecium, superficial, solitary, globose to long cylindrical, wall pseudoparanchymatous, outer wall brown, thick, inner wall hyaline thin and compressed, with numerous setae over entire surface, setae septate.

Ascus: Bitunicate, numerous, cylindric, straight, bearing 4-8 uniseriate spores, diverging from the centre. Pseudoparaphyses, persistent, septate.

Ascospores: Ascospores oblong with broadly rounded ends, not conspicuously constricted and without enlarged cells with thin yellowish-brown, glabrous walls with a thin sheath when young and without sheath when mature, spores with two vertical septa at right angles dividing the entire spores into four equal quadrants with 5-8, often 7, transverse septa.

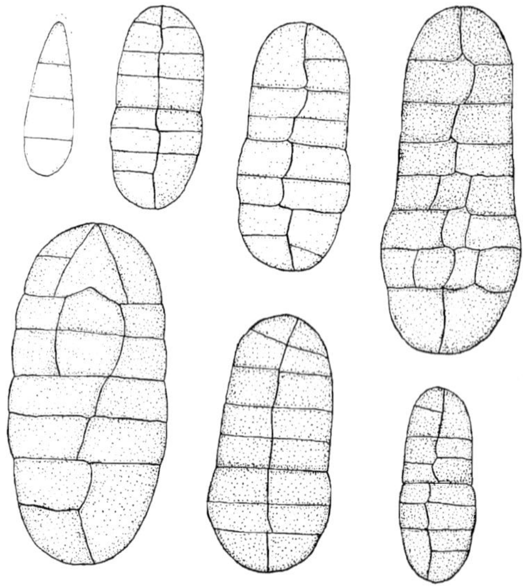

Ascospores with cross and vertical septa

Anamorph: ***Drechslera verticillata***

Synnemata: Dark-brown, densely covered by setae and conidiophores.

Conidiophores: Conidiophores median variable in size, yellowish-brown, simple, straight, bearing conidia through apical pores, geniculate and marked by a brown.

Conidia: Cylindrical with broadly rounded ends, straight median, yellowish-brown except for hyaline zone, around brown scar, included in contour of basal cell, 5-7 septate, cells subequal, longer than width.

Cytochalasins: Cytochalasin A, B, F, T, Z_1, Z_2 and Z_3

Diagnostic features: Perithecium glabrous, globose to cylindrical cell pseudoparanchymatous with numerous setae over entire surface, bitunicate ascus, cylindrical, stroma cylindrical erect, dark-brown, conidiophores geniculate, conidia straight or slightly curved, obclavate, 7-12 pseudoseptate, golden-brown.

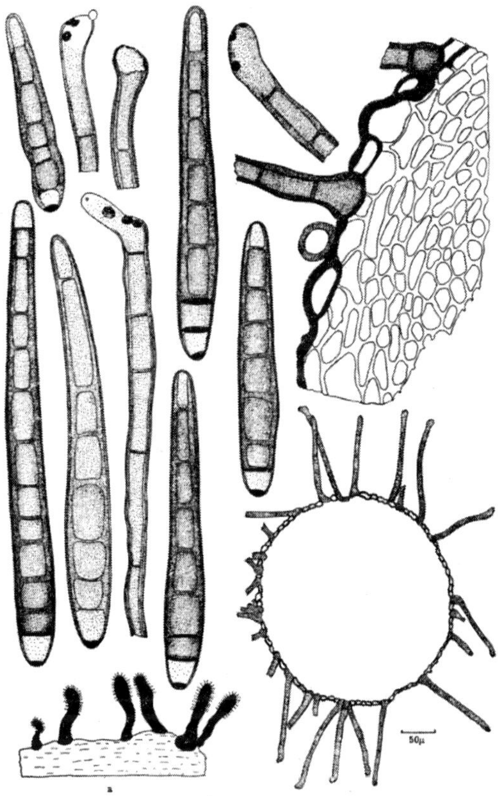

Conidia, part of conidiophore arising from Sclerotia

58. ***Pyricularia grisea*** Sacc

Colonies: Effuse, thinly hairy, grey, grayish-brown, olivaceous-brown

Mycelium: Immersed

Chlamydospores: Sometimes formed in culture

Stroma: None

Setae and hyphopodia: Absent

Conidiophores: Macronematous, mononematous, slender, thin-walled, mostly unbranched, straight or flexuous, geniculate towards apex, pale-brown, smooth

Conidiogenous cell: Polyblastic, integrated, terminal, sympodial, cylindrical, geniculate, denticulate, thin walled, cut off as a rule by a septum to form a separating cell.

Conidia: Solitary, dry, acroleurogenous, simple, obpyriform, obtunicate or obclavate, hyaline to pale-olivaceous-brown, smooth, 1-3 septate, hilum often protuberant.

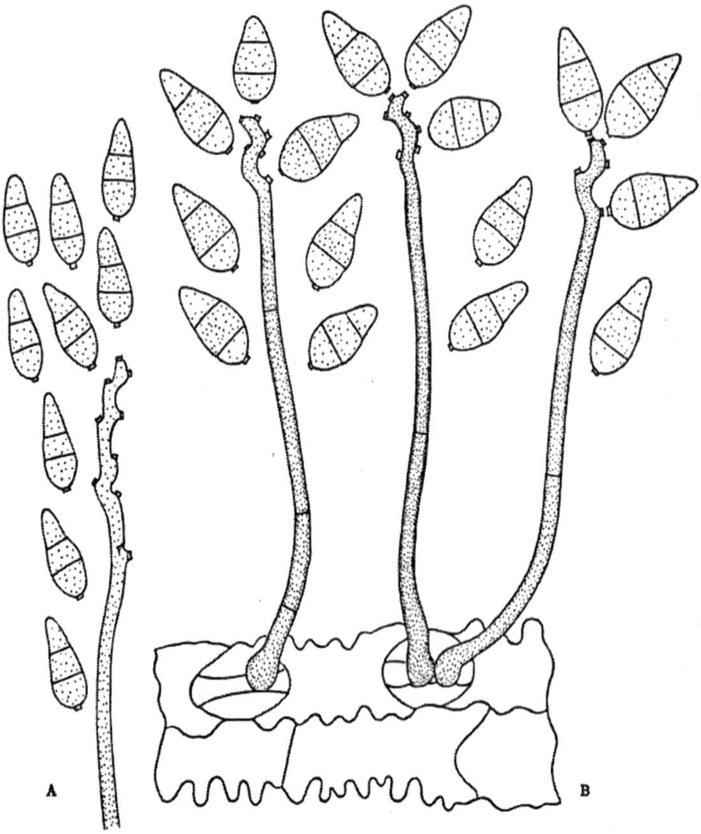

Conidiophore and 3-celled conidia

Teleomorph: ***Magnoparthe grisea***

Mycelium: Septate, branched, olivaceous.

Perithecia: Groups or single, stroma not apparent, long beaked, base globose, 60-300μm in diam, beak 60-150 x 100-1200 μm.

Asci: Cylindrical to subclavate, nonpersistent, arising from the base of the perithecium, unitunicate, thin walled except at the tip, 55-90 x 7-10 μm.

Ascospores: Fusiform, 3-septate, slightly curved, two middle cells are pigmented, 17-24 x 4-7 μm.

Cytochalasins: Pyrichalasin H

Diagnostic features: Conidiophores macronematous, unbranched geniculate towards apex, conidiogenous cells polyblastic, sympodial, denticulate, conidia solitary, acroleurogenous, 1-3 septate with protruberant hilum. *Magnoparthe grisea* is the teleomorph.

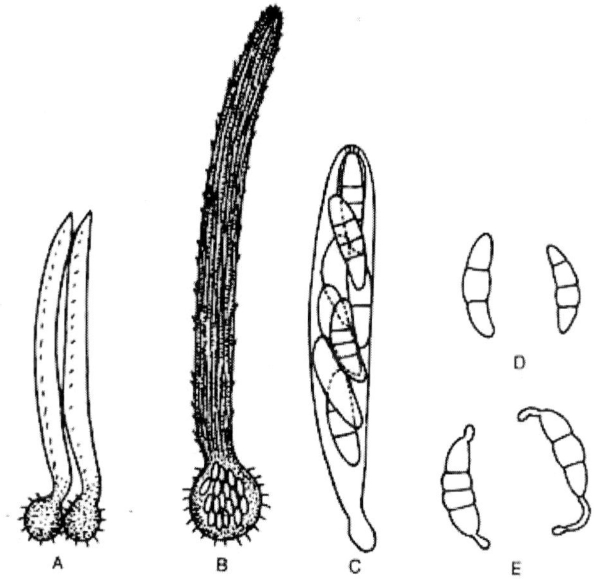

(A) Perithecia (B) perithecium much enlarged (C) single ascus containing ascospores (D) ascospores (E) germinating ascospores

59. ***Rhinocladiella*** Nannfeldt

Colonies: effuse, hairy or felted, grey, brown, olivaccous or black, often slow growing.

Mycelium: partly immersed, partly superficial.

Stroma: none. Setae and hyphopodia absent.

Conidiophores: macronematous or semi-macronematous, mononematous, unbranched or loosely branched, straight or flexuous, pale to mid dark brown or olivaceous brown, smooth, verruculose or echinulate.

Conidiogenous cells: polyblastic, integrated, terminal sometimes becoming intercalary, sympodial, cylindrical, cicatrized, scars small, ramo-conidia often present.

Conidia: solitary, dry, acropleurogenous, simple, ellipsoidal, cylindrical, clavate or fusiform, colorless to pale brown or olivaceous brown, smooth or verruculose, almost always 0-septate, 1 or more septa very rarely formed.

Cytochalasins: Cytochalasin E and F

Diagnostic features: Conidiophores macronematous or semi-macronematous mononematous, unbranched, scars small, ramo-conidia often present. Conidia solitary, acropleurogenous, one celled, pale dark brown or olivaceous brown.

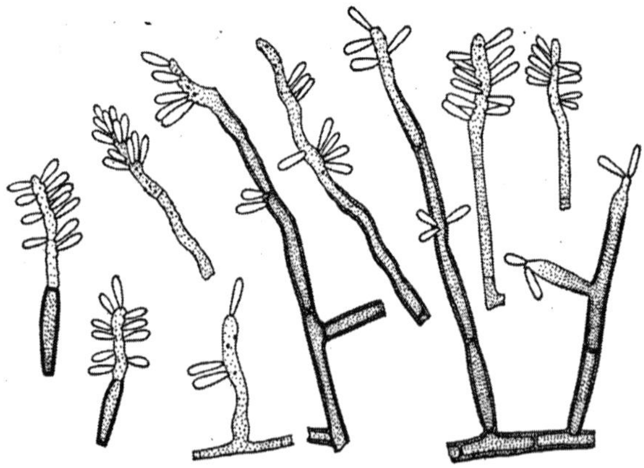

Conidiophores bearing with conidia

60. ***Rosellinia necatrix*** Harting

Colonies: Effuse, dark chocolate to blackish brown, with numerous erect, dark brown synnemata clearly visible to the naked eye.

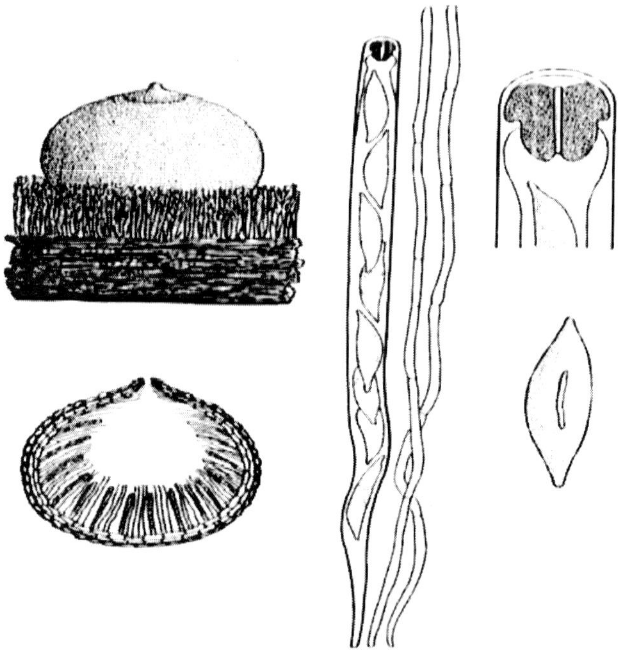

(a) Perithecium (b) cross section of perithecium showing arrangement of asci, ostiole (c) ascus apex (d) ascospores

Mycelium: Partly superficial, partly immersed.

Stroma: None.

Setae: Hyphopodia absent.

Ascomata: Single or gregarious borne on a felt like subculumn of dark hyphae, ostiolate globose or broadly or pyriform. Ostiolar neck papillate or comical.

Ascus: Unitunicate, cylindrical, long talked with am amyloid special apparatus, liming inside of perithecial wall, 8-spored.

Ascospores: One celled, dark brown, ellipsoid, often laterally compressed with a long longitudinal germ slit.

Anamorph: ***Dematophora*** Hartig

Conidiophores: Macronematous, synnematous, component threads flexuous, intertwined, repeatedly branched, often dichotomously towards the apex, splaying out to form a head, pale to mid brown or olivaceous brown.

Conidiogenous: Cells polyblastic, integrated and terminal on branches or discrete, sympodial, geniculate, denticulate with short, thin-walled separating cells which corresponds to a frill at the base of each conidium.

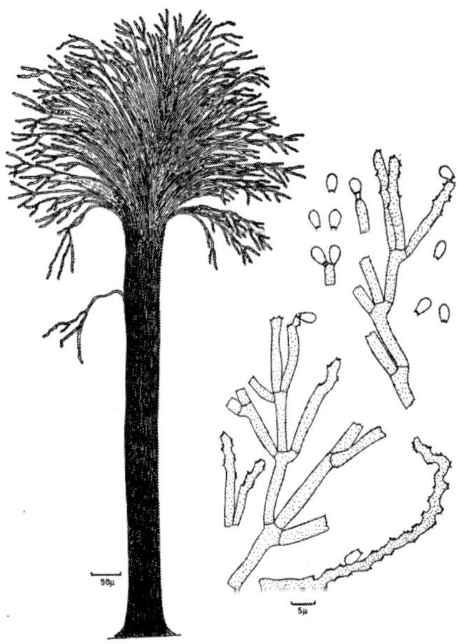

Group of conidiaphores and conidia

Conidia: Solitary, dry, acropleurogenous, simple, ellipsoidal or obovoid, colourless to pale olivaceous brown, 0-septate, smooth.

Cytochalasins: Cytochalasin C, Chaetoglobosin O and Rosellichalasin

Diagnostic features: Ascostomata perithecium ostiolate ascus unitunicate, cylindrical with amyloid apparatus, 8-spored, ascospores one celled, dark-brown, ellipsoid, conidiophores synnematous, conidiogenous cells polyblastic, integrated, sympodial, geniculate, denticulate with short thin walled separating cells which break across the middle leaving minute fill or collar at each geniculation which cross ponds to a fill at the base of each conidium acropleu-rogenous, ellipsoidal or obovoid (0-septate).

61. ***Spicaria elegans*** Corda

Synonymy= *Hormodendrum elegans*

Mariannaea elegans

Paecilomyces elegans

Penicillium elegans

Colonies: Reaching 2.5-3.5 cm in one week at 25ºC on MEA.

Conidiophores: Erect, branched, bearing 6 or more phialides at the tip of the branches;

Phialides: Slender, awl shaped,

Conidia: Producing opening asymmetrical; conidia fusiform, developing obliquely from phialidic opening that results in asymmetrical shaped conidia, 4-6 x 1.5-2.5 µm, cohere in imbricate chains;

Chlamydospores : Catenulate.

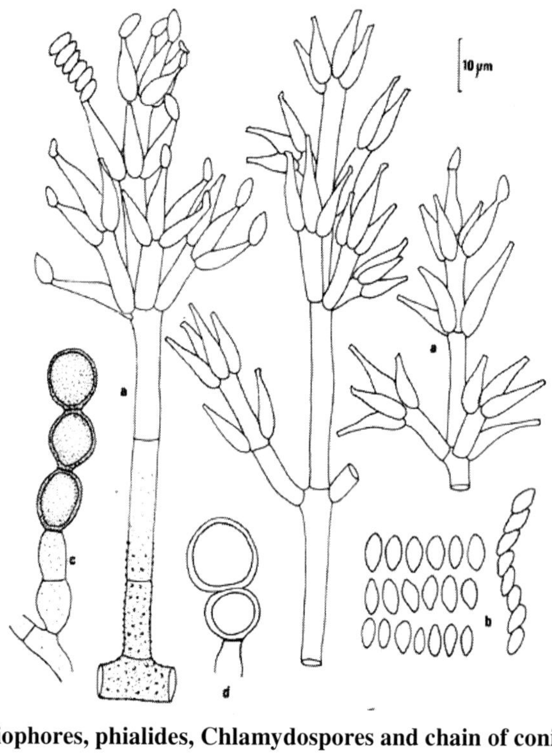

Conidiophores, phialides, Chlamydospores and chain of conidia

Cytochalasins: Cytochalasin F, Z7 to Z15, Z21-Z23, Aspochalasin M,N,O, P and Q, Spicochalasins.

Diagnostic features: Conidiophores branched bearing six or more phialides conidia fusiform developing obliquely forming a Imbricate chain. Chlamydospores are produced in a chain.

62. ***Stachybotrys chartarum*** Ehrend

Synonymys = *Oospora chartarum*

= *Stilbospora chartarum*

Colonies: Reaching 1.4cm diam in five days on OA at 25°C, covered with a dark powdery mass of conidia

Reverse: Uncoloured

Conidiophores : Simple or irregularly branched, usually about 100 μm, sometimes up to 1000 μm long and 3-6 μm wide, hyaline at the base and ± dark-olivavceous, roughened towards the apex bearing cluster of 4-10 phialides

Phialides: Obovate or ellipsoidal, hyaline, later becoming olivaceous, 9-14 x 4-6 μm

Conidia: Aggregating into slimy masses, ellipsoidal, at first hyaline, at maturity dark olivaceous-grey, ± opaque, almost smooth walled to coarsely roughened with warts and ridges, 7-12 x 4-6 μm.

Cytochalasins: Alachalasins A-G

Diagnostic features: Conidiophores elongate, hyaline at the base and dark-olivaceous at the apex, with 4-10 phialides, conidia aggregating into slimy masses, ellipsoid, olivaceous-grey at maturity.

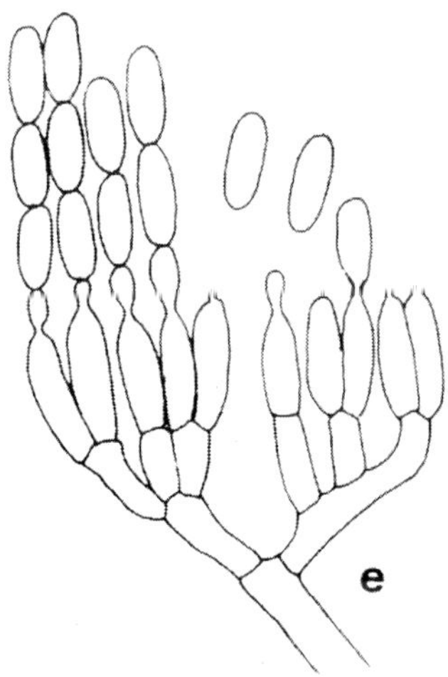

Conidiophores with phialides and conidial chain in slimy mass

63. *Trichoderma gamsii* Samuels & Druzhin

Colonies: Grown on PDA in intermittent light forming conidia sporadically within 1 weak

Mycelium: White covering the 9-cm-diam Petri plate

Conidia: Tending to form in 3 broad, often obscure, concentric rings; conidia form in a dense, No pigment diffusing through the agar, no distinctive odour.

Conidiophores: More or less uniformly branched for a short distance, solitary phialides common, internodes between branches typically long, occasionally short and branches crowded; often branches arising from swollen nodes on the conidiophore, occasionally conidiophores terminating in a long extension; extension sterile for a long distance below the tip, a single phialide forming at the tip.

Phialides: Lageniform, at most slightly swollen in the middle, straight; solitary or terminating branches in pairs or in appressed heads of several, (5.2-) 8.5-10(-18.5) μm long, (1.5-)2.2-3.0(-4) μm at the widest point, (1.0-)1.5-2.2 (-2.7) μm at the base, L/W = (1.4-)2.4-4.2(-8.4) (n = 300); arising from a (1.0-) 1.8-3.0(-5.0) μm wide cell. Intercalary phialides common.

Conidia: Ellipsoidal, oblong, or narrowly ovoidal, smooth.

Cytochalasins: Aspochalasin D, J and I

Diagnostic features: Colonies white covering entre plate producing conidia with one weak, conidiophores branched branches arising from swollen nodes on the conidiophore, phialides lageniform, conidia **e**llipsoidal, oblong, or ovoid, smooth walled one celled.

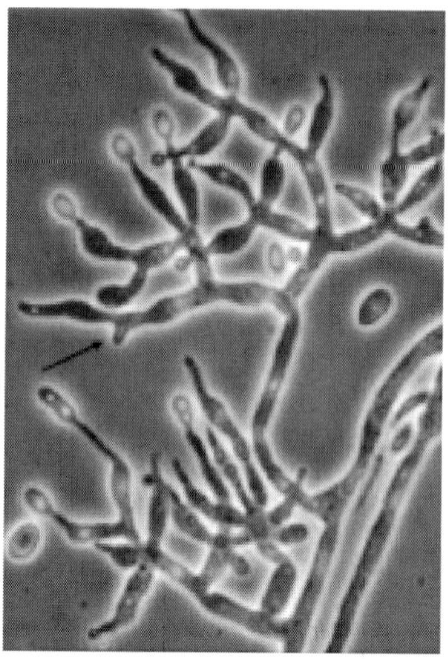

Conidiophores and conidia

64. ***Xylaria hypoxylon*** Grev.

Colonies: Dark brown to black with antler-like white tips

Ascomata: Perithecium cylindrical or flattened with dimensions of 3–8 centimetres (1.2–3.1 in) tall × 2–8 mm thick. The erect ascocarps are often twisted or bent, and typically sparsely branched, often in a shape resembling a stag's antlers

Asci: 100 × 8 μm, cylindrical, 8-spored.

Ascospores: Kidney-shaped, black, and smooth, the asexual spores (mitospores) are ellipsoid in shape, smooth, and hyaline.

Cytochalasins: Cytochalasin C, D**,** N,O,P,Q and R

Diagnostic features: Ascomata brown dark, erect stroma, perithecium immersed in stroma with a short neck lined periphyses, Ascospores kidney shaped, black, smooth produces anamorph produces hyaline, 1-celled, covers the fungal fruit body.

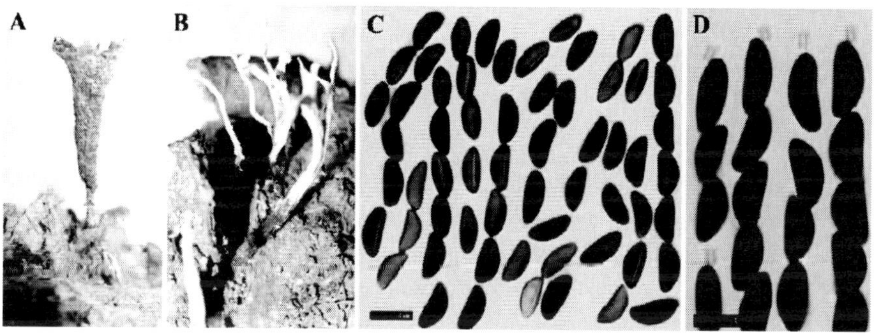

(A) Apothecium (B) cross section of apothecia (C and D) ascospores

65. ***Xylaria obovata*** Berk

Stromata: With fertile part subglobose, irregular-cylindrical to reniform, with rounded fertile apex, 9-17 mm total length x 8-8.5 mm cm diam, short-stipitate. Stipe smooth, 3-6 mm in length x 2-3 mm in diam. External surface black, smooth to wrinkled. Internally white to cream coloured, occasionally becoming hollow

Perithecia: Black, completely immersed, up to 1 mm diam, ostioles papillate, black.

Asci: Eight-spored, cylindrical, stipitate, 135-176 μm total length and 7-9 μm broad, the spore-bearing part in 95-120 μm long, with apical ring bluing in Melzer's iodine reagent, 6-10 μm high to 4-5μm broad.

Ascospores: Ellipsoid, inequilateral, grayish-black, unicellular, smooth, (25-)27-35 x 7-9 μm, germ slit conspicuous, oblique, wavy, less than ascospore length.

Diagnostic features: Stroma irregular cylindrical to reniform, with rounded fertile apex, stipitate, ascus cylindrical apical ring bluing in Melzer's iodine reagent, ascospores ellipsoid, single celled, germ slit conspicuous.

Stalked perithecia

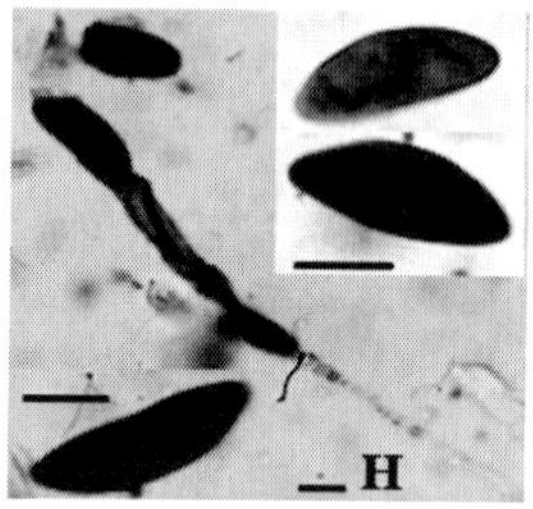

Ascospores

66. ***Xylaria striata*** **Pat.**

Stromata: Gregarious on soil, several arising from common base, 44–120 mm long, fertile portion cylindrical, 9–18(–22) × (2–)2.5–3(–4) mm diam., with acute sterile apices; stipe root-like, 70–100 mm long, stromatal surface more or less roughened from whitish to pale-peeling outer layer and minute wrinkles.

Perithecia: Almost immersed, 0.38–0.56 mm diam.; ostioles papillate to slightly papillate.

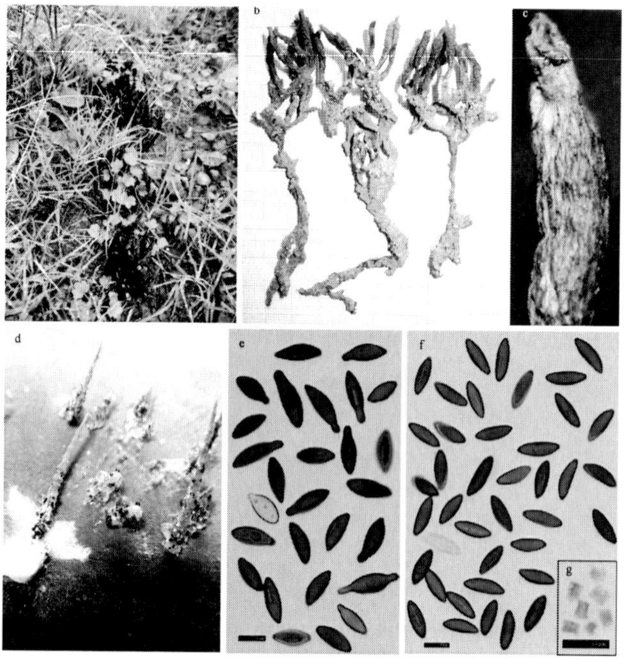

Ascocarp and ascopores

Asci: Degenerated and not seen; apical ring bluing in Melzer's iodine reagent, rectangular, 3–3.5(–4) × 2(– 3) μm.

Ascospores: Ascospores (13–)15–19(–22.5) × 5–7(–9) μm, fusiform to ellipsoid, often apiculate at one end when produced on stromata under natural conditions but not apiculate when produced on stromata under laboratory conditions, with straight about 1/3 of spore-length

Cytochalasin: Xylastriasan A

Diagnostic future: Stromata gregarious, several arising from base, perithecia immersed, 0.38–papillate ostioles, Ascospores fusiform to ellipsoid, apiculate, one-celled.

67. ***Xylaria*** Hill ex Schranck

Stromata: Erect, up to 10cm high, cylindrical, clavate or slender, simple or branched, terete to ellipsoidal in cross-section, corky or fleshy; exterior dark, composed of thick-walled cells, often roughened or covered with matted, dark brown, hairs; dark brown to dark gray or black, tip often whitish and conidiogenous; interior white or sometimes brownish, composed of hyaline, interwoven hyphae, center hollow in some species.

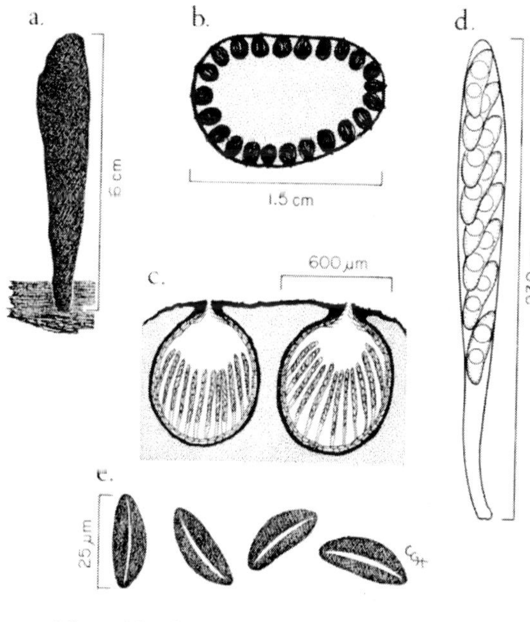

(a) Stroma with perithecia (b) & (c) Cross sectinal view of perithecium (c) Ascospores (d) Ascus with ascospores

Ascomata: Ostilate perirhecium, immersed in a stroma composed entirely of fungal tissue. Immersed, forming a layer on periphery of stroma, with slightly protruding ostiolar papillae, obpyriform to ellipsoidal, often laterally compressed by mutual pressure, ostiole lined with periphyses. Perithecial walls dark and flattened on exterior, subhyaline on interior, paraphyses filamentous.

Asci: Unitunicate, forming a layer along sides and base of perithecium, cylindrical to clavate, with distinct, amyloid apical apparatus, 4-8 spored.

Ascospores: 1-celled, dark brown, oval to ellipsoid, sometimes inaequilateral, with a longitudinal germ slit or a lateral germ pore.

Cytochalasins: Cytochalasin H, H2, Z27, Z28 and Chaetoglobosin O

Diagnostic features: Ascomata immersed in a stroma which is erect, cylindrical to clavate producing ostiolate perithecia, ostiole lined with periphyses, asci cylindrical, unitunicate, 8 spored, ascospores 1-celled, dark brown, oval to ellipsoid. *Nodulisporium* is the Anamorph.

68. *Xylaria arbuscula* Sacc.

Colonies: Grown three weeks at 20°C in diffuse daylight on OA 5 cm diam., with pronounced concentric rings of slightly raised, yellow-tan to orange-tan mycelium alternating with broader bands of appressed white mycelium.

Stromata: Erect, scattered to gregarious, simple to ramified, 5-24 mm total height, the fertile head 3-11 mm high x 1.5-2.5 mm diam, cylindrical to slightly fusiform or flattened, hard-textured, with a pointed sterile apex, long-stipitate to nearly sessile, the stipes well-defined, 1-15 mm high black, glabrous to hairy. Stromatal surface silvery grey to blackish, without visible perithecial contours, roughened by persistent grey to dull black carbonaceous elongated plaques and undulating lines furrowing the surface of the underlying stromatal crust; crust carbonaceous, 80-120 μm thick, interior white, spongy, solid.

Apothecia

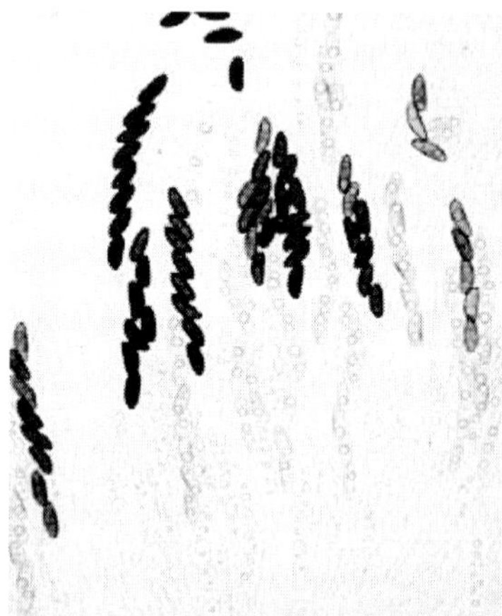

Ascospores

Perithecia: Fully immersed, subglobose to obovoid, 0.4-0.5 mm diam.

Ostioles: Hemispherical to discoid, inconspicuous, ca. 80 μm diam.

Asci: Cylindrical, 8-spored, 190-210 µm total length, the spore-bearing parts 82-110 µm long x 7-8 µm broad, the stipes 64-72 µm long, with apical ring tubular with flared upper rim, 3.5 µm high x 2.5 µm broad, bluing in Melzer's reagent. Paraphyses filiform, copious.

Ascospores: 12-15 x 5.5-6 µm, ellipsoid-inequilateral with narrowly rounded ends, uniseriate to often irregularly biseriate in upper part of the ascus, medium brown (olive brown when fresh), smooth, biguttulate, with a conspicuous straight germ slit 3/4 spore-length on the ventral side.

Anamorph: Not observed.

Diagnostic features: Stromata erect, scattered to gregarious, perithecia fully immersed, subglobose, ostiole inconspicuous, cylindrical ascus with tubular, apical ring, ascospores one celled, brown, biguttulate with straight germ slit.

69. ***Xylaria psidii*** J.D. Rogers & Hemmes

Colonies: Pink, mycelia with red media on malt extract medium; blackish-white stromata developed faster on rice agar medium.

Stromata: Upright, 5–6 cm long, branched or unbranched, single or in clusters and extended to the fertile cylindrical or conical apex.

Perithecia: Embedded, stipes about 1.5 mm thick, twisted, externally brown and internally white coloured, while fertile part is black. Texture hard, asci eight spored arranged in uni or biseriate manner.

Ascospores: Black-brown, unicellular, ellipsoid, inequilateral, smooth with acute ends and 4–5 × 7.3–11 µm in size with straight germ slit throughout spore-length.

Cytochalasins: Cytochalasin B

Diagnostic features: Colonies: Pink, blackish-white stromata, stromata upright, 5–6 cm long, branched or unbranched, perithecia embedded, stipes short and thick, ascospores black-brown, unicellular, ellipsoid, straight germ slit.

Ascocarp on a dead wood

70. *Zygosporium masonii* Hughes

Colony: Effuse or compact, brown, blackish-brown or black

Conidiophores: Up to 100µm long, about 2 µ thick, with chains of up to 6 integrated vesicles and a hyaline, sterile apical region which terminates in a small knob.

Vesicles: 7-12 µ long, 4-5 µ thick in the broadest part.

Conidia: Ellipsoidal,, hyaline 6-8 x 3-4 µ.

Cytochalasins: Cytochalasin D, Zygosporin A, D, E, F, G.

Diagnostic features: Conidiophore with up to 6- integrated vesicles, hyaline, vesicular conidiogenous cells, non-celled, hyaline.

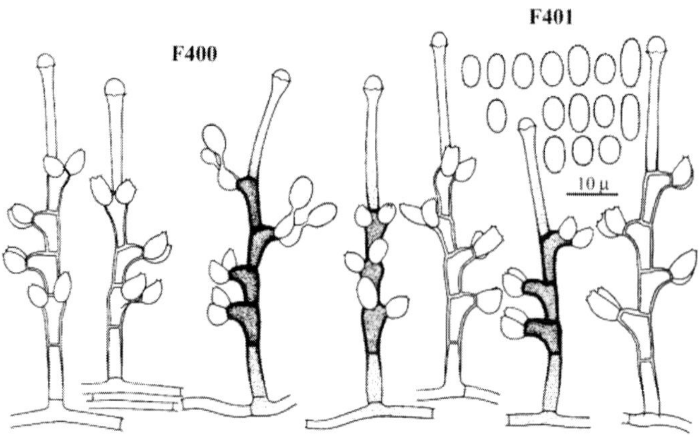

Conidiophores and Conidia

■■■■

Chapter-4

Isolation and Detection of Cytochalasins

In view of diversity of fungi producing cytochalasins, they are being grown on diverse media. Some of the media commonly employed for, this purpose are PDA (Potato-200g; dextrose-20g; agar-20gr and distilled water- 1 lit), Malt extract medium (Malt extract-20g; peptone-1g; dextrose-20g; agar-20g and distilled water- 1 lit), Rice medium (Rice-40g and distilled water-100ml), Yeast extract medium (Glucose-2%, peptone -1%and yeast extract-0.5% inartificial seawater) and Soy sauce-sucrose medium (Soy sauce -50ml; sucrose-50mg and distilled water-1lit). They are grown on solid or liquid media at specific pH and incubation temperature. The fungi grown for 10-15 days are extracted for cytochalasins. Both the culture broth and mycelium are employed for extraction of different cytochalasins depending on the fungus.

4.1. Cytochalasins

(i) Cytochalasin A and B

Analytical and preparative TLC of cytochalasins has been carried out on silica gel by most workers. Examples of solvent systems are given below. Detection can be effected with iodine vapour, ethanolic sulphuric acid, vanillin-phosphoric acid and Dragendorff reagent, or p-anisaldehyde followed by examination under long-wave UV light (360nm).

The TLC of cytochalasins A and B, together with other cytochalasins in several solvent systems, are characterized. TLC on silica gel G of several derivatives of cytochalasin B was carried out by Rothweiler and Tamm (1970). Chloroform containing 0.5-20% of methanol was mostly used as the solvent system. Detection was effected with iodine vapour.

(ii) Cytochalasins C and D

Aldrige and Turner (1969) separated cytochalasins C and D on silica gel G using chloroform-methanol-formic acid (90:5:5). The plates were sprayed with 5% ethanolic sulphuric acid and heated to ca. 110°C for a few minutes. Cytochalasin C gave an orange and cytochalasin D a yellow fluorescence under UV light.

TLC of cytochalasin D was carried out on silica gel GF_{254} with benzene: ethylacetate (7:3) or chloroform : methanol (9:1) (or both) and the spots were rendered visible by spraying with concentrated sulphuric acid followed by heating at 180°C (Hayakawa *et al.*, 1973).

Padhye *et al.* (1975) published R_f values of seven cytochalasins (including C and D) in eleven solvent systems using three reagents such as ethanolic sulphuric acid, vanillin-phosphoric acid and Dragendroff's reagent for detection of cytochalasins **(Table 4.1).** Cytochalasin C and D had almost identical R_F values in most of the solvent systems used. They could be readily distinguished from each other by spraying with ethanolic sulphuric acid and observing their fluorescence under UV light. Cytochalasin C gives a dull orange colour and cytochalasin D a weak yellow colour. Durackova *et al.*, (1976) gave additional information on the TLC behaviour of cytochalasins A, B and D in eight solvent systems and their detection with p-anisaldehyde **(Table 4.2).**

Table 4.1. R_f x 100 values of cytochalasins on silica gel F plates[2]

Cytochalasin	Solvent system*										
	1	**2**	**3**	**4**	**5**	**6**	**7**	**8**	**9**	**10**	**11**
A	63	53	62	39	36	0	93	67	79	95	56
B	58	38	53	36	30	0	92	63	78	95	47
C	46	34	44	23	26	0	90	59	78	92	30
D	45	32	42	19t**	20t	0	89	58	76	93	34
H	63	55	64	46	50	0	87	59	80	92	60
J	44	38	56	34	35	0	75	55	72	90	46
E	79	57	63	48	40	0	90	62	82	96	71t

* Solvent systems : 1 = chloroform-methanol (95:5); 2 = chloroform-methanol formic acid (95:5:5); 3 = chloroform diethylamine (90:10); 4 = diisopropyl ether-ethyl acetate (90:10); 5 = cyclohexane-ethyl acetate-diethylamine (60:30:10); 6 = benzene-chloroform (50:50); 7 = n-butanol-formic acid-2ater (80:10:10); 8 = benzene-methanol (70:30); 9 isopropanol; 10 = acetone; 11 = benzene-acetone (70:30). ** t = Tailing.

(iii) Cytochalasin E

The procedure was developed for the detection of cytochalasin E by Demain *et al.* (1976). Spores from a soil cultures maintained at 4°C were inoculated on to Czapek Dox medium, enriched with 0.5% yeast extract and 0.5% casamino acids (Difco Lab.), in a 120-ml dilution bottle. The incubation lasted for 4-6 days at 30°C. Spores were suspended in 100 ml of 0.01% (w/v) sodium lauryl sulphate, and 10ml of the spore suspension were added to each 2.8-1, wide-mouthed Fernbach flask. The Fernbach flasks had received 300 g of pearled barley and

150 ml of warm water. They were then autoclaved, cooled and incubated on a rotary shaker at 200 rpm (ca. 5.08 cm diameter orbit) for 12 days at 30°C.

Table 4.2. R_f values of different cytochalasins on (a) silica gel plates and (b) silufol sheets

Cytochalasin	Solvent system*								Detection (Spray reagent)	
	A	B	C	D	E	F	G	H	1	2
A (a)	41	58	25	85	17	33	46	99	Pale beige	Blue
(b)	34	42	23	83	20	31	44	95		
B (a)	28	46	13	78	8	16	24	98	Violet	Blue
(b)	34	34	13	74	7	14	26	94		
D (a)	19	24	10	75	0	8	14	90	Beige	Orange
(b)	32	27	10	74	4	9	20	85		

*Solvent systems : A = benzene-methanol-acetic acid (24:2:1); B = toluene-ethyl acetate-90% formic acid (6:3:1); C = benzene-ethanol (95:5); D = chloroform-methanol (4:1); E = chloroform-isobutyl methyl ketone (4:1); F = chloroform-acetate (9:1); G = chloroform-acetic acid-diethyl ether (17:1:3); H = n-butanol-acetic acid-water (4:1:4, upper layer). *Detection; 1 = p-anisaldehyde ; 2 = UV at 366 nm after p-anisaldehyde.

The barley was extracted twice, each with 500ml of methylene chloride per flask. After the first addition of methylene chloride, the suspension was allowed to soak over night. The flask contents were then filtered through glass-wool and the grains were blended in a warning blender for 60 sec with the second 500-ml portion of methylene chloride. The homogenate was filtered through cheese cloth and the solids were discarded. After combining the two extracts, anhydrous magnesium sulphate was added in order to remove trace amounts of water; the suspension was then filtered on a Buchner funnel with No. 595 filter-paper (Schleicher & Schuell) and with either celite filter aid No. 503 (Johns-Manville) or Hyflo Super Cel. After evaporating the filtrate to dryness in a flask evaporator, the oily residue was precipitated by pouring it slowly into ice-cooled light petroleum and placing it over night in an ice-bath. The crude toxin fraction was recovered by filtration under vacuum on a medium-pore, sintered-glass filter and allowed to dry in the hood before being weighed. To remove trace amounts of light petroleum before determination of the weight of the fraction, the vial was stored overnight under negative pressure. From 1 kg of barley, approximately 35 mg of cytochalasin E and 80 mg of combined tremorgens were isolated in pure form. This yield represents an improvement over the original production of 10 mg of cytochalasin E and 11 mg of tremorgens per kilogram of glutinous rice. The tremorgens formed from this strain usually occur in a 3:1 ratio of tryptoquivaline to tryptoquivalone (Demain *et al.*, 1976).

(iv) Cytochalasin G

A large-scale fermentation of *Z. masonii* was used to produce cytochalasin D (Hayakawa *et al.*, 1973). In 500 ml flask containing 100 ml of a pre-seed medium (30 g of glucose, 20g of peptone and 5g of sodium chloride in 1 l of tap water, pH adjusted to 6 prior to sterilization) was employed to grow at 28°C for 96 h on a reciprocating shaker.

The culture filtrate of initial pH 3.4, was adjusted to pH 7.0 and extracted with chloroform (3 x 3 l). The solvent was removed to give a brown semi-solid residue (2.2 g), which was combined with similar extracts to give a total of 7.47 g. The extract was adsorbed on silica gel (75 g) by evaporation of an acetone solution and placed on the top of a silica gel column (300 g) made up in benzene. Elution with benzene containing increasing proportion of chloroform gave gummy fractions (3 g), which were discarded. Elution with chloroform gave a series of solid fractions, the earlier of which were a single compound and combined (2.0 g) and recrystallised from acetone to give cytochalasin C.

From a shake-flask fermentation, cytochalasin D was isolated as follows. A 25 g amount of celite was added to culture broth and filtered to obtain 55g of mycelium (including the celite) and 860 ml of the filtrate. The mycelium was extracted with methanol (2 x 200 ml) and the extract was evaporated in vacuo. The residue was dissolved in ethyl acetate, washed with water, 2 N sodium carbonate solution and water, dried over sodium sulphate and evaporated at 50-60°C in vacuo. The residue gave cytochalasin D.

The producing strain of *Nigrosabulum* sp. (Cameron *et al.*, 1974) was grown as a surface culture for 28 days in a medium adjusted to pH 5.5 and containing (g l^{-1}) cerelose (50), $NaNO_3$ (2), KH_2PO_4 (1), $MgSO_4.7H_2O$ (0.5), KCl (0.5), $FeSO_4$ (0.01), oxoid yeast extract (1) and 0.1% of a minor element concentrate.

Pseudoeurotium zonatum was grown as a surface culture in a medium containing (g per litre of demineralized water) cerelose (50), $NaNO_3$ (2), KH_2PO_4 (1), $MgSO_4.7H_2O$ (0.5), KCl (0.5), yeast extract (1) and 1 ml of minor element solution pH 5.3-5.7 (non-adjusted). The sterilized medium was inoculated with a spore suspension and the flasks, each containing 1 : l of the medium, were incubated for 25 days at 27°C.

The culture filtrate was extracted at pH 6.8 with chloroform (2 x 18 l) to afford an amorphous solid (5 g), part of which (3.8 g) was dissolved in the two-phase system toluene-methanol-chloroform-water (40.5:40.5:9.5:9.5) (100 ml). The mixture was placed in an equal portions in the first four separators of Gallenkamp automatic counter current apparatus programmed to perform 120 solvent transfers with a partition period of 1 min and a settling time of 10 min. Additional lower-solvent phase (12.5 ml) was placed to each of the remaining separators and upper-solvent phase was placed in the reservoir. At completion the contents of separators 56-80 were pooled and concentrated in vacuo to ca. 50 ml, then extracted with ethyl acetate (3 x 100 ml) to afford a solid (0.81 g), which was recrystallized from acetone-ethyl acetate. Two further crystallizations of the product from methanol yielded cytochalasin G (350 mg)(Cameron *et al.*, 1974).

Probst and Tamm (1981) isolated Cytochalasin G from both the mycelium and the medium of *Pseudoeurotium zonatum*, were extracted twice with ethyl acetate. The organic layer was dried over sodium sulphate and evaporated in vacuo at 40°C. The dark-brown extract was chromatographed on silica gel (100:1) using increasing amount of methanol in methylene chloride. From the fractions containing 1.25% of methanol, cytochalasin G was isolated by crystallization from acetone-hexane.

(v) Cytochalasin H and J

Phomopsis paspalli was cultivated in large conical flasks in 2% potato-dextrose-agar medium for 3 weeks (Patwardhan *et al.*, 1974). The entire material including the medium, was agitated with an amount of 95% ethanol sufficient to cover the entire material and the mixture was left at 20°C for 4 days. The aqueous alcoholic extract was drained off and evaporated under reduced pressure. The residue (65 g) was mixed with sodium sulphate (300 g) and extracted by cold percolation, with the following sequence of solvents to furnish various extracts; light petroleum (124 mg) (gum), diethyl ether (585 mg) (yellow sticky mass), ethyl acetate (1.7 g) (yellow solid) and acetone (618 mg, gum).

TLC and preparative TLC were carried out on silica gel layers containing 15% gypsum. TLC of the diethyl ether extract in chloroform-methanol (:1) and spraying with concentrated sulphuric acid gave four spots with R_f 0.56 (yellow, minor) and 0.32 (red, major). The red spots correspond to kodo-cytochalasin-1 (cytochalasin H) and kodo-cytochalasin-2 (cytochalasin J).

Extraction of the mycelium resulted a mixture of cytochalasins N,O,P,Q and R on TLC plates. The compound giving a red (N), yellow (O, P, Q), or blue (R) colouration after spraying with an p-anisaldehyde-acetic acid : sulphuric acid (1 : 98 : 1) mixture followed by heating at 110°C for 2-3 min.

4.2. Zygosporins

TLC was used in the isolation and purification of zygosporins from a culture of *Zygosporium masonii*. The culture filtrate was extracted with ethyl acetate, washed and dried extract was evaporated to about one-third of its volume and the separated product was filtered off. The filtrate was evaporated in *Vacuo* to give crude cytochalasin D (zygosporin **A**) and a paste **(A)**.

Fraction 1, eluted with benzene, benzene-chloroform (9:1. 7:3 and 9:5) and chloroform, contained an oil.

Fraction 2, eluted from the silica gel column with chloroform, was crystallized from ethyl acetate to give a crystalline product (C) and a paste (D). The latter was chromatographed on alumina to give an oil and a paste (E), eluted with chloroform-methanol.

Fraction 3, eluted from the silica gel column with chloroform-methanol (9:1), was dissolved in light petroleum and the precipitate (F) was collected. The crystalline product C was separated into cytochalasin D (R 0.40) and zygosporin E (R_f 0.48) by TLC with ethyl acetate as the solvent. The paste E was re-chromatographed on silica gel to give an amorphous powder, which was separated into zygosporin G (R 0.35) and zygosporin F (R_f 0.28) by TLC using toluene-methanol (10 : 1) as the solvent system. The precipitate F was separated into cytochalasin D (R_f 0.50) and zygosporin D (R_f 0.40) by TLC using chloroform-methanol (10:l) as the solvent system. In addition the isolation of zygosporins, TLC has been used to characterize, degradation of products and derivatives of the four zygosporins. R_f values of zygosporins on silica gel plates developed with chloroform-methanol (9:l) are R_f 0.50 for zygosporin A (cytochalasin D), 0.40 for zygosporin D, 0.55 for zygosporin E and 0.57 for zygosporin F and G.

4.3. Chaetoglobosins

Udagawa *et al.* (1979) extracted chaetoglobosin A, B, C, D and E by dissecting agar surface of 14-day-old cultures grown on oatmeal agar or potato dextrose agar slants at 23-25°C. Each *Chaetomium* spp. was grown in 1-1 Roux flasks on a medium composed of 200 g of polished rice and 20 ml of sterile water per litre flask. After inoculation, the flasks were incubated in a static condition at 23-25°C for 21 days.

Production of 19-0-acetyl derivatives of chaetoglobosins, A, B and D by *C. globosum* (strain Lederle H-124) was as follows. *Chaetomium globosum* was grown in shake cultures in a medium containing (g l^{-1}) molasses (20), glucose (10) and bactopeptone (5) in demineralized water; pH adjusted to 7.5 with sodium hydroxide before sterilization. The sterilized medium was inoculated with a spore suspension. For chaetoglobosin C and chaetoglobosin K-producing *Penicillium aurantio-virens and Diplodia macrospora* were grown on shredded wheat medium was cultured on potato dextrose.

To produce chaetoglobosins K and L, *Diplodia macrospora* was grown in standing cultures in ten 1-1 Erlenmeyer flasks. Each flask contained 40 g of shredded wheat, 80 ml of Difco mycological broth (pH 4.8), 16 g of sucrose and 1.6 g of yeast extract. The flasks were sterilized and inoculated with a suspension of spores of *D. macrospora* and incubated at 27°C for 22 days.

At the end of incubation period, the cultures were harvested and extracted with methylene chloride twice at room temperature and the combined solution was evaporated. The residue was extracted with ethyl acetate twice at room temperature and the solution was evaporated. The extracts were analysed by TLC for detection of chaetoglobosin.

The mycelium of *P. aurantio-virens* was extracted with chloroform and subjected to a series of elutions on a silica gel column. Toxicity was found in the ethylacetate fraction (oral, 1-day-old cockerels). A second column employing gradient elution (chloroform-ethyl acetate) was used for further purification. The crude toxin was crystallized from ethyl acetate to give small needles of Chaetoglobosin C.

Chaetoglobosin A and B were isolated from 14 days old submerged culture of *C. globosum* by adding equal amount of ethanol and kept overnight at 4°C. The ethanol was evaporated in vacuo at 40-50°C and the aqueous layer was extracted four times with methylene chloride. The organic layer was dried over sodium sulphate and evaporated in vacuo at 40°C. The brown residue was dissolved in benzene and kept for 3 days at room temperature. The crystallized mixture of Chaetoglobosin A and 19-0-acetylchaetoglobosin A was separated by TLC on silica gel using increasing amounts of methanol in methylene chloride. The benzene solution was evaporated in vacuo at 40°C and the residue chromatographed on silica gel (100:1) in the same manner to get 19-0-Acetylchaetoglobosin A, chaetoglobosin A, 70-80 mg for chaetoglobosin C, 19-0-acetylchaetoglobosin D and 19-0-acetylchaetoglobosin B.

For the separation of Chaetoglobosins K and L (Probst and Tamm, 1982) on TLC, CH_2Cl_2 : methanol (95:5) and toluene-ethylacetate-formic acid (5:4:1) served as solvents. The fractions from the column chromatography containing the chaetoglobosins K and L were purified by preparative TLC on silica gel with toluene-ethyl acetate-formic acid (5:4:1).

Chaetoglobosins A-E were chromatographed on silica gel F_{254} using benzene-ethylacetate (1:1) and benzene-chloroform-methanol (10:10:3) as the solvents. Metabolites were detected by UV irradiation at 254 and 365 and by spraying with Ehrlich's reagent (5% in dilute hydrochloric acid) and colouration after heating (Udgawa *et al.*, 1979).

The R_f value of Chaetoglobosin K was 0.53-0.56 on silica gel 60 TLC plates developed with toluene-ethylacetate-formic acid (5:4:1) and it was observed as a dark-spot under shortwave UV light (Cutler *et al.*, 1980).

Decreasing R_f values of the five Chaetoglobosins on the same sorbent and methylene chloride-methanol (95:5), showing increasing polarity from left to right: 19-0-acetylchaetoglobosin A > chaetoglobosin C > 19-0 acetylchaeto-globosin B > 19-0-acetylchaetoglobosin D > chaetoglobosin A (Probst and Tamm, 1981).

4.4. Aspochalasins

Two hundred ml medium containing (g l^{-1}) glucose (50), L(+)-tartaric acid (3), diammonium D-tartrate (3), $(NH_4)_2HPO_4$ (0.4), K_2CO_3 (0.4), magnesium hydroxide carbonate (0.3), $FeSO_4.7H_2O$ (0.05), $ZnSO_4.7H_2O$ (0.05) and demineralized water; pH 3.7 before and after sterilization was inoculated with 2 ml of a conidial suspension (9 x 10^7 ml^{-1}) and incubated on a rotary shaker at 27°C for 48 h. The culture filtrate was extracted repeatedly with about 1/20 volume of ethyl acetate (Heberle *et al.*, 1974).

TLC of the aspochalasins was performed on Kieselgel 60-Fertigplatten F_{254} and the spots were rendered visible by spraying with 50% sulphuric acid and heating at 200°C, with iodine vapour or fluorescence under UV light. R_f values of 0.35, 0.27 and 0.54 were obtained for aspochalasins C, D and B respectively (in ethyl acetate, blue fluorescence). In chloroform-methyl acetate (4:1), aspochalasins A and B had R_f values of 0.53 and 0.26, respectively. TLC data for derivatives and degradation products of aspochalasins can also be recorded.

4.5. Deoxaphomin, proxiphomin and protophomin

Phoma sp., strain S 298, was grown in static culture in a medium containing (g l^{-1}) glucose (20), peptone (2), yeast extract (2), malt extract (2), KH_2PO_4 (2), $MgSO_4.7H_2O$ (2) and deminaralized water at 18°C for 11 days.

The culture medium was separated from the mycelium by filtration and the mycelium was washed three times with water. Ten liters of the filtrate were extracted four times with 8 l of ethyl acetate and the extract was washed three times with mycelial washings. The extract was dried over calcium chloride, filtered and evaporated in vacuo. The dark-brown crude extract was dissolved in methylene chloride methanol (9:1) and filtered through alumina (activity II). From this partially purified extract, phomin (cytochalasin B) was obtained by fraction crystallized alumina or Kieselgel. Phomin was eluted from alumina (activity II) with methylene chloride-methanol (98:2). Dehydrophomin, which moved slightly more quickly, was eluted with methylene chloride or methylene chloride-methanol (99:1).

Deoxaphomin was isolated from 1 liter of homogenized culture and extracted four times with 1 liter of ethyl acetate. A mixture of silicone oil and propylene glycol was used as an anti-foaming agent. The extract was evaporated and the residue was separated between light petroleum and methanol-water (9:1). The non-polar fraction was stored and the polar fraction was purified by filtration through alumina in methylene chloride-methanol (9:1). Methylene chloride was added to the extract and phomin was precipitated quantitatively. Crystallization from acetone gave phomin, which, on TLC (methylene chloride-methanol, 95:5), showed a polar strongly UV-active impurity.

Repeated crystallization of crude phomin from acetone yielded pure phomin. The mother liquors from both crystallizations were combined (ca. 1g) and deoxaphomin was separated on four preparative plates (20 x 20 cm, 2 mm) with chloroform-acetone (3:1). The crude extract was further purified using four preparative separations (chloroform : acetone (3:1), twice with chloroform-acetone-formic acid (90:5:5), chloroform : acetone (3:1). The substance was extracted with chloroform (1:1) and the extracts were checked for purity by TLC (chloroform-acetone, 3:1; chloroform-acetone-formic acid, 90:5:5). Amorphous deoxaphomin was obtained (Rothweiler and Tamm, 1970).

Proxiphomin and protophomin were extracted from the residue after isolation of phomin and deoxaphomin (Binder and Tamm, 1973). The extract was chromatographed on Kieselgel and eluted with methylene chloride-methanol (9:1), contained several non-polar components. Fractions 7-9 were combined and chromatographed again on Kieselgel. From fractions 17-26 crude proxiphimin resulted and fractions 60-68 (methylene chloride-methanol, 98:2) contained protophomin.

Impure proxiphomin was purified using preparative TLC (20 x 20 cm, 1 mm) with methylene chloride-methanol and chloroform-acetone (4:1) (TLC with methylene chloride-methanol, 98:2, and methylene chloride-ethyl acetate, 9:1).

4.6. Spicochalasin

Spicochalasin was isolated from *Spicaria elegans* (Zhenjian *et al.,* 2009). The fungus was grown in a liquid medium (PDA) for 8 days at 28°C. The whole broth was exhibited three times with ethyl acetate and concentrated under reduced pressure. The extract was separated into 5 fractions by silica gel chromatography and eluted with gradient of methylene chloride/methyl alcohol and further purified with reverse phase chromatography eluting with methanol: water (7:3).

4.7. Scoparasin

Scoparasin is elaborated by an endophytic fungus *E. scoparia* (Pongcharoen *et al.,* 2007). It is grown on PDA broth at 25°C from 5 days at room temperature for 4 weeks. At the end of incubation period, the filtrate was extracted three times with ethyl acetate and reduced to a brown gum which is subjected to separation by column chromatography with a gradient system of increasing polarity to get scoparasin.

4.8. Periconiasin

Periconiasin is elaborated by *Periconia* sp (Zhang *et al.,* 2013). The fungus was grown in 1 liter Erlenmeyer conical flask containing PDA growth for 10 days at 25°C. At the end of incubation period, cultures were harvested. The filtrate thus obtained were reduced and applied to Amberlite XAD-16 macroporous adsorbent resin column by eluting in H_2O and ethyl acetate, successively. The extract was reduced by evaporation and subjected to column chromatography to get periconiasin.

Some of the methods of separation, solvents and spray reagents for detection of cytochalasins are précised in **Table 4. 3.**

Table 4.3. Isolation and detection of different cytochalasins

Name of cytochalasin	Substratum/ medium	Solvent system	Spray reagent	Rf value	Reference
Cytochalasin A & B	Semi synthetic liquid medium	Chloroform-Acetone (93:7) Toluene-Ethylacetone-formic acid	50% ethanolic H_2SO_4	0.29 0.68	Rothweiler and Tamm (1970).
Cytochalasin C & D		Chloroform-methanol-formic acid (90:5:5)	5% ethanolic sulphuric acid and heat at 110ºC for few mints.	0.40	Hayakawa *et al.* (1973)
Cytochalasin E	PDA liquid medium	Acetone-hexane (2:3)	H2SO4 spraying and heat	0.25	Qiao *et al.* (2011)
Cytochalasin H&J		Chloroform-methanol (9:1)	Concentrated sulphuric acid	0.56	Patwardhan *et al.* (1974)
Cytochalasin K,L & M		using reversed-phase chromatography.			
Cytochalasin N,O,P,Q,R	Malt extract medium	ethyl acetate	anisaldehyde-acetic acidsulphuric acid (1 : 98 : 1)		Edwards and Maitland, (1989)
Cytochalasin W	Wheat kernel medium	CH_2Cl_2-*ipr*OH		0.6	Evidento *et al.* (1996)
Cytochalasin X, Y, Z	Malt extract medium	Ether: EtOAC (3:2)	HPLC: Flow rate 1:0 ml/min, H2O/acetonitrile; 90:10-25:75 in14 min, held at 25:75 for 10 min, 25:75-0:100 in 2 min, held at 0:100 for 4 min, then 0:100-90:10 in 2 min, and held at 90:10 for 8 min. retention time13.8, 12.3, and 15.1 min, respectively.	-	Feng *et al.*(2002).
Cytochalasin Z1	-	$CHCl_3$-iso-PrOH (9:1)	10% H2SO4 in methanol and 5% Pospho molybdic	0.32	Evidente *et al.*(2002)

Name of cytochalasin	Substratum/ medium	Solvent system	Spray reagent	Rf value	Reference
Cytochalasin Z2	-	EtoAC-n-Hexane (7:3)	acid in methanol followed by heating at 110C for 10 min	0.50	Evidente *et al.* (2002)
Cytochalasin Z3	-	$CHCl_3$-iso-PrOH (93:7)		0.60	Evidente *et al.* (2002)
Cytochalasin Z4	Wheat kernels autoclaved	$CHCl_3$: PrOH (9:1)	Reverse phase EtOH-H_2O (6:4)	0.58	Evidente *et al.*, (2003)
Cytochalasin Z5		$CHCl_3$: PrOH (9:1)	Reverse phase EtOH-H_2O (6:4)	0.54	Evidente *et al.* (2003)
Cytochalasin Z6		$CHCl_3$: PrOH (9:1)	Reverse phase EtOH-n-hexane (7:3)	0.48	Evidente *et al.* (2003)
Cytochalasin Z16-17	PDA	Ethyl acetate : MeOH (1:1, v/v)	HPLC with MeOH/H_2O (64:36, v/v)		Zhang *et al.* (2010)
Cytochalasin Z21-23	PDA	Petroleum ether/CHCl3/ MeOH	HPLC (75% MeOH; 4ml/min)		Wang *et al.* (2011)
Cytochalasin Z24	Potato glucose liquid medium	Petroleum ether/$CHCl_3$/ MeOH (2:1:1)	HPLC (60% MeOH/H_2O)		Zheng *et al.*(2013)
Cytochalasin Z27-Z28	Rice medium	$CHCl_3$: MeOH (1:5)	Column run with 60-90% MeOH/H_2O at flow rate .0Ml/min for 30 min		Zhang *et al.* (2014)
Chaeto-globosin A-E		benzene-ethylacetate (1:1) and benzene-chl-oroform-metha-nol (10:10:3)	Ehrlich's reagent (5% in dilute hydrochloride acid and colouration after heating		Udgawa *et al.* (1979)
Chaeto-globosin M & N	potato-dextrose-agar	petroleum ether/ ethyl acetate 1: 1		0.55	Burlot *et al.* (2003)
Chaeto-globosin Z	potato-dextrose-agar/T-top medium	CH_2Cl_2–MeOH (100:1–10:1)			

Name of cytochalasin	Substratum/ medium	Solvent system	Spray reagent	Rf value	Reference
Zygosporin D		Chloroform-methanol (10:1)		0.50	Minato and Matsumoto (1970)
Zygosporin E		Ethyl acetate		0.48	Minato and Matsumoto (1970)
Zygosporin F		Toluene : methanol (10:1)		0.28	Minato and Matsumoto (1970)
Zygosporin G		Toluene : methanol (10:1)		0.35	Minato and Matsumoto (1970)
Penochalasin A	Yeast extract	CH_2C1_2-MeOH (1 9 : 1)	10% H_2S0_4 followed by heating.		Numata *et al.* (1996)
Alachalasins	PDA	n-hexane–CH_2C1_2-MeOH	Used HPLC		Zhang *et al.* (2008)
Spicochalasin	soybean flour	$CHCl_3$/MeOH (95:5)	HPLC		Zhenjian *et al.* (2009)
Aspochalasin	PDA	CH_2Cl_2: MeOH (1:1)	silica gel column CH_2Cl_2/MeOH (94:6)		Zhou *et al.*(2004)
Aspochalasin F & G	Czapek Yeast extract agar	n-hexane: ethyl acetate (4:1)	HPLC: mobile phase: n-hexane: EtOAC (4:1) Flow rate: 7ml/min Detection at 240 nm.		Fang *et al.* (1997)
Pyrichalasin H	Soy sauce-sucrose medium	$CHCl_3$/MeOH (10:1)		0.56	Manabu and Tsuneo (1991)
Rosellichalasin	Malt extract medium	Benzene : Acetone (9:1)	Column run with methanol		Kimura *et al.* (1989)
Flavichalasins	PDA	CH_2Cl_2:MeOH	Column run with petroleum ether: Ethyl acetate (20:1) HPLC= 65% MeCN in H_2O 2ml/min		Wei *et al.* (2017)

■■■■

Chapter-5

Biosynthetic Pathways and Chemical Synthetic Strategies

Molecular diversity of cytochalasin family members such as cytochalasin, chaetoglobosins and aspochalasins are derived from three parts (1) a pyrrolidinone consisting of distinct amino acid, (2) a cyclohexane with functionalities derived from epoxide (3) and a macrocycle including 11- and 13-membered carbocycles, a lactone, and a carbonate with various oxidative decorations.

I. BIOSYNTHESIS OF CYTOCHALASINS

The biosynthetic origins of cytochalasin molecular skeleton were established by a number of feeding experiments with isotope-labeled precursors such as acetate and amino acids (Tanenbaum *et al.,* 2010). Although involvement of the intramolecular Diels–Alder reaction (IMDA) was proposed for construction of the characteristic perhydroisoindol one with macrocyclic system in the cytochalasin biosynthesis, it was difficult to define the reaction sequence in the late stage modifications (Probst and Tamm, 1981). In the biosynthetic study of the structurally related chaetoglobosins, Oikawa *et al.* (1992) succeeded in the determination of the origin of oxygen atoms and obtained a series of less oxidized analogs including prochaetoglobosin I by treatment of the producer strain with the specific P450 inhibitors .These results enabled us to propose IMDA of the pyrrolinone with linear polyketide chain to give prochaetoglobosin I. Schumann and Hertweck (2007) using an RNA silencing method, identified the first cytochalasin biosynthetic gene cluster consisting of a polyketide synthase, non-ribosomal peptide synthetase (PKS–NRPS) hybrid and oxidative modification enzymes which are expected in this proposal.

5.1. Biosynthesis of cytochalasin E and K

Qiao *et al.* (2011) based on isotope labeling studies identified the *ccsA* gene cluster involved in biosynthesis of cytochalasin E and K (**Fig. 5.1**). The domain organization of the PKS-NRPS enclosed by *ccsA* is similar to that of other reported PKS-NRPS genes including Che A in *P. expansum*, Ten S in *Beauveria bassiana* and APdA in *Aspergillus nidulans*. A dissociated ER (encoded by gene *ccsC*) downstream of *ccsA* will function *in trans* with the PKS module of *ccsA* to synthesize a reduced *β*-keto octaketide backbone, analogous to the roles played by TenC and ApdC during the biosynthesis of tenillin and aspyridone respectively. Subsequently, one molecule of phenylalanine is selectively activated by the A domain of the NRPS module and transferred to the phosphopantetheinyl arm of T domain. C domain then catalyzes the condensation between the nucleophilic amino group of phenylalanine and the electrophilic carbonyl of the upstream nascent octaketide chain to yield the T domain-tethered amide intermediate.

The R domain at the end of the NRPS module catalyzes the reductive release of the nascent aminoacyl-thioester to give an aminoaldehyde intermediate, which can readily undergo an intramolecular Knoevenagel condensation to yield the 3,5-disubstituted 3- pyrrolin-2-one. The terminal R domains of some PKS-NRPSs was reported to off load the polyketide-amino acid intermediates via a Dieckmann condensation to afford a tetramic acid moiety. Under these conditions, the released tetramic acid intermediate from CcsA would have to undergo further ketoreduction and dehydration to form the proposed 3-pyrrolin-2-one dienophile.

The intramolecular Diels-Alder *endo* cycloaddition of the pentaene intermediate was to form between the terminal diene of the polyketide chain and the 3-pyrrolin-2- one dienophile to afford the 11-membered carbocycle-fused perhydroisoindolone scaffold. The Diels-Alder cycloaddition in cytochalasin biosynthesis is predicted to occur after the release of the hybrid polyketide-amino acid molecule from the PKS-NRPS. For this purpose solanapyrone and spinosyn A biosynthesis are required for post-PKS-NRPS Diels-Alder reaction in cytochalasin biosynthesis.

Multiple sequence alignment of the primary protein sequences of CcsB and previously characterized BVMOs (Baeyer-Villiger Monooxygenase) indicated that CcsB contains the two intact conserved Rossmann fold motifs, GxGxxG and GxGxxA, as well as the so-called BVMO signature. CcsB is responsible for the consecutive Baeyer-Villiger insertion of two oxygen atoms into the 11-membered carbocyclic intermediate to generate the unique vinyl carbonate moiety in cytochalasin E and K. Which occurs in two steps. One step is oxidation step is consistent cytochalasin synthesis and the second Baeyer oxidation occurs on an acrylate moiety. Thus CcF and its homologous may play role in post-PKS-NRPS biosynthesis step.

The identification of *ccs* gene cluster not only allows for continuous investigation of the molecular basis of the structural diversity generated by fungal PKS-NRPSs, but also opens the door for genome mining of cytochalasin gene cluster and application of molecular engineering to generate novel cytochalasin derivatives.

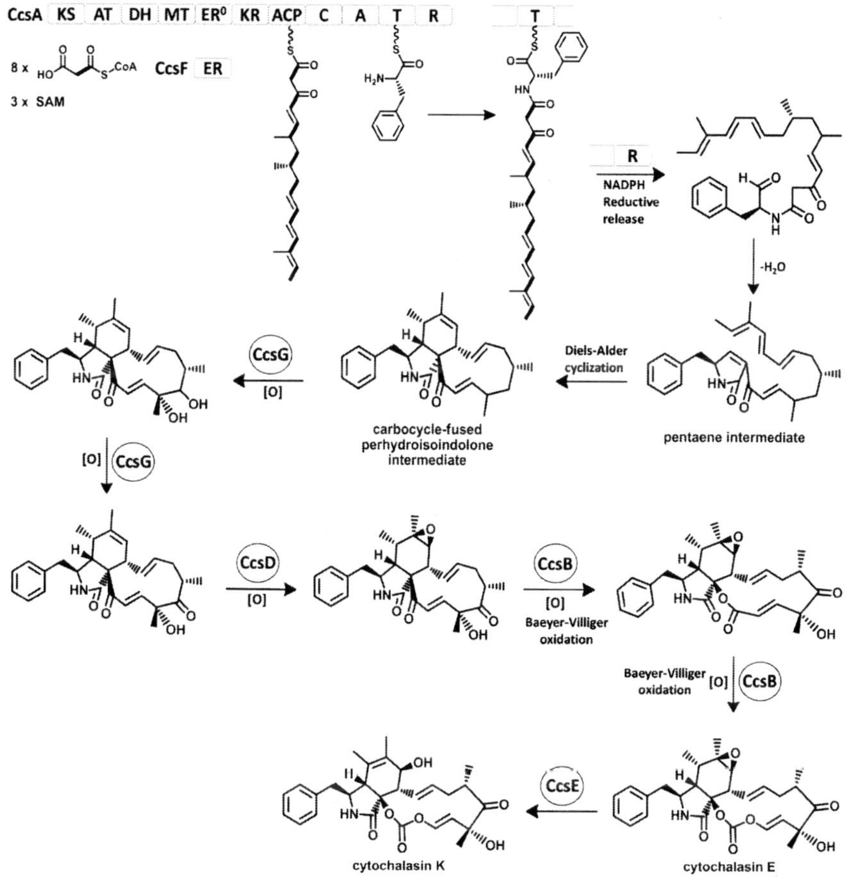

Fig. 5.1. Biosynthetic pathway for cytochalasin E and K.

Binder *et al* (1970) elucidated the biosynthesis of cytochalasin which results with the fusion of one unit of phenylalanine, nine units of acetate-malonate and two units of methionine. Labelling experiments revealed that nine acetate units are directly incorporated, in cytochalasin while long chain fatty acids were not incorporated. Robert and Tamm (1975) demonstrated the synthesis of cytochalasin B by an enzymatic Bayer-villiger type oxygen insertion by direct conversion of labeled deoxaphomin by *Phoma* sps. They also demonstrated that stereo specifically labeled precursor, phenylalanine, was primary precursor of Cytochalasin D. Further, they recorded the rapid incorporation of D and L isomers of phenylalanine in equal proportion through phenylpyruvic acid.

5.2. Biosynthesis of cytochalasin B

The complex structures of the cytochalasins suggest a mixed biosynthetic pathway involving amino acid and malonyl building blocks. The first biosynthetic studies were performed by Thomas and Whitehead, (1986) in the context of cytochalasin B production by *Phoma* sp. Cultures were supplemented with a series of ^{14}C- and ^{3}H-labelled precursors, and the obtained radioactive cytochalasin B was investigated by chemical degradation experiments (Binder *et al.,* 1970). The alternating distribution of the radioactivity measured in the different degradation products pointed towards a polyketide-like biosynthetic pathway (Staunton and Weissman, 2001). Furthermore, since malonate is decarboxylated at (one of the two possible) C-1 during polyketide chain elongation, the radioactivity measured for [1-^{14}C]-malonate incorporation into cytochalasin B was reduced to 50% in comparison to activities obtained in [2-^{14}C]- malonate feeding experiments. Notably, propionate (or methylmalonate) does not account for methyl branches. However, feeding experiments with labelled propionate revealed two possible pathways for this precursor to enter the cytochalasin B pathway. [2-^{14}C]-Propionate caused an alternating distribution of radioactivity identical to that of [1-^{14}C]-labelled acetate and malonate, which is in accord with the well known enzyme-catalyzed oxidative fission of propionate to acetate (α-oxidation). However, feeding [1-^{14}C]-propionate to the cytochalasin B producer resulted in an activity solely located within the β-lactam portion of the metabolite, suggesting an alternative route of propionate into cytochalasin biosynthesis via the shikimic acid pathway **(Fig. 5.2).** Thus, propionate is converted into phosphoenolpyruvate that in turn is condensed with shikimic acid eventually to form C-1-labelled phenylalanine or other aromatic amino acids (Vederas *et al.,* 1975). Incorporation of propionate through this (second) pathway results in cytochalasin B, which is exclusively labelled at the C-4 position (labelled C-1 of [1-^{14}C]- propionate was removed in the course of the α-oxidation pathway and failed in labelling remaining portions of cytochalasin B.

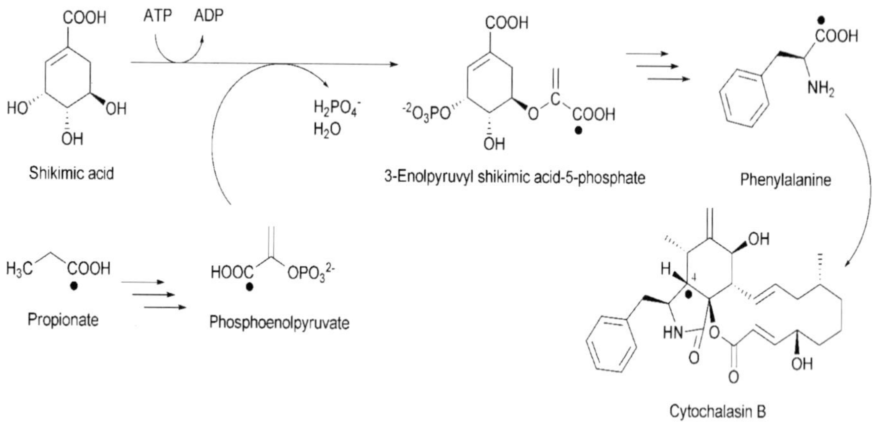

Fig. 5.2. Alternative route for propionate to enter cytochalasin B biosynthesis via the shikimic acid pathway. Incorporation of [1-^{14}C]-propionate results in cytochalasin B labelled at C-4.

The total synthesis of cytochalasins has been of considerable interest for several years. Both inter-and intramolecular Diels-Alder reactions have been used to assemble the isoindolane components stereoselectively. The stereogenic contents around the macrocycle can be introduced either before or after the Diels-Alder reactions. The general synthetic route towards the formation of cytochalasins can be divided in to two strategies.

1. Simultaneous formation of the six-member isoindolane ring and the large macrocycle in a later stage Diels-Alder cyclization (Merifield and Thomas 1999,).
2. Initial formation of isoindolane core (through the use of an intramolecular Diels- Alder reaction) and subsequent appendices of the macrocycle.

Although the Diels-Alder strategy is mostly used to generate the eleven membered ring of the cytochalasins. The other reaction types have been performed with the purpose such as ring expansion (Vedejs *et al.,* 1990), ring fragmentation (Clark and Fuchs, 1979), Pd-assisted macrocyclizations (Trost and Brickner, 1983) or Reformatsky reaction (Vedejs and Ahmad, 1988).

The first total synthesis of cytochalasin B was reported by Stork *et al.* (1978) which consisted of the simultaneous formation of the isoindolone core and the macrocyclic fused ring in a late-stage intramolecular Diels-Alder condensation. Haidle and Myers (2004) have developed an enantioselective and convergent synthetic route in which the isoindolone core is synthesized first by a Diels-Alder reaction, with the macrocyclic appendage. It is also possible to synthesize cytochalasins B and L employing this strategy by using common Tricyclic precursors and a late-stage macrocyclization reaction (namely, through an intramolecular Horner-Wadsworth-Emmons olefination ring closure (Kelly, 1991 and Nicolaou *et al.,* 1997).

Dodecahydrocytochalasin B, which resulted from the catalytic hydrogenation of cytochalasin B with platinum oxide in glacial acetic acid, was oxidized by chromium trioxide in sulfulric acid (Kuhn-Roth oxidation) to give three equivalents of acetic acid [C(5) and C(11), C(6) and C(12). C(16) and 16-methyl]. Acetic acid was degraded further with sodium azide in polyphosphoric acid (Schmidt procedure) to carbon dioxide (isolated as barium carbonate) and methylamine (isolated as picrate). After treatment of di-O-acetylcytocahalasin B with ozone and catalytic hydrogenation with palladium of the ozonide, formaldehyde was obtained (isolated as formaldimedone) [C(12)]. Ozonolysis of cytochalasin B and subsequent treatment of the products with sodium borohydride yielded the γ-lactams I and II, and 3-methyloctane-1,7,8-triol. Kuhn Roth oxidation of γ-lactam II using a modified procedure allowed the isolation of one equivalent each of acetic acid [C915) and C(11)] and benzoic acid [C(10) and 10-phenyl]. Catalytic hydrogenation of γ-lactam led to the hexahydro derivative. Kuhn-Roth oxidation of the latter gave one equivalent of acetic acid [C(5) and C(11)] which again was degraded to carbon dioxide and methylamine.

5.3. Biosynthesis of cytochalasin D

The structural similarity of the carbocyclic cytochalasin D and the macrolide cytochalasin B, and the fact that their relative and absolute configurations are identical in the region of identical structural elements, suggested similar biogenetic pathways. Biosynthetic experiments with cytochalasin D using growing cultures of *Z. masoaii* have shown that the same precursors are incorporated to a similar degree as in cytochalasin B.

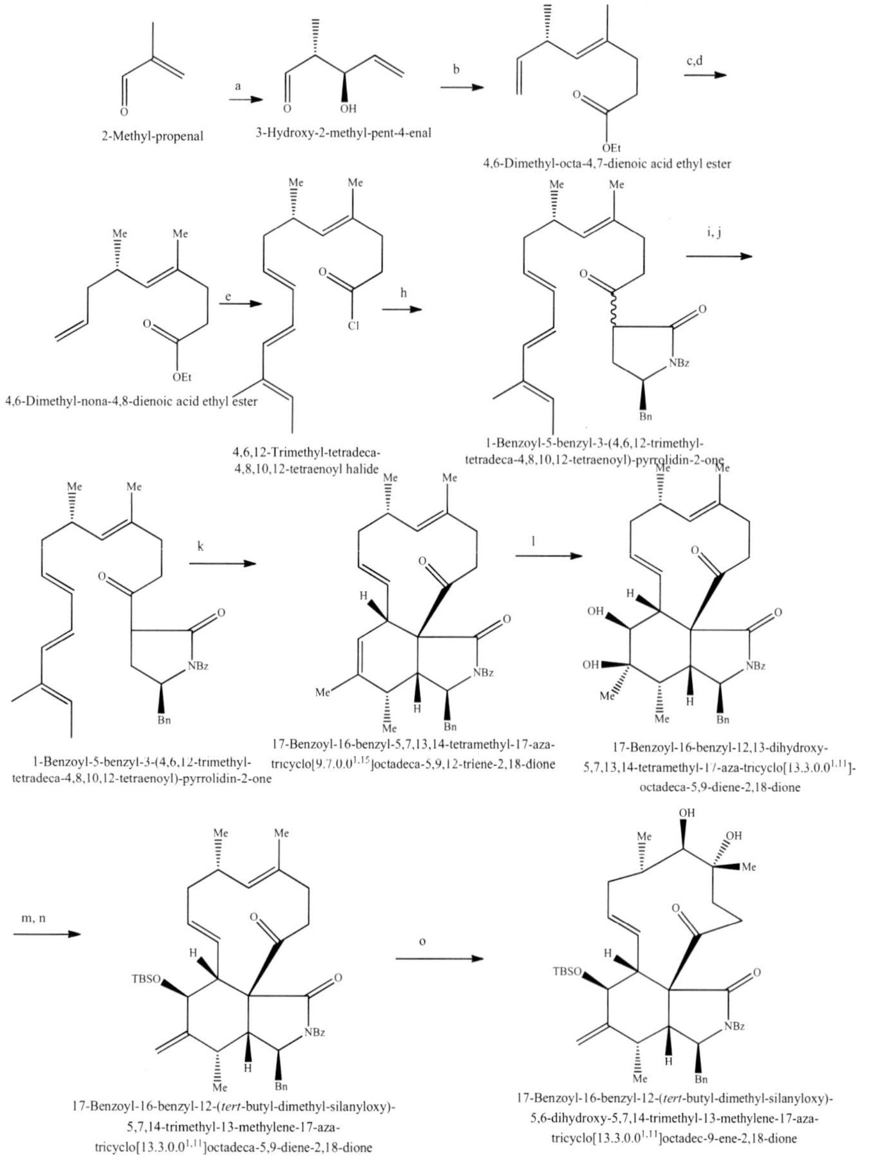

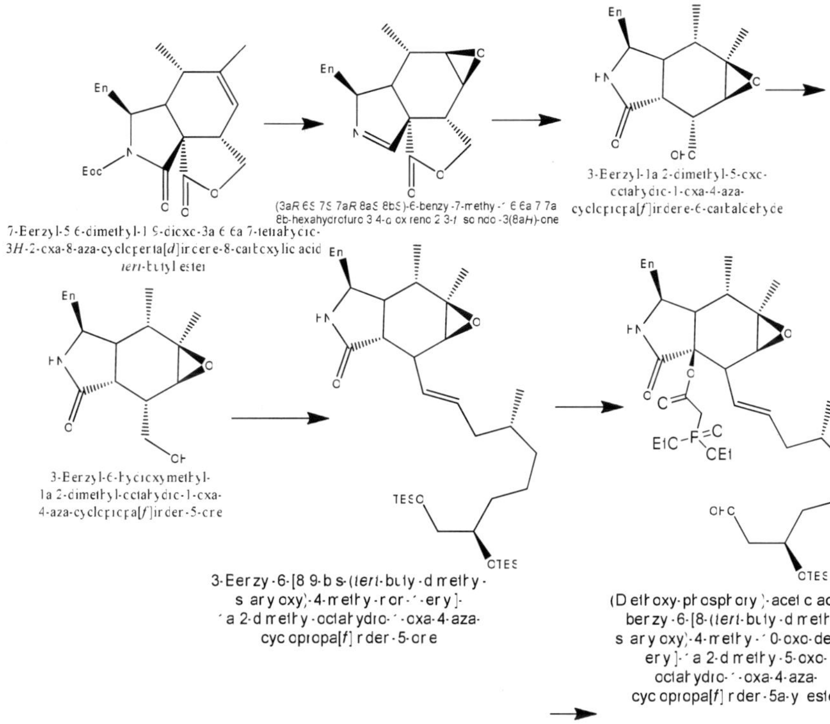

Fig. 5.3. Synthesis of Cytochalasin D by (Merifield and Thomas, 1999)
Reagents and conditions: (a) (E)-But-2-etnyldi-isopinocampheylborane,-78°C, 3h, then H_2O_2. b)MeC(OEt)3, propanoic acid, 140-170°C. (c) 9-bore-bivyvlo (3.3.1) nonane then H_2O_2. (d)$(COCl)_2$, DMSO. (e) dienylphoshonate Li, THF, hexamethylphosphoramide. (f) NaOH, EtOH, H_2O. (g) CO(imidazole)$_2$, THF. (h) (5R)$^{-5}$-Phenylmethylpyrrolidinone lithium enolate. (i) LiN(SiMe3)$_2$, PhSeCl. (j)m-chloroperoxybenzoic acid, H_2O_2, 50°C. (k) toluene, 80°C. (l) OsO4 (m) ^{t}BuMe2SiOtf, 2,6-lutidine. (n) $SOCl_2$, Et_3N. (o) OSO_4.

Cytochalasin D, carbocyclic cytochalasin, is an isomer of cytochalasin H, differing from the latter given a carbonyl function at C-17 and the opposite configuration at C-18. It was isolated from the microorganism *Metarrhizum anisopliae*, and its first total synthesis was published by Vedejs and Reid (1984). These authors had previously reported the advantages of using *N*-acylpyrrolinones for Diels-Alder reactions in the cytochalasin area (Vedejs and Gadwood, 1978).

Merifield and Thomas (1999), who had previously achieved the total synthesis of cytochalasin H, developed another synthetic route for the total synthesis of cytochalasin D (**Fig. 5.3**). In this strategy consisted of employing the same approach used for cytochalasin H -an intramolecular Diels-Alder reaction to form the reduced isoindolone ring and the macrocycle simultaneously (Merifield and Thomas, 1990). Through this method, only the stereogenic centers at C-16 and C-3 were present before the cyclization. The remaining parts of the molecule were introduced either during the Diels-Alder reaction itself or later in the synthesis using the conformational preference of the macrocyclic ring to control the stereochemistry.

5.4. Biosynthesis of Chaetoglobosin

Molecular investigations of the chaetoglobosin producer *Penicillium expansum* led to the discovery of the first cytochalasin biosynthesis gene cluster **(Fig. 5.4).**

Fig. 5.4. Organization of the chaetoglobosin (che) biosynthetic gene cluster. Predicted functions of gene products: CheA: PKS–NRPS; CheB: enoyl reductase; CheD, CheE, CheG: oxygenases; CheC, CheF: regulators

The *che* gene locus codes for a polyketide synthase (PKS) and nonribosomal peptide synthetase (NRPS). PKS–NRPS hybrid synthetase with a stand-alone enoyl reductase, as well as various oxidoreductases that are likely to be involved in tailoring processes. The identity of the chaetoglobosin synthetase (CheA) could only be verified through an RNA-silencing approach (Schuemann and Hertweck, 2007). CheA comprises all domains described for typical fungal PKS–NRPS hybrids and is closely related to pyrrolinone synthetases, like those involved in the fusarin C (Song *et al.*, 2004), equisetin (Sims *et al.*, 2005), tenellin (Halo *et al.*, 2008), aspyridone (Bergmann *et al.*, 2007) and cyclopiazonic acid pathways (Seshime *et al.*, 2009). The *N*-terminus of CheA belongs to the family of highly reducing PKS, such as lovastatin nonaketide synthase (Kennedy *et al.*, 1999) which lack a functional enoyl reductase (ER) domain. Thus, CheA probably depends on the enoylreductase activity of a trans-acting monofunctional protein encoded by another gene (cheB) present in the che gene cluster. During chaetoglobosin A biosynthesis **(Fig. 5.5),** CheA likely catalyzes the stepwise assembly of a nonaketide from an acetyl-CoA starter unit and eight malonyl-CoA extenders that are loaded on to the acyl carrier protein (ACP) by means of the acyl transferase (AT) domain. After each elongation, ketoreductase (KR), dehydratase (DH) and enoylreductase (ER) domains mediate β-keto processing steps. At this point it would be important to note that the code for the degree of β-keto processing after each elongation cycle is fully enigmatic for iterative fungal PKS (Hertweck, 2009). After completion of the complex nonaketide chain tethered to the ACP domain, an activated tryptophan bound to the NRPS portion of CheA would attack the polyketide chain to yield an amide. Two mechanisms for the subsequent off-loading process are conceivable. Based on sequence similarities, the *C*-terminal domain might function as a reductase (Oppermann *et al.*, 2003). In this case, an amido aldehyde would be liberated that could undergo a Knoevenagel condensation to form a pyrrolinone. Alternatively, the terminal domain could function as a Die-ckmann cyclase, thus directly releasing a tetramic acid derivative, as shown for the equisetin and tenellin pathways (Halo *et al.*, 2008). Irrespective of the route involved, a pyrrolinone would represent the predicted dienophile for the proposed Diels–Alder [4 + 2] cycloaddition. The requirement of a "Diels–Alderase" or an entropic trap is still under debate (Oikawa and Tokiwano, 2004; Stocking and Williams, 2003), and one may even consider a non-concerted cycloaddition leading to the tricyclic ring system. However, Diels–Alder reactions have been implicated in the formation of many natural products, e.g. for solanapyrone from *Alternaria solani* (Kelly, 2008) and lovastatin from *Aspergillus terreus* (Auclair *et al.*, 2000). Furthermore, the harsh reaction conditions required to obtain the cytochalasin macrocycle in chemical synthesis suggest the involvement of an enzyme (Thomas, 1991). Eventually, after the cytochalasin framework has been established, a sequence of oxidative tailoring reactions putatively catalyzed by CheD, CheE and CheG would yield the final products, chaetoglobosins A and C (Schuemann and Hertweck, 2006)

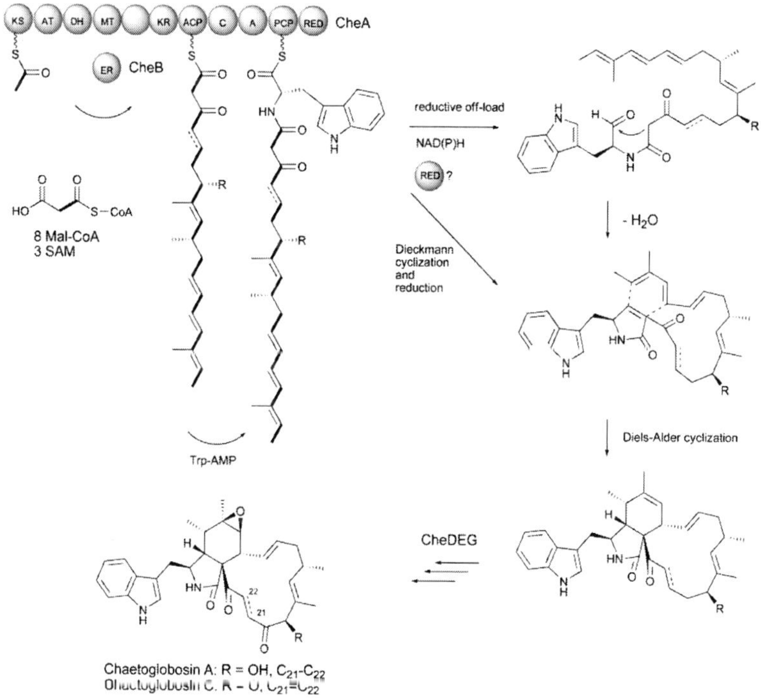

Fig. 5.5. Architecture of the CheA synthetase and model for chaetoglobosin A biosynthesis. Two alternative routes for off-loading and cyclization of the polyketide–amino acid hybrid are shown

To date, the *che* gene cluster from *P. expansum* represents the only known example for a biosynthetic gene cluster coding for the production of a cytochalasin. Yet, it should be noted that Amnuaykanjanasin *et al.* (2005) cloned and sequenced a putative PKS–NRPS hybrid gene from *Xylaria* sp. BCC 1067, a producer of 19,20-epoxycytochalasin Q. However, the involvement of the encoded synthetase in cytochalasin biosynthesis remains to be proven. The large diversity of cytochalasins results not only from the variety of polyketide chains assembled by the requisite PKS part, but also from the range of amino acids incorporated by the NRPS portion. The latter would be a primary determinant for the sub-group of cytochalasins: phenylalanine for 104 tryptophan most cytochalasins and zygosporins, mainly for cytochalasins and chaetoglobosins, tyrosine for pyrichalasins and leucine for aspochalasins. The identification and analysis of cytochalasin biosynthesis genes and their heterologous expression could permit the application of molecular engineering and combinatorial biosynthesis methods. As a consequence, the diversity of the highly complex cytochalasins could be further increased through biological and biochemical approaches, thus being a useful option to complement conventional chemical methods.

II. CHEMICAL SYNTHESIS OF CYTOCHALASINS

5.5. Chemical synthesis of cytochalasin B

The first total synthesis of the 14-membered macrolactone cytochalasin B was reported by Stork and Nakamura (1983). **Scheme 2 (Fig. 5.6)** involves an intramolecular [4 + 2] cycloaddition as key step. As an extension of a preceding formal total synthesis of cytochalasin B (Stork *et al.*, 1978), a triene **1**, derived from (+)-citrollenol, was esterified with N-acetylhydroxypyrrolone **2** using Mukaiyama's method. The subsequent intramolecular Diels–Alder reaction required harsh conditions (180-190°C for 6 days) and provided the macrocycle **4** in only moderate yields (30%, endo/exo 4 : 1). Various subsequent steps were needed for functional group conversions. In a similar fashion, Thomas and co-workers synthesized the 11-membered cytochalasins H, D, G and O through late-stage intramolecular [4 + 2]-cycloadditions (Merifield and Thomas, 1999).

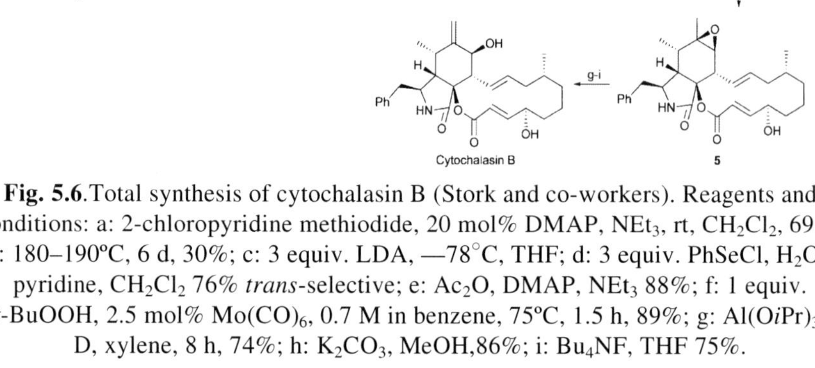

Fig. 5.6. Total synthesis of cytochalasin B (Stork and co-workers). Reagents and conditions: a: 2-chloropyridine methiodide, 20 mol% DMAP, NEt_3, rt, CH_2Cl_2, 69%; b: 180–190°C, 6 d, 30%; c: 3 equiv. LDA, —78°C, THF; d: 3 equiv. PhSeCl, H_2O_2, pyridine, CH_2Cl_2 76% *trans*-selective; e: Ac_2O, DMAP, NEt_3 88%; f: 1 equiv. *t*-BuOOH, 2.5 mol% $Mo(CO)_6$, 0.7 M in benzene, 75°C, 1.5 h, 89%; g: $Al(OiPr)_3$, D, xylene, 8 h, 74%; h: K_2CO_3, MeOH,86%; i: Bu_4NF, THF 75%.

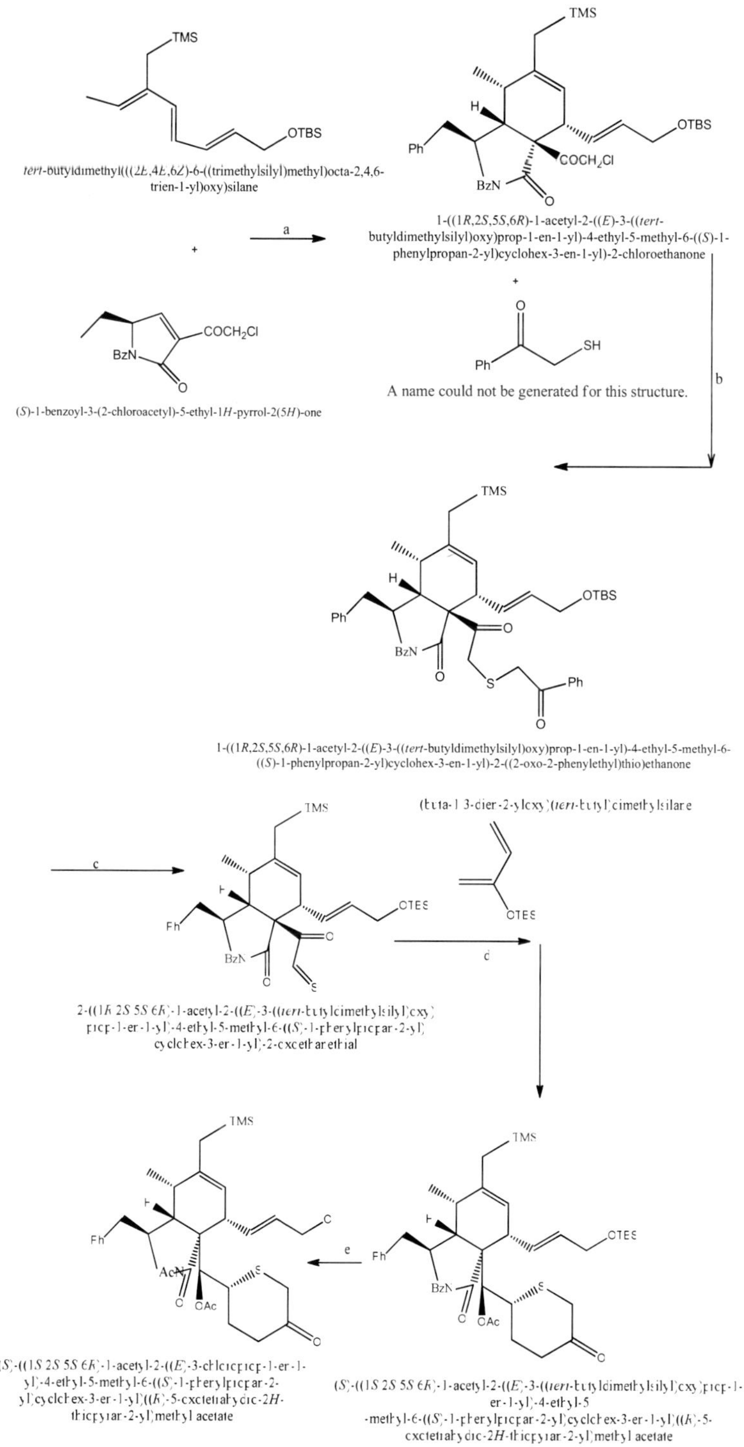
TMS
OTBS
tert-butyldimethyl(((2E,4E,6Z)-6-((trimethylsilyl)methyl)octa-2,4,6-trien-1-yl)oxy)silane
+
COCH2Cl
BzN
O
(S)-1-benzoyl-3-(2-chloroacetyl)-5-ethyl-1H-pyrrol-2(5H)-one
a
H
Ph
COCH2Cl
1-((1R,2S,5S,6R)-1-acetyl-2-((E)-3-((tert-butyldimethylsilyl)oxy)prop-1-en-1-yl)-4-ethyl-5-methyl-6-((S)-1-phenylpropan-2-yl)cyclohex-3-en-1-yl)-2-chloroethanone
+
SH
A name could not be generated for this structure.
b
S
1-((1R,2S,5S,6R)-1-acetyl-2-((E)-3-((tert-butyldimethylsilyl)oxy)prop-1-en-1-yl)-4-ethyl-5-methyl-6-((S)-1-phenylpropan-2-yl)cyclohex-3-en-1-yl)-2-((2-oxo-2-phenylethyl)thio)ethanone
c
d
e
OAc
AcN

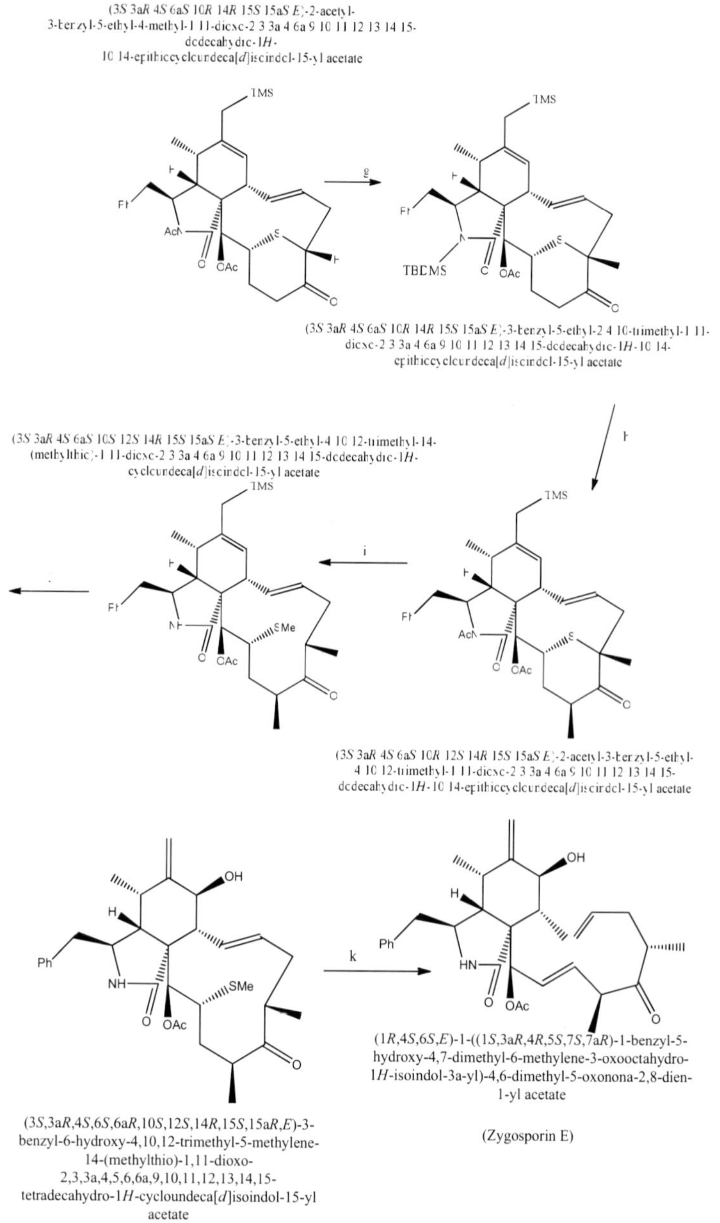

Fig. 5.7. Synthesis of zygosporin E

Reagents and conditions: a: rt, 93%; b: K_2CO_3 88%; c: hn, D; d: 1. hn; 2. DBU, 0°C, 60%, 3. DIBAl-H, 60%; 4. Ac_2O; e: 1. Et_3NHF, 2. Bu_3P, CCl_4, 40% over 3 steps; f: NaI, K_2CO_3 71%; g: 1. K_2CO_3, —15°C; 2. TBDMSCl; 3. LHMDS, MeI; h: 1. LHMDS, MeI; 2. Et_3NHF, 60%; 3. Ac_2O, 88%; i: 1. Me_3OBF_4, TMS-allyl, Rieke zinc, MeOH, 87%; 2. K_2CO_3; j: 1. PhSeSe(Me)$PhBF_4$; 2. $NaIO_4$, H_2O, 85% over 2 steps; k:1. mCPBA, 75%; 2. 135°C,70%.

5.6. Biosynthesis of Zygosporin E

An alternative strategy for the formation of 11-membered carbacyclic cytochalasins was developed by Vedejs *et al.* (1990). For the total synthesis of zygosporin E, they established a high-yielding (93%) endo-selective intermolecular Diels–Alder reaction of a doubly activated dienophile, obtained from N-benzoyl-(5R)-(phenylmethyl)-2-pyrrolidinone, with a trimethylsilyl-substituted triene. The resulting isoindolone served as the starting point for a sulfurmediated route that allows the stereospecific introduction of ring substituents at a late stage of the synthesis. A short lived thioaldehyde, photo-chemically generated from a phenacyl sulfide, was trapped by a hetero-Diels–Alder reaction with 2-(tert-butyldimethylsiloxy)-1,3-butadiene. Then, transformation of the protected allylic alcohol into the allylic chloride provided the substrate for a base-mediated macrocyclization. The resulting sulfur-bridged carbocycle could be methylated at C-16 and C-18 with high stereospecificity, which could be rationalized by the rigid constitution of the polycyclic ring system. To cleave the C-16–S bond, S-methylation with Me_3OBF_4 and reduction using Rieke zinc proved to be appropriate. A selenoxide rearrangement of the allylic silane finally provided the corresponding allylic alcohol, and oxidative sulfur elimination furnished zygosporin E (**Fig. 5.7**).

5.7.1. Synthesis of Aspochalasin B

Since the biomimetic route involving a Diels–Alder macrocyclization proved to be a bottleneck, alternative approaches were tested, in which the isoindolone cores are prepared first. In this context, Trost *et al.* (1989) developed a straight forward Pd-catalyzed synthesis to 11-membered cytochalasins, such as (-) aspochalasin B (Trost and Gowland, 1979). Based on a protocol of Tamm (1978), the isoindolone subunit is formed by an intermolecular Diels–Alder cycloaddition of 4-methylsorbinol with an alkylidene malonic ester, derived from a Cbz-protected isoleucine building block (Schmidlin *et al.,* 1983). After lactone ring opening and oxidation, the aldehyde side chain was extended through alkylation and Claisen rearrangement, and the methyl ester was transformed into the sulfone required for the palladium catalyzed macrocyclization. Finally, the resulting enol ether was converted into the α-hydroxy ketone, and sulfoxide elimination yielded (-)-aspochalasin B (**Fig. 5.8**).

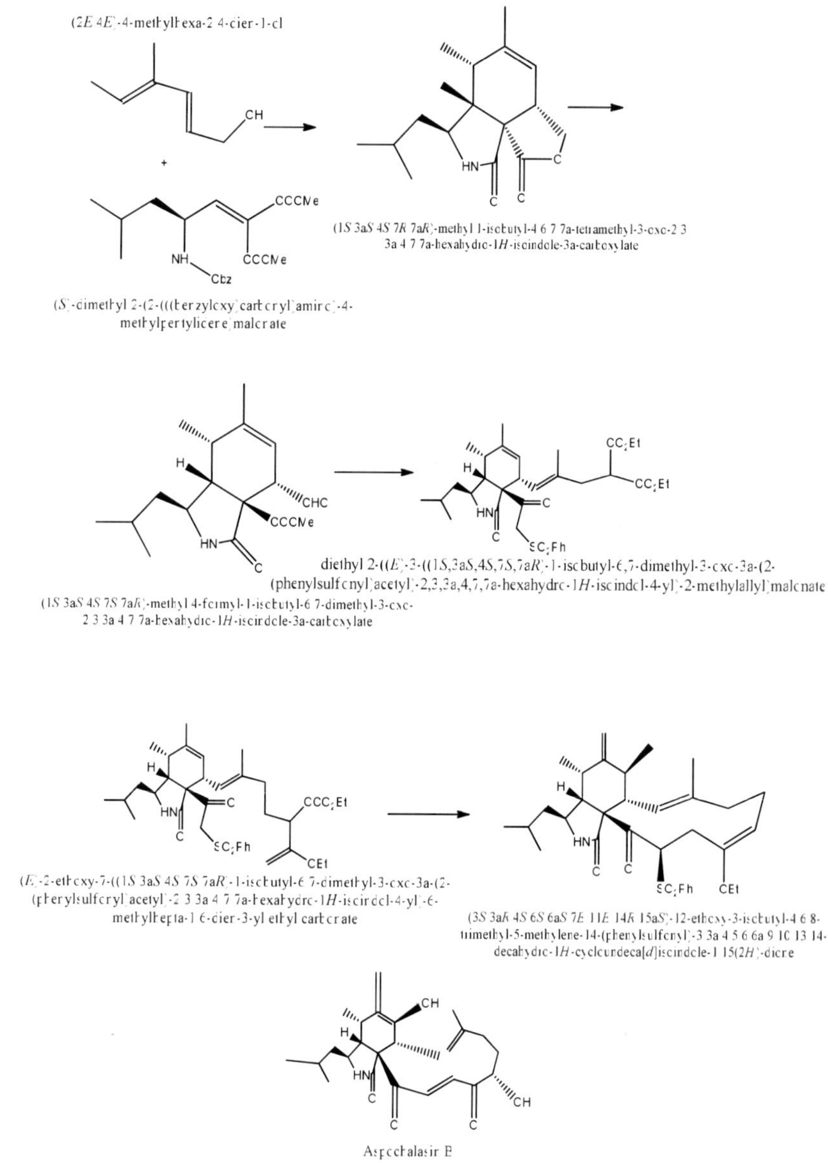

Fig. 5.8. Total synthesis of ( )-aspochalasin B (Trost et al. 1989). **Reagents and conditions**: a: xylene, BHT, 130 C, 33–40%; b: KOH, H_2O– MeOH, toluene, rt, $NaHSO_4$; c: CH_2N_2, Et_2O, MeOH, 94% over 2 steps; d: $(COCl)_2$, DMSO, CH_2Cl_2, Et_3N, 78 C, 90–100%; e: $[CH_2]C(Me)]_2$ CuLi, THF, 78 C 58%; f: n-BuLi, $MeSO_2Ph$, THF, HMPA, 0 C 80%; g: $(EtO)_2C]CHCO_2$ Et, PPTS, rt, 68%; h: 1. 1 N aqueous KOH, THF, rt, HCl; 2. toluene, D; 3. CH_2N_2, Et_2O, rt, 71% over 3 steps; i: 1. n-BuLi/DiBAl-H, THF, 0 C; 2. PCC, CH_2Cl_2, rt, 47% over 2 steps; j: 1. $[CH_2]C(OEt)_2]CuLi$, THF, Et_2O, 78 C; 2. $ClCO_2Me$, pyridine, 0 C; k: 10% $(Ph_3P)_4Pd$, 10% dppp, THF, rt, D, 49%; m: 1. 2 eq. $MeCOO_2H$, K_2CO_3, CH_2Cl_2, 2. $BnNMe_3F$, THF–CH_2Cl_2, rt.

5.7.2. Chemcal synthesis of (-)-Aspochalasin B

A different approach to cytochalasin natural products was investigated by Trost *et al.* (1989) and resulted in the total synthesis of the leucine-derived [11]cytochalasin, (-)-aspochalasin B. The strategy consisted in synthesizing an isoindolone part and the utilization of a palladium-catalyzed formation of the 11-membered carbocycle (**Fig. 5.9**).

As starting material for this synthesis, the Cbz-protected leucine ester-Benzyloxycarbonylamino-4-methyl-pentanoic acid methyl ester was used. Reduction of the ester group and condensation with malonic acid dimethyl ester gave olefin 2-(2- Benzyloxycarbonylamino-4-methyl-pentyl)- malonic acid-dimethyl ester, which was reacted with diene 4-methyl-hexa-2, 4-dien-1-ol in a Diels-Alder reaction to yield tricycle 3-Isobutyl-4,5-dimethyl-1,9-dioxo-3a,4,6a,7,8,9hexahydro-3H-2-aza-cyclopenta [*d*]indene-2-carboxylic-acid benzyl ester. The lactone ring was hydrolyzed and the resulting carboxylic acid converted to the methyl ester using diazomethane (4 Hydroxymethyl-1-isobutyl-6,7-dimethyl-3-oxo-1,2,3,4,7,7a-hexahydro-isoindole-3a-carboxylic acid methyl ester). The following Swern oxidation proceeded in excellent yield and delivered aldehyde 1166. Addition of dipropenylcuprate (-(1-hydroxy-2-methyl-allyl)-1-isobutyl-6,7-dimethyl-3-oxo-1,2,3,4,7,71-hexahydro-isoindole-3a-carboxylic acid methyl ester), introduction of a sulfone unit (7a-(2-Benzenesulfonyl-acetyl)-7-(1-hydroxy-2-methyl-allyl)-3-isobutyl-4,5-dimethyl-2,3,3a,4,7,7a-hexahydro-soindol-one), and treatment with ethyl b,b-diethoxyacrylate (2-Ethoxycarbonyl-but-2-enedioic acid diethyl ester) to undergo condensation/rearrangement allowed for the preparation of 2-{3-[3a-(2-Benzenesulfonyl-acetyl-)-1-isobutyl-6,7-dimethyl-3-oxo-2,3,3a,4,7,7a-hexahydro-1*H*-isoindol-4-yl]-2-methyl-allyl}-malonic acid diethyl ester, which upon decarboxylation, methyl esterification, chemoselective reduction and oxidation yielded the corresponding aldehyde 5-[3a-(2-Benzenesulfonyl-acetyl)-1-isobutyl-6,7-dimethyl-3-oxo-2,3,3a,4,7,7a-hexahydro-1*H*-isoindol-4-yl]-4-methyl-pent-4-enal. For the introduction of another C2 unit, again a cuprate was the reagent of choice, leading to an enol ether alcohol, which was trapped with methyl chloroformate (Carbonic acid 5-[3a-(2-benzenesulfonyl-acetyl)-1-isobutyl-6,7-dimethyl-3-oxo-2,3,3a,4,7,7a-hexahydro-1*H*-isoindol-4-yl]-1-(1-ethoxy-vinyl)-4-methyl-pent-4-enal ester methyl ester). Next, the key step of this total synthesis was a Pd-catalyzed macrocyclization, which proceeded via a syn-p-allyl palladium complex with a remarkable diastereoselectivity and delivered a single isomer 7a-(2-Benzenesulfonyl-4-cyclopropyl-4-ethoxy-butyryl-7-(2,4,dimethyl-pent-1-wnyl)-3-isobutyl-4,5-dimethyl-2,3,3a,4,7,7a-hexahydro-isoindol-1-one with a (Z)-configured enol ether double bond. Treatment with peracetic acid gave an enol ether epoxide, which was immediately hydrolyzed to hydroxy ketone 3-Benzenesulfonyl-6-hydroxy-16-isobutyl-9,13,14-trimethyl-17-aza-tricyclo [9,7,0,$0^{1.15}$] octadeca-9,12-diene-2,5,18-trione. The second macrocyclic double bond was finally introduced by sulfone elimination, hence yielding (-)-aspochalasin B in 19 overall steps from simple starting materials.

2-Benzyloxycarbonylamino-4-methyl-pentanoic acid methyl ester

↓

2-(2- Benzyloxycarbonylamino-4-methyl-pentyl)- malonic acid- dimethyl ester

↓

4-methyl-hexa-2, 4-dien-1-ol

↓

3-Isobutyl-4,5-dimethyl-1,9-dioxo-3a,4,6a,7,8,9-hexahydro-3H-2-aza-cyclopenta [*d*]indene-2-carboxylic-acid benzyl ester

↓

4-Hydroxymethyl-1-isobutyl-6,7-dimethyl-3-oxo-1,2,3,4,7,7a-hexahydro-isoindole-3a-carboxylic acid methyl ester

↓

4-formyl-1-isobutyl-6,7-dimethyl-3-oxo-1,2,3,4,7,7a-hexahydro-isoindole-3a-carboxylic acid methyl ester

↓

4-(1-hydroxy-2-methyl-allyl)-1-isobutyl-6,7-dimethyl-3-oxo-1,2,3,4,7,71-hexahydro-isoindole-3a-carboxylic acid methyl ester

↓

7a-(2-Benzenesulfonyl-acetyl)-7-(1-hydroxy-2-methyl-allyl)-3-isobutyl-4,5-dimethyl-2,3,3a,4,7,7a-hexahydro-soindol-one

↓

2-Ethoxycarbonyl-but-2-enedioic acid diethyl ester

↓

2-{3-[3a-(2-Benzenesulfonyl-acetyl-)-1-isobutyl-6,7-dimethyl-3-oxo-2,3,3a,4,7,7a-hexahydro-1*H*-isoindol-4-yl]-2-methyl-allyl}-malonic acid diethyl ester

↓

5-[3a-(2-Benzenesulfonyl-acetyl)-1-isobutyl-6,7-dimethyl-3-oxo-2,3,3a,4,7,7a-hexahydro-1*H*-isoindol-4-yl]-4-methyl-pent-4-enoic methyl ester

↓

5-[3a-(2-Benzenesulfonyl-acetyl)-1-isobutyl-6,7-dimethyl-3-oxo-2,3,3a,4,7,7a-hexahydro-1*H*-isoindol-4-yl]-4-methyl-pent-4-enal

↓

Carbonic acid 5-[3a-(2-benzenesulfonyl-acetyl)-1-isobutyl-6,7-dimethyl-3-oxo-2,3,3a,4,7,7a-hexahydro-1*H*-isoindol-4-yl]-1-(1-ethoxy-vinyl)-4-methyl-pent-4-enal ester methyl ester

↓

7a-(2-Benzenesulfonyl-4-cyclopropyl-4-ethoxy-butyryl-7-(2,4,dimethyl-pent-1-wnyl)-3-isobutyl-4,5-dimethyl-2,3,3a,4,7,7a-hexahydro-isoindol-1-one

↓

3-Benzenesulfonyl-6-hydroxy-16-isobutyl-9,13,14-trimethyl-17-aza-tricyclo [9,7,0,0$^{1.15}$] octadeca-9,12-diene-2,5,18-trione

↓

6-Hydroxy-16-isobutyl-9,13,14-trimethyl-17-aza-tricyclo [9,7,0,0$^{1.15}$] octadeca-3,9,12-triene-2,5,18-trione

Fig 5.9. Chemical synthesis of Aspochalasin B

5.8. Chemical Synthesis of Cytochalasin H

Cytochalasin H, an isomer of cytochalasin D, was synthesized by Thomas and Whitehead (1986) utilizing an intramolecular Diels-Alder reaction to build up both the isoindolone and macrocycle moieties in a single step. The formation of the Diels-Alder precursor 1-benzoyl-5-benzyl-4-phenylselanyl-3-[4,6,12-trimethyl-4-(2-trimethylsilanyl ethoxymethoxy)-tetradeca-8, 10, 12-trinoyl]-pyrrolidin-2-one together with the following steps for the completion of the synthesis of cytochalasin H are outlined in **Fig.5.10.**

Diels-Alder precursor 1-benzoyl-5-benzyl-4-phenylselanyl-3-[4,6,12-trimethyl-4-(2-trimethylsilanyl-ethoxymethoxy)-tetradeca-8, 10, 12-trinoyl]-pyrrolidin-2-one was accessible from aldehyde 3-(*tert*-Butyl-dimethyl-silanyloxy)-2-methyl-propionaldehyde, dienylphosphonate (4-methyl-hexa-2,4-dienyl)-phosphonic acid diethyl ester, and benzoyl pyrrolidinone 1-benzoyl-5-benzyl-pyrrolidin-2-one in 18 steps. Elimination of the phenylselenyl group generated a double bond, which was subjected to an intramolecular, diastereoselective Diels-Alder reaction at elevated temperature to yield tricycle 17-Benzoyl-16-benzyl-5,7,13,14-tetramethyl-5-(2-trimethylsilanyl-ethoxymethoxy)-17-aza-tricyclo-[13.3.0.0$^{1.11}$] octadeca-9,12-diene-2,18-dione. After deprotection of the amine moiety(16-Benzyl-5,7,13,14-tetramethyl-5-(2-trimethylsilanyl-ethoxymethoxy)-17-aza tricycle [13.3.0.0$^{1.11}$]-octadeca-9,12-diene-2, 18-dione), the macrocyclic double bond is introduced by a-deprotonation of the ketone, and subsequent phenylselenylation followed by elimination (16-Benzyl-5,7,13,14-tetramethyl-5-(2-trimethylsilanyl-ethoxymethoxy)-17-aza-tricyclo[13.3.0. 0$^{1.11}$]-octadeca-3,9,12-triene-2, 18-dione). Reduction with sodium borohydrate and O-acetylation gave compound acetic acid 16-benzyl-5,7,13,14-tetramethyl-18-oxo-5-(2-trimethyl-silanyl-ethoxymethoxy)-17-aza tricycle [13.3.0. $0^{1.11}$]-octadeca-3,9,12-trien-2-yl, ester. The final functionalization at the cyclohexene ring to create an exo-double bond was realized by epoxidation to acetic acid 16-benzyl-[12,13]-epoxy-5,7,14-trimethyl-13-methylene-18-oxo-5-(2-trimethyl-silanylethoxymethoxy)-17-azatricyclo[13.3. 0. $0^{1.11}$]-octadeca-3,9-dien-2-yl ester (unfortunately, epoxidation occurred mainly at the unwanted double bond present in the macrocycle. However, separation of the regioisomers was possible), followed by elimination and ring opening to acetic acid 16-benzyl-12-hydroxy-5,7,14-trimethyl-13-methylene-18-oxo-5-(2-trimethylsilanyl-ethoxymethoxy)-17-aza-tricyclo [13.3.0. $0^{1.11}$]-octadeca-3,9-dien-2-yl ester**,** which, after SEM-deprotection, finally yielded cytochalasin H.

In addition, the same research group conducted an alternative approach to cytochalasin H synthesis. For this purpose, they prepared a phenylalanine analogue of cytochalasin G and treated it with a Methyl-Grignard reagent to yield a key intermediate for the cytochalasin H synthesis (Sauter *et al.,* 1989).

3-(*tert*-Butyl-dimethyl-silanyloxy)-2-methyl-propionaldehyde

↓

(4-methyl-hexa-2,4-dienyl)-phosphonic acid diethyl ester

↓

1-benzoyl-5-benzyl-pyrrolidin-2-one

↓

1-benzoyl-5-benzyl-4-phenylselanyl-3-[4,6,12-trimethyl-4-(2-trimethylsilanyl-ethoxymethoxy)-tetradeca-8, 10, 12-trinoyl]-pyrrolidin-2-one

↓

17-Benzoyl-16-benzyl-5,7,13,14-tetramethyl-5-(2-trimethylsilanyl-ethoxymethoxy)-17-aza-tricyclo-[13.3.0.$0^{1.11}$] octadeca-9,12-diene-2,18-dione

↓

16-Benzyl-5,7,13,14-tetramethyl-5-(2-trimethylsilanyl-ethoxymethoxy)-17-aza-tricyclo[13.3.0. $0^{1.11}$]-octadeca-9,12-diene-2, 18-dione

↓

16-Benzyl-5,7,13,14-tetramethyl-5-(2-trimethylsilanyl-ethoxymethoxy)-17-aza-tricyclo[13.3.0. $0^{1.11}$]-octadeca-3,9,12-triene-2, 18-dione

↓

Acetic acid 16-benzyl-5,7,13,14-tetramethyl-18-oxo-5-(2-trimethyl-silanyl-ethoxymethoxy)-17-aza-tricyclo[13.3.0. $0^{1.11}$]-octadeca-3,9,12-trien-2-yl, ester

↓

Acetic acid 16-benzyl-[12,13]-epoxy-5,7,14-trimethyl-13-methylene-18-oxo-5-(2-trimethyl-silanyl-ethoxymethoxy)-17-aza-tricyclo[13.3.0. $0^{1.11}$]-octadeca-3,9-dien-2-yl ester

↓

Acetic acid 16-benzyl-12-hydroxy-5,7,14-trimethyl-13-methylene-18-oxo-5-(2-trimethylsilanyl-ethoxymethoxy)-17-aza-tricyclo[13.3.0. $0^{1.11}$]-octadeca-3,9-dien-2-yl ester

↓

Acetic acid 16-benzyl-5,12-dihydroxy-5,7,14-trimethyl-13-methylene-18-oxo-17-aza-tricyclo[13.3.0. $0^{1.11}$]-octadeca-3,9-dien-2-yl ester

Fig. 5.10. Chemical synthesis of cytochalasin H

5.9. Chemical synthesis of Cytochalasins D and O

Merifield and Thomas (1990) reported a total synthesis of cytochalasin D. Some years later, they also achieved the total synthesis of cytochalasin O by functionalization of a late-stage intermediate of cytochalasin D (1991). These syntheses proceeded similarly to the above-described synthesis of cytochalasins H and G, in utilizing a Diels-Alder reaction to form the tricyclic core. Therefore, only the total synthesis of cytochalasin O, structurally characterized by the presence of a syn-diol moiety, is described in the following scheme starting from Diels-Alder product 17-Benzoyl-16-benzyl-5,7,13,14-tetramethyl-17-aza-tricyclo -[13,3,0, $0^{1.11}$] octadeca-5,9,12-triene-2,18-dione (**Fig. 5.11**).

After oxidation with an excess of osmium tetraoxide to tetraol 17-Benzoyl-16-benzyl-5,6,12,13-tetrahdroxy5-7-13-14-tetramethyl-17-aza-tricyclo[13,3,0, $0^{1.11}$]octadeca-9-ene-2,18-dione, this was protected to give bis-acetonide (3a*R*,6a*R*,9s,9a*R*,10*S*,10a*R*,13a*S*,13b*R*,17*S*,17*S*,17a*R,E*)-8-benzoyl-9-benzyl-2,2, 3a,10,10a,12,12,17—octamethyl-4,5,9,9a,10,10a,13b,16,17,17adecahydro-3a*H*-[1,3]dioxolo[4',5':7,8]cycloundeca [4,5-*f*]isoindole-6,7(8*H*,a3aH)-dione. As a result of the previously developed protocol, a phenylselenyl mediated elimination, as well as N-deprotection was carried out. The resulting a,b-unsaturated ketone was reduced to the corresponding alcohol with sodium borohydride in excellent yield and afterwards acetylated to a compound (3a*R*,4*E*,6*S*,6a*R*,9*S*,9a*R*,10*S*,10a*R*,13a*S*,13b*R*,14*E*,17*S*,17a*R*)-8-benzoyl-9-benzyl 2,2,3a,10,10a,12,12,17—octamethyl-oxa-6,7,8,9,9a,10,10a,131,13b,16,17,17a-decahydro-3a*H*-[1,3]dioxolo[4',5':7,8]cycloundeca [1,2-*d*]dioxolo[4,5-*f*]isoindol-6-yl acetate. One acetonide functionality was removed selectively, and the resulting diol was converted into the α-hydroxyketone (1*E*,4*S*,5*R*,6*R*,7*E*,

9*S*,9a*R*,12*S*,12a*R*,13*S*,13a*R*,16a*S*,16b*R*)-11-benzoyl-12-benzyl-5,6-dihydroxy-4,6,13,13a,15,15-hexamethyl-10-oxo-4,5,6,9,10,11,12,12a,13,13a,16a,16b-dodecahydro-3*H*-cycloundeca[*d*][1,3]dioxolo[4,5-*f*]isoindole-9-yl acetate. Finally, cytochalasin O was obtained through deprotection of the remaining diol in good yield. By comparison of several properties of this product with an authentic sample, the absolute stereochemistry at the cyclohexane ring of cytochalasin O was finally confirmed as a result of the total synthesis that was carried out.

17-Benzoyl-16-benzyl-5,7,13,14-tetramethyl-17-aza-tricyclo-[13,3,0, $0^{1.11}$]octadeca-5,9,12-triene-2,18-dione

↓

17-Benzoyl-16-benzyl-5,6,12,13-tetrahdroxy5-7-13-14-tetramethyl-17-aza-tricyclo[13,3,0,$0^{1.11}$]octadeca-9-ene-2,18-dione

↓

(3a*R*,6a*R*,9s,9a*R*,10*S*,10a*R*,13a*S*,13b*R*,17*S*,17*S*,17a*R*,*E*)-8-benzoyl-9-benzyl-2,2,3a,10,10a,12,12,17—octamethyl-4,5,9,9a,10,10a,13b,16,17,17a-decahydro-3a*H*-[1,3]dioxolo[4',5':7,8]cycloundeca [4,5-*f*]isoindole-6,7(8H,a3aH)-dione

↓

(3a*R*,4*E*,6*S*,6a*R*,9*S*, 9a*R*,10*S*,10a*R*,13a*S*,13b*R*,14*E*,17*S*,17a*R*)-8-benzoyl-9-benzyl-2,2,3a,10,10a,12,12,17—octamethyl-oxa-6,7,8,9,9a,10,10a,131,13b,16,17,17a-decahydro-3a*H*-[1,3]dioxolo[4',5':7,8]cycloundeca [1,2-*d*]dioxolo[4,5-*f*]isoindol-6-yl acetate

↓

(1*E*,4*S*,5*R*,6*R*,7*E*,9*S*,9a*R*,12*S*,12a*R*,13*S*,13a*R*,16a*S*,16b*R*)-11-benzoyl-12-benzyl-5,6-dihydroxy-4,6,13,13a,15,15-hexamethyl-10-oxo-4,5,6,9,10,11,12,12a,13,13a,16a,16b-dodecahydro-3*H*-cycloundeca[*d*][1,3]dioxolo[4,5-*f*]isoindole-9-yl acetate

↓

Acetic acid 17-benzoyl-16-benzyl-5,6,12,13-tetrahydroxy-5,7,13,14-tetramethyl-18-oxo-17-aza-tricyclo[13,3,0,$0^{1.11}$]-octadeca-3,9-dien-2yl-ester

Fig. 5.11. Chemical synthesis of cytochalasin O

Chapter-6

Biological Activities

6.1. Cytotoxicity

Cytochalasins exhibit variety toxicities, such as cytotoxicity, mutagencity and teratogenecity. Cytophysiological effects including metabolism and macro-molecular synthesis, interacts with contractile system and alters membrane structure and function. Some of the prominent toxic effects include the following

1. Inhibition of cytological cleavage but not nuclear division
2. Formation of polynucleate cells
3. Inhibition of cell movement
4. Nuclear extrusion leading to isolation of viable nuclei from cytoplasm
5. Potential mycotoxins
6. Mitotic reduction
7. Inhibition of phagocytosis and pinocytosis
8. Alteration of surface adhesion properties and microfilaments
9. Dispersion, synthesis and uptake of chemistry of membrane compounds.

Cytochalasins are not only diverse in their structure and producing organisms, they also exhibit diverse biological activities which are described below. The cytotoxicity of cytochalasins, Chaetoglobosins and zygosporins are précised in **Table 6.1**.

Table 6.1. Cytotoxicity and acute toxicity of different cytochalasins

Toxin	Cytotoxicity ID_{50} (μg/ml)	Species	Sex	Age or Weight	Route	LD_{50} (mg/kg)
Cytochalasin A	3.2	--	--	--	--	--
Cytochalasin B	1.0-3.2	--	--	--	--	--
Cytochalasin C	<0.32	Mouse (Swiss albino, DS)	M	18-20g	i.p.	>100[b]

Toxin	Cytotoxicity ID_{50} (μg/ml)	Species	Sex	Age or Weight	Route	LD_{50} (mg/kg)
Cytochalasin D	0.32-1.0	Mouse (Swiss albino, DS)	M	18-20g	i.p.	10[b]
	0.89	Mouse	--	--	s.c.	1.85
		Mouse (Swiss Webster)	F	32 g	i.p.	2.1
		Mouse (BALB/c)	F	22 g	i.p	2.2
		Mouse (C578L/6J)	F	29 g	i.p.	2.2
Cytochalasin E	--	Mouse (Carworth)	M	21 g	i.p.	4.6
			F	20 g	i.p.	4.9
		Guinea pig (Hartley)	--	Weanling	i.p.	0.5-1.5
		Rat (Fisher)	M.F	1 day; 5g	i.p.	0.98
			M	69 g	i.p.	2.6
			F	71 g	i.p.	2.61
			M	70 g	p.o	9.1
			M	80 g	i.t.	1.3
Cytochlasin H	--	Chicken (DeKalb)	M	1 day	p.o.	12.5
Kodo-cytochalasin-1	--	Mouse	--	--	--	<2[b]
Zygosporin D	0.4	Mouse (Swiss albino, Ds)	M	18-20 g	i.p.	10[b]
Zygosporin E	0.4	--	--	--	--	--
Zygosporin F	$\geq$10	--	--	--	--	--
Zygosporin G	0.68	Mouse (Swiss albino, Ds)	M	18-20 g	i.p.	750[b]
Chaetoglobosin A	3.2-10	Mouse (DDD)	M	5 weeks	s.c.	6.5
			F	5 weeks	s.c.	17.8
		Mouse (DDD)	M.F	5 weeks	Po.	>400
		Rat (Wistar)	M.F	5 weeks	i.p.	2.0[b]
			M.F	5 weeks	p.o.	>400
Chaetoglobosin B	3.2-10	--	--	--	--	--

Toxin	Cytotoxicity ID_{50} (µg/ml)	Species	Sex	Age or Weight	Route	LD_{50} (mg/kg)
Chaetoglobosin C	10-32	--	--	--	--	--
Chaetoglobosin D	3.2	--	--	--	--	--
Chaetoglobosin E	3.2	--	--	--	--	--
Chaetoglobosin F	10	--	--	--	--	--
Chaetoglobosin G	3c	--	--	--	--	--
Chaetoglobosin J	3c	--	--	--	--	--

a. The concentration of toxin producing an actual or approximated inhibition of growth in Hela cell cultures, b. LD_{100} values, c. The degree of inhibition was not stated.

Cytochalasins have been used extensively to understand the role of F-actin in different aspects of cellular function (Peterson and Mitchison, 2002). Cytochalasin A inhibits glucose transport, actin polymerization and blocks the formation of microtubule. They inhibit cell division and inhibit HIV-1 protease (Trendowski, 2014). Inhibition of polymerization of brain tubulin and muscle action by cytochalasin A has also been reported. Cytochalasin A (10-20µg/ml) inhibits growth and sugar uptake in *Saccharomyces* strain.

It inhibits cell division by interfering with the formation of contractile microfilaments, inhibits cell movement and glucose transport and induces nuclear extrusion and interferes with actin polymerization (Vassiliev *et al.,* 2014). Inhibition of actin filament polmerization is through binding to the fast-growing (barbed) end of F-actin filaments. An alternative could involve capping proteins. By doing so, Cytochalasin B not only inhibits actin polymerization but also consecutive processes such as filament network build-up. This inhibition can affect all three major steps of actin polymerization. They are

1. **Nucleation:** A core of minimal 3 actin monomers is formed.
2. **Elongation:** The core is used for elongation by addition of actin monomers.
3. **Steady state/Annealing:** A balance between polymerisation and depolymerisation is reached (steady-state). The F-actin filament stops growing and two barbed ends fuse to create one filament.

Cytochalasin B reversibly blocked motility and eliminated most of the phalloidin-stainable F- actin from the leading lamella. The loss of F-actin was nearly complete within 2-3 min of Cytochalasin B application and was largely reversed within 5-6 min of its removal. F-actin networks receded from the entire peripheral margin of the lamella forming a band devoid of F-actin (Forscher and Smith, 1988). Bonder and Moosker (1986) have observed that cytochalasin B slows down the actin but does not prevent monomer addition at the bonded end of actin filament.

Cytochalasin B is a good suppressor of smooth muscle cells (SMC) contraction and migration *in vitro*. Although perivascular infusion of cytochalasin D inhibited collar induced SMC migration from media to intima *in vivo* as well, the intimal hyperplasia was not reduced due to concomitant development of an inflammatory response (Richard and Hidde, 2001).

It also interferes with cell division but not with nuclear division, and cause extrusion of cell nuclei, blocks cellular movements and inhibits phagocytosis. Phagocytosis inhibition is reverse, in the ciliated protozoan *Tetrahymena*, but cell division in heat-synchronized populations remain unaffected. It also causes starvation in *Tetrahymena* which results in the formation of small cells. Mortensen and Larsson (2003) have made detailed investigations on the effect of cytochalasins on actin.

Collings *et al.* (1995) reported that cytochalasins rearrange the cortical actin of *Nitella sp* into short stable rods. Goddette and Frieden (1986) discussed the mechanism of action of cytochalasins D on polymerization of actin. Carlier *et al.*, (1986) recorded the inhibitory activity of Cytochalasin D varied with the concentration of ATP-actin. Sampath and Pollard (1991) recorded almost equal inhibition of ATP-actin or ADP-actin by the cytochalasin D. Tian *et al.* (2012) reported that cytochalasin D affect at micromolar in ventricular myocytes.

Cytochalasins, C, D and Q were cytotoxic against tumor cell lines PC-3 M, NCI-H460, SF-268, MCF-7, and MDA-MB-231. Cytochalasin D presented the highest cytotoxicity against the five cell lines with an IC_{50} range of 0.22-1.44 lM, and selectivity for line NCIH460 with an IC_{50}value of 0.22 lM. Pongcharoen *et al.* (2007) established that cytochalasin D isolated from *Xylaria sp.* BCC 9653, was cytotoxic towards African green monkey kidney fibroblast (Vero) cells with an IC_{50} value of 0.19 lM. Cytochalasin Q is also cytotoxic against KB and BC-1 cell lines, with IC_{50} values of 3 and 1 lg/mL (Chinworrungsee *et al.* 2001).

Cytochalasins are microfilament directed agents and has anticancer activity. Cytochalasins preferentially damage malignant cells and minimal effects on normal epithelial cells and immune system. Cytochalasins influence the end stages of mitosis suggesting that such agents could be combined with microtubule directed agents to elicit a profound synergetic effect on malignant cells. Cytochalasins could be used to supplement current chemotherapeutic measures to improve efficacy rates as well as decrease the prevalence of drug resistance in the clinical setting.

Goddette and Frieden (1986) tried to explain the mechanism of actin polymerization under the influence of cytochalasin D. Cytochalasin D induces an increase in the initial rate of polymerization and decrease in the final stages of reaction. In this process Mg induces dimmers formation and conformational changes which is associated with ATP hydrolysis. Phomopsin elaborated by *Phomopsis leptostromifarmin* is potent microtubule depolymerizer. Cytochalasin H produced by *Phomopsis paspali* which infects a millet *Psrpalam scrobiculatum* is reported to act on central nervous system (Patwardhan *et al.*,

1974). Howard *et al.* (1981) tried to correlate the binding of cytochalasins to human polymorpholonuclear leucocytes and induction of morphological alternations and inhibition of cell motility.

Dozens of cytochalasins with distinct cytotoxic or anti-HIV activities have been isolated from the arthropod-derived fungus *Chaetomium globosum* (Chen *et al.,* 2015). Chaetoglobosins preferentially induce apoptosis in chronic lymphocytic leukemia (CLL) cells over normal B cells. Inhibits actin polymerization and leads to the accumulation of cells in G_2 phase (Knudsen *et al.*, 2014)

The Ustiloxin and Phomopsin families of natural products isolated from *Ustilaginoidea virens* and *Phomopsis leptostromiformis* are potent microtubule depolymerizers. Cytochalasin H derivatives regulate plant growth (Cox *et al.,* 1983) and cytochalasin E inhibits angiogenesis (Udagawa *et al.,* 2000).

6.2. Mutagenecity and teratogenecity

Cytochalasin B is reported to cause long multicentric chromosomes. Human lymphocytes exposed to cytochalasin H developed a variety of chromosomal aberrations including tetraploidy and premature chromosomal condensation, extension and constriction. Cytochalasins also exhibit teratogenicity. Cytochalasin B caused neural tube closure in chicken embryo which includes everted neural tubes, absence of optic vessels and general retardation of neurulation. Increased foetal resonation was also noted. Exencephally was the predominant defect along with other abnormalities including hypognathia and coxial skeletal defects of mycotoxins.

6.3. Locomotion and cytoplasmic streaming

Cytochalasin B is a rapid and reversible inhibitor of cellular locomotion and ruffling activity at low concentration but did not affect flagellar or ciliary movement at higher concentrations (Carter, 1967). Locomotion was totally inhibited in human, horse and rabbit leukocytes after 1 hr exposure to cytochalasin B at 2.5 μg/ml concentration. Inhibition of locomotion by cytochalasin B was not dependent on mucopolysaccharide synthesis. Cytoplasmic streaming in plant cells was also inhibited. No ultrastructural changes in microfilaments of *Nitella* cells were observed during inhibition of streaming.

Cytochalasins B and D inhibit cytoplasmic streaming in characean algal cells at concentrations of about 10 gM, without disrupting the subcortical actin bundles. Cytochalasin D reversibly reconstructs the more labile cortical actin strands into short, stable rods (Collings *et al.* 1995). Microtubule disassembly potentiates Cytochalasin D effects on both cytoplasmic streaming and the structure of the cortical actin cytoskeleton (Collings *et al.* 1996). Microtubule disassembly affected the disruption of subcortical actin bundles in prolonged exposure to low concentrations of Cytochalasin D (Foissner and Wasteneys, 2000). Cytochalasins reported wide range of effects on intra cellular motility and actin filament, recognisation of actin skeleton in internodal cells of *Nitella pseudoflabellata* (Foissner and Wasteneys, 2007). Cytochalasins disruptive effects on subcortical actin bundles are greatly enhanced by the concurrent disassembly of microtubules.

Tominaga *et al.* (1997) reported that combined propyzaminde-cytochalasin B treatments caused, microtubule disassembly and the formation of meshworks instead of parallel actin bundles, preventing the recovery of reverse-fountain streaming after removal of cytochalasin B.

6.4. Cytophysiological effects

Interestingly large number of cytophysiological processes are altered by cytochalasins. Most of the information is based on *in vitro* system and their relevance to animal toxicology can only be summerised. Subacute cytochalasin exposure may produce a spectrum of subtle but profound effects in animals. Cytochalasin action is rapid with effects noted within the first few minutes after exposure to the toxin and even the reversibility of action is also fast after removing cytochalasin suggesting non-covalent interaction of toxin to cellular receptor molecules.

Cytochalasin H is an effective inhibitor of the cytoskeletal reorganization that occurs in platelets following activation by adenosine diphosphate (ADP). Cytochalasin H prevented the normal incorporation of actin, alpha-actinin and actin-binding protein into the cytoskeleton that occurred following ADP activation, and incorporation of myosin was markedly reduced. Aggregation was only partially inhibited but, more dramatically, while the rate of disaggregation following addition of certain agents that interfere with fibrinogen binding to glycoprotein IIb/IIa on the surface of platelets was markedly increased. It also inhibited the ADP-induced shape change.

Cytochalasin D at 25–100 μg/ml concentrations blocked mitosis in living cells of *Oedogonium* (Sampson *et al.*, 1996).

Certain morphological alterations following cytochalasin treatment in susceptible cell models was recorded. These features appear to be a result of the cessation of movements at the cell surface as a consequence of alterations in the microfilament-plasma membrane complex and include such effects as generalized cellular contraction, retraction of surface microvilli, zeiosis (blebbing) and arborization. These responses are often differentially manifested in various cell types, but are invariably dose-dependent and reversible. Cytochalasin B induced morphological alterations in chicken embryo glial and nerve cells, not affected in the absence of glucose.

The ED_{50} for cytochalasin D induced zeisosis varied from 0.005 to 0.35 μg cytochalasin D per milliliter in a range of cell types after a 60-min exposure. In many cases, apical aggregation was completed after 3 hrs exposure. Cytochalasin D was qualitatively a more potent inducer of zeiosis in HEp-2 cells than cytochalasin B by at least a factor of four (Janet and Gabriel, 1983). Cytochalasin

E produced zeiotic knobs blebs and the retraction of the cytoplasm to the cell center (arborization) in mouse fibroblasts. In similar experiments, cytochalasin D and cytochalasin E were approximately 5 and 50 times more potent than cytochalasin B respectively.

The process of enucleation in a wide variety of cytochalasin B-treated cells may also be considered a variant of zeiosis. It differed from zeosis primarily on the basis that higher concentrations of cytochalasin B (10 to 20 μg/ml) were required. Spontaneous enucleation of a cell population rarely exceeded 3% and was the consequence of the displacement of nuclei to the cell periphery followed by protrusion of the nuclei from the cell surface and eventual segregation of the nuclei into pockets attached to the main cell body by thin cytoplasmic strands. The nearly total enucleation of a cell population may be obtained by moderate centrifugation of cell at this stage. Epithelial cells were more resistant to enucleation than other cell types. As with zeosis, disruption of the cortical cytoskeletal microfilament network and blebbing or herniations of the cell surface as a result of hydrostatic pressure are thought to be responsible for nuclear extrusion. Cytochalasin B and dihydro-cytochalasin B were equally effective in enucleating cells of a mouse mammary tumor line.

6.5. Attachment and adhesion

The cytochalasins affect the attachment of suspended cells to a solid surface and the substrate adhesivity of these cells once attached. Attachment and the concomitant process of cell spreading is mediated by the cellular skeletal and contractile systems. Once a cell comes to rest on the substrate, microspikes or filopodial extensions arise from the cortex around the circumference of the base and contact with and attach to the substratum. Cytochalasin B and D prevented the invasion of cells by fungal pathogen *Cryptosporidium parvum* up to 70% without affecting attachment (Chen and Russo, 2000). Cytochalasins are reported to convert rolling of neutrophils to stationary adhesion on P-selectin (Sheikh and Nash, 1998). Cytoplasmic spreading and filopodial extension were inhibited in HeLa cells treated with 0.5μg cytochalasin D per milliliter. The cells remained rounded and were covered with numerous zeiotic blebs under these conditions. Cytochalasin B also alters the distribution of microtubules, normally, microtubules are restricted to the neurite shaft and a central growth cone domain (Forscher and Smith, 1988). Bylund *et al.* (2004) reported that cytochalasin B triggers novel pertussis toxin sensitive pathway in TNF-alpha primed neutrophils.

Different cell lines exhibited variable sensitivities to the effects of cytochalasins on attachment which generally required higher doses than those used for inhibition of cytokinesis or locomotion. Attachment and spreading of L cells and Vero cells was not remarkably altered by cytochalasin D levels capable of inducing nuclear extrusion, whereas approximately 50% inhibition of attachment was noted at 0.25 μg cytochalasin D per milliliter for HeLa and HEp-2 cells. Attachment of Ehrlich ascites tumor cells was not affected by 10 μg cytochalasin B per milliliter, whereas decreased attachment was reported with similar concentrations of cytochalasin A and E at higher levels (100 μg/ml). However, the order of potency was CA>CB>CE. Reduced attachment of frog embryo endodermal cells to glass surfaces has also been reported for cytochalasin H. Once cells are attached to the substrate, they may become dislodged by cytochalasin treatment, possibly because of toxin-induced concentration and the resultant detachment of the cells. In some instances, adhesion may be promoted by cytochalasin D.

6.6. Cell sorting, aggregation and disaggregation

The ability of cytochalasin B to inhibit ruffled membrane activity and cellular locomotion of vertebrate cells in monolayer culture prompted its use to study the necessity for this kind of active cellular locomotion in cell sorting in heterotypic cell aggregates. Cell sorting was inhibited in chick embryo heart, pigmented retina aggregates but a remarkable degree of sorting did occur in neural retina-pigmented aggregates. Cytochalasin B at 5 or 10 μg/ml is sufficient to inhibit completely locomotion of these cell types in monolayer culture. The effect of cytochalasin B on specific sorting during reaggregation of embryonic chick heart and neural retina cells was studied by Maslow and Mayhew (1972).

Inhibition of aggregation by cytochalasin B in embryonic corneal epithelium was attributed primarily to membrane immobilization preventing normal apposition of the membrane rather than to the disruption of desmosomes which serve as adhesive cellular organelles. Aggregation was promoted in trypsinized suspended BHK21 cells after treatment with 5μg cytochalasin B milliliter for 45 min. The formation of MCF-7 aggregates and a decrease in stainability were observed on treatment with the conditioned medium of neutrophils stimulated with cytochalasin B (Satoru *et al.*, 2005)

The disaggregation of previously formed cell aggregates has been reported in certain amphibians. Cytochalasin B caused the dose-dependent disaggregation of frog gastrulae. Cytochalasin H produced a complete and reversible disaggregation of frog embryo endodermal cells within 1.5 hr.

Series of events in platelet activation involving the formation of microfilament-containing pseudopodia followed by adhesion of cells to the substrate and finally aggregation of the cells into a clump held together by the pseudopodial mesh. Activation was inhibited by 10μg cytochalasin B per ml. Pseudopodia in activated platelets disappeared after cytochalasin B treatment.

Cytochalasin B inhibits ruffled membrane activity and cellular locomotion of vertebrate cells in monolayer culture suggests its possible employment as model to study its necessity in cell sorting in heterotypic cell aggregates. Cell sorting was inhibited in chick embryo heart-pigmented retina aggregates but a remarkable degree of sorting did occur in neural retina-pigmented retina aggregates. Cytochalasin B at 5 or 10 μg/ml concentration caused total inhibition locomotion of these cell types in monolayer culture.

Cytochalasin B and D were found to inhibit monosaccharide transport, while cytochalasin D failed to inhibit monosaccharide transport. Both agents caused an aggregation of filamentous actin into dispersed felt-like masses enveloping large cystic vacuoles (Brett and Godman, 1984).

Cytochalasin D causes changes in the cytoskeleton which contributes to the stability of the aggregates that form due to the inhibition of GpIIb-IIIa-fibrinogen complex formation (May *et al.*, 1998).

6.7. Morphogenesis

Cytochalasin B possess significant activity in altering morphogenesis and could serve as a mechanistic basis for explaining the teratogenicity of this toxin. Cytochalasin B was effective in blocking axonal growth in cultured chicken embryo dorsal root ganglia. With the possible exception of bronchial branching in lung rudiments, the effects of cytochalasin B have been reversible. In addition, a loss of visible microfilament structure, possibly into a condensed or collapsed state, has been strongly correlated with altered development patterns. Lastly, neither glucose uptake nor protein or polysaccharide synthesis was required for the activity of cytochalasin B or for the recovery of these systems from cytochalasin B-induced damage. Cytochalasin B prevented entry of male sperm into the eggs of Pacific Oyster (Ledu and McComble, 2003). The affect of cytochalasins on morphogenesis and fertilization in different experimental plants and animals are summarized in **Table 6.2**

Table 6.2. Effect of cytochalasins on fertilization and morphogenesis

Biological system	Dosage/duration	Response
Brown alga eggs and embryos (*Pelvetia fastigiata*)	100 μm CB/ml; 30 min	Inhibition of photopolarization in eggs
	4 μg CB/ml; 14 days	Inhibition of apical hair formation
Polyporus biennis	20μg CB/ml; 3 hr	Irregular hyphae growth; inhibition of initiation of clamp connection formation
Neurospora crassa	50μg CD/ml; time not given	Stimulation of hyphae branching
Surf clam eggs (*Spisula solidissima*)	0.5-5μg CB/ml; 10 min	Prevention of polyspermy block after fertilization
Echiuroid worm eggs and sperm (*Urechis caupo*)	1-2μM CB; 30 sec-10min	Inhibition of sperm penetration and formation of sperm penetration cone
Marine snail embryos (*Ilyanassa obsoleta*)	1-10μg CB/ml; 1.5-4hr	Prevention of polar lobe formation

Biological system	Dosage/duration	Response
Squid embryos (*Loligo pealei)*	0.2-2.0μgCB/ml; 10-30min	Blastodisc formation and polar body extrusion inhibited in precleavage embryos; generalized anomalies of organ placement and differentiation in 3- to 4- day embryos
Ascidian larvae (*ten species*)	0.25-10μg CB/ml; 90 sec-4hr	Inhibition of tail resportion in metamorphosis via disruption of apical filaments and prevention of caudal epidermis contraction
Frog embryos (*Xenopus laevis*)	7.5 μg CB/ml; 4-10min	Regression of furrow formation during cleavag and interference with cell junction development
Newt embryos (*Triturus alpestris*)	2-10μg CB/ml; 0-100min	Regression of first cleavage furrow
Chicken embryos	10μg CB/ml	Inhibition of neurulation and separation of neural folds with concomitant disturbances of microfilament arrays
Chicken embryos	5μg CB/ml; 3hr	Prevention of nuclear migration in neural cells. Inhibition of estrogen-induced tubular gland formation and regression of presently existing gland cells and associated microfilaments
Chicken embryo spinal cord and dorsal root ganglia cell cultures	7-10μg CB/ml; 0-24hr	Cessation of axon elongation and alteration of microfilament net work
Chicken oviduct explants	5 μg CB/ml; 3 hr	Prevention of nuclear migration in neutral cells
	30-50 μg CB/ml 6 h4	Inhibition of estrogen induced tubular gland formation and regression of presently existing gland cells and associated microfilament
Chicken embryo duodenal explants	0.1-10 μg CB/ml; 15 min - 6 hr	Formation of elongated branched microvilli and alteration of their spatial distribution
Chicken embryo duodenal explants	1.0μg CB/ml; 24-36hr	Prevention of previllous ridge formation and disruption of microfilament bundles
Chicken embryo cardiac muscle	20-100μM CB; 45min-4hr	Disruption of preexisting myofibrils and inhibition of new myofibril formation; loss of Z-bands and disarrayed myosin filaments

Biological system	Dosage/duration	Response
Chicken embryo limb bud mesodermal cells	2.5μg CB/ml; 24-72hr	Irreversible inhibition of chondrogenesis. Independent of reduction in protein synthesis or glucose uptake
Mouse oocytes	5μg CB/ml; 5-16hr	Arrest of oocyte meiosis evidenced by inhibition of polar body emission and loss of homologous chromosome separation
Mouse eggs and spermatozoa	5-10 μg CB/ml; 1-7hr	Reduction of fertilization rate; induction of triploidy in fertilized eggs
Preimplantation mouse embryos	0.25-10μg CB/ml; 0-24hr	Inhibition of morula to blastocyst transformation; decrease of trophoblast outgrowths noted within 3 to 5min
Mouse embryos	10μg CB/ml; 3-8.5hr	Induction of tetraploidy and 2n/4n mosaic embryos
Mouse embryos	10μg CB/ml; 2hr	Abnormal structure and development of neuroepithelial cells
Mouse embryo salivary glad epithelium cultures	7-10μg CB/ml; 18-72hr	Loss of normal branching morphogenesis noted within 6 to 10 hr; microfilament disorganization
Mouse skeletal muscle explants	0.25-2μg CD/ml; 0-7 days	Increased contractility, retraction and segmental constriction of myotubes after short exposures; prolonged treatment resulted in formation of leptomeric bodies and myofibril disorientation

6.8. Endocytosis

The endocytosis is a process in which extracellular fluid particles are incorporated by the cell via invaginations of plasmamembrane and the resultant formation of endocytic vacuoles or vesicles. There are three types of endocytosis which are characterized by the size of the particles ingested or the vesicles formed.

i. **Phagocytosis**: Ingestion of particles more than 1 μm in diameter and is performed by macrophages and polymorphonuclear leucocytes and produces vesicles 200 – 700 nm in diameter.

ii. **Macropincytosis**: It is also noted in macrophages and certain specialized epithelial cells and produces vesicles 200 – 700 nm in diameter.

iii. **Micropinocytosis:** It is responsible for protein transport in vascular endothelial cells and forms microvesicles of 60-70 nm in diameter. Phagocytosis of bacteria was inhibited by cytochalasin B, while attachment of bacteria to phagocytes was not affected. Cytochalasin E prevented ingestion of *E. coli* by guinea polymorphonuclear leucocytes. Sucrose uptake and macropinocytosis were inhibited by the presence of cytochalasin B. on the other hand, micropinocytosis was insensitive. Cytochalasin D is a specific inhibitor for macropinocytosis by binding to actin filaments, thus it can interfere with actin polymerization and assembly (Peterson and Mitchison, 2002). Inhibition of A-MLV (Amphotropic Murine Leukemia Virus) infection was also true of F-actin depolymerization with cytochalasin D (Rasmussen and Frederik, 2015)

6.9. Exocytosis

Exocytosis is a process by which cells secrete substances into the extracellular space and is reported to be inhibited by cytochalasins. The secretion of hormones, neurotransmitters and proteins including lysosomal and exocrine enzymes and immunologically active products, has been extensively studied with regard to the mechanisms of cytochalasin action. Actual disruption of the microfilament web located beneath the plasmamembrane (cortical band) was reported in pancreatic beta cells in the presence of cytochalasin B. Cytochalasin A was examined for its effects on antigen-stimulated histamine release in human lungs and leukocytes and in rat peritoneal mast cells. Cytochalasin A unlike, cytochalasin B and D was ineffective in enhancing complement-induced histamine release in human leukocytes. Cytochalasin B and E were equally effective in this system. However, cytochalasin A was a more potent inhibitor of histamine release in rat mast cells and a more effective stimulator of histamine release in serum-sensitized human lungs than cytochalasin B at equivalent concentrations. Serum stimulated prostaglandin secretion in mouse 3T3 cells was approximately three times more effectively inhibited by cytochalasin B than cytochalasin A.

Treatment of neutrophil with cytochalasin D confirmed that ILT-4 (immunoglobulin-like transcript 4) up-regulation was mediated through granule exocytosis since it enhanced ILT4 up-regulation on neutrophils stimulated with fMLF (either N-formyl-methionyl-leucyl phenylalanine) (Jeremy *et al.*, 2013). Some of the important effects of cytochalasins on exocytosis are précised in **Table 6.3.**

Table 6.3: Effect of cytochalasin on Exocytosis

Biological system	Secretory product	concentration	Response
Human leukocytes	Lysosomal enzymes (β-glucuronidase marker)	5μg CB/ml; 15 min	An approximate 35% increase in enzyme release after zymosan particle challenge
Rabbit leukocytes	Lysosomal enzymes (β-glucuronidase marker)	0.5-10μg CB/ml; 15-30 min	Enhanced release of enzymes following stimulation with retinol or rabbit globulin coated latex particles and antigen antibody complexes; enzymes not released from isolated secretory granules
Mouse peritoneal macrophages	Lysosomal enzymes (β-glucuronidase and (β-galactosidase markers)	2-10μg CB/ml; 5-72hr	Time and dose-dependent release of enzymes; effect noted after 24-hr exposure to 10mg CB/ml
Guinea pig peritoneal leukocytes	Lysosomal enzymes (β-glucuronidase marker)	2μg CB/ml; 30min	An approximate 45% increase in enzyme release without particle challenge
Human blood platelets	Lysosomal enzymes (β-glucuronidase and N-acetyl glucosaminidase)	1-50μg CB/ml; 10 min	Enzyme release stimulated by CB only after collagen induction
Mouse peritoneal macrophages (Thioglycollate-stimulated)	Elastase and lysozyme	4μg CB/ml; 24-72hr	An approximate 60% inhibition of elastase secretion after 48hr; no effect on lysozyme
Rat parotid tissue slices	α–Amylase	52μM CB; 1 hr	Stimulated α-amylase release at low cyclic AMP levels (<0.3mM); and high cyclic AMP (1.0mM) induced amylase release
Water mold *Achlya ambisexualis*	Cellulase	2-20μM CA; 1hr	Reversible dose-dependent inhibition of cellulase secretion; variable effects with CB
Rabbit and human fibroblast cultures	Collagenase	1-10μg CB/ml;; 2-8 days	Enhanced collagenase activity in medium

Biological system	Secretory product	concentration	Response
Rat liver slices	Plasma proteins (albumin)	5-50 μM CB; 3 hr	No significant inhibition of plasma protein release
Rat pancreas pieces	Chymotrypsinogen and lipase	0.1-10μg CB/ml; 1hr	Optimum stimulation of enzyme release at 0.5-2 μg CB/ml
Guinea pig hypogastric nerve-vas diferens preparation	Dopamine-β-hydroxylase	3μg CB/ml; 75mins	Blockage of electrically induced enzyme release; little effect on basal enzyme secretion
Rat lymph node cells	Lymphotoxin	0.05-5μg CB/ml; 3days	Dose-dependent inhibition of ovalbumin-stimulated lymphotoxin production
Mouse plasmacytoma cells	Ig	100μM CB; 4hr	No effect on Ig release
Mouse spleen cells	Ig antibodies	3-300μg CB/ml; 18hr	No effect on antibody secretion toward sheep erythrocytes
Mouse spleen cells and viral-infected BHK cells	Interferon	0.6-12.5μg CB/ml or 0.12-10μg CD/ml; 8hr	Complete inhibition of interferon production at 10μg CB/ml or 0.37μg CD/ml

6.10. Immunological functions

Several immunological important processes are altered under the influence of cytochalasins. However, mechanism of action mainly is unknown although membrane surface properties and other factors controlling cellular aggregation are undoubtedly implicated. The reversible inhibition of chemotoxis by cytochalasin B has been reported for human/ rabbit leucocytes. Cytochalasins strictly regulate immunoregulatory factors such as interleukin-2, prostaglandin-E2, insulin-like growth factor-binding protein-1 and macrophage inflammatory protein-2 (Lewitt and Baxter, 1991 and Isowa *et al.,* 2000).

A rapid increase in Ca+2 uptake in lectin-stimulated lymphocytes after treatment with 10 μM cytochalasin A, B and E. No effects were noted with the cytochalasins alone. It was suggested that the increased Ca+2- flux might potentiate the lymphocyte response by activating actomyosin contraction.

Cytochalasin congeners were found to enhance lectin-induced DNA synthesis in lymphocytes. Cytochalasin A, D and E were approximately ten times more effective than cytochalasin B, whereas dihydro- Cytochalasin B and γ-lactone cytochalasin B were inactive.

Rosette formation was inhibited by cytochalasin B with thymus-derived peripheral blood lymphocytes (T cells), but not with thymocytes. The potency series CA>CE>CD>CB was obtained, with values ranging from 77 to 23% for inhibition of cytolysis at equimolar (10μM) concentrations. Cytochalasin A and B both inhibited cytolysis by preventing binding of the K cells to target cells.

The reversible inhibition of chemotaxis by Cytochalasin B (2 to 4μg/ml) has been reported for human and rabbit leukocytes. This effect was presumably because of the inhibition of leukocyte locomotion.

6.11. Viral infectibility

There appears to be a dichotomy between the effects of cytochalasins on RNA and DNA virus infectibility or the sensitivity of the host to virus infection as measured by plaque formation. Cytochalasin B (2 μg/ml) promoted the production of C-particles in mouse L-cells. In a departure from the increased infectivity normally observed after cytochalasin treatment of RNA virus host cells, recent findings have indicated a dose-dependent inhibition (10 to 25 μg cytochalasin B per milliliter) of the release of Kirsten murine sarcoma-leukemia virus from rat kidney cells. Cytochalasin B (54μg/ml) was reported to decrease the antiviral activity of murine interferon in mouse L929 cells.

Cytochalasin B and D (4 to 6 μg/ml) produced a 5-fold enhancement of the poliovirus infectivity in Vero cells, although cytochalasin D was more effective than cytochalasin B at suboptimal concentrations. Maximal infectibility (3-fold increase) of Vero cells with parainfluenza virus obtained at 1 μg cytochalasin B per milliliter.

Cytochalasin B enhanced RNA virus infectivity could be a consequence of a reduction in host protein synthesis, thereby allowing for more efficient translation of viral mRNAs.

Both cytochalasin B and dihydrocytochalasin B totally inhibited Ca^{+2} dependent blood clot retractions in platelet rich human plasma. Cytochalasin B caused osmotic fragility and increased deformability of human erythrocytes water movement across toad bladders was promoted by the action of vasopressin or cyclic AMP. These also caused profound disruption of fluid balance mechanisms in rodents by cytochalasin F or Chaetoglobosin A. Inhibition of fluid transport was also accompanied by alternations in the electrical properties of the membrane by cytochalasin B. Neurotransmission was also blocked through cat cervical sympathetic ganglia by high concentration of cytochalasin B. Low levels of cytochalasin B caused a rapid nearly total loss of end plate activity at nerve muscle contracts in cultured chicken embryo muscle cells and spinal cord explants. Cytochalasin B also caused a reversible dose-dependent blockage of MSH or cyclic AMP induced darkening in frog skin with a drastic reduction in microfilaments of the dendritic process of melanocytes.

Cytochalasin B (10 μg/ml) prevented the migration of the virions from the factories into the cytoplasm and also disrupted the virus-containing membrane structures. Herpes simplex virus (HSV) infectibility of human embryonic lung cells was reduced by 60 to 100%, depending upon cytochalasin B concentration and exposure time. Fucose and glucosamine uptake by lung cells were nearly total inhibited by 10 μg cytochalasin B per milliliter without affecting protein synthesis. Cytochalasin B treatment also reduced much of the HSV-induced nuclear and membrane cytopathology in the infected lung cells.

Cytochalasin B and D reversibly enhanced the infectivity of isolated HSV-DNA from three- to six fold in rabbit kidney cells, un-derlining the role that viral proteins play in modulating host cell functions. Cytochalasin D was reported to depolymerise the cellular actin and microtubules, in order to study the effect of Japanese Encephalitis Virus infection in the cell (Magdline, 2015).

6.12. Blood clot retractions

Cytochalasin B antagonizes or impairs any mechanism leading to blood platelet aggregation, either during the phases of activation and shape change or following the dense-granule release reaction and stimulation of the prostaglandin-thromboxane system (Ressom *et al.,* 2012).

Cytochalasins interfere with the polymerisation of actin. Cytoachalasin H is reported to inhibit platelet cytoskeletal reorganisation and disorganise platelet function (Natarajan *et al.,* 2000). Inhibition of fibrinolysis was observed when clot retraction was prevented by cytochalasin D or abciximab, or when platelet membranes were tested (Carrieri *et al.,* 2010).

The inhibition of clot retraction by cytochalasin D reduced clot strength and enhanced clot sensitivity to lysis (Collet *et al.,* 2001). Neutrophils triggered with TNF-ὰ mobilize receptor storing organelles and concomitantly but in independently, the cells are primed to respond to the microfilament disrupting toxin cytochalasin B (Bylund *et al.,* 2004).

Cytochalasin B reduced the release of antimicrobial peptides (eg. Apolipophorin 3) and enzymes (eg. Serine protease) from phorbol 12-myristate 13-acetate (PMA) stimulated haemocytes (Nessa *et al.,* 2011). Haemocytes are sensitive to cytochalasin B.

Cytochalasin B inhibit platelet aggregation and clot retraction. Cytochalasin B also inhibits the uptake of particles during neuetrophil mediated phagocytosis (Mitchell *et al.,* 2008).

Both cytochalasin B and dihydro-cytochalasin B (50 μm) totally inhibited Ca^{+2} dependent blood retractions in platelet-rich human plasma. Cytochalasin D and E effects fibrinogen receptor movement and cytoskeletal reorganization in fully spread, surface activated platelets (Olorundare *et al.,* 1992). Cytochalasin E was potent and selective inhibitor of bovine capillary endothelial (BCE) cell proliferation (Udagawa *et al.*, 2000).

Cytochalasin D can inhibit the rapid polymerization of actin in human platelets after thrombin stimulation and induce rapid depolymerization of filamentous actin in platelets (Casella *et al.,* 1981). Morris and Tannenbaum (1980) reported no net depolymerization of actin in spreading fibroblasts treated with cytochalasin D, despite the well known morphological effects of the drug on these cells.

Cytochalasin E inhibited angiogenesis induced by basic fibroblast growth factor by 40% to 50% and inhibited the growth of Lewis lung tumors by approximately 72%. It is also potent antiangiogenic agent that may hold promise for the treatment of cancer and other types of pathologic angiogenesis (Udagawa *et al.,* 2000). The mechanisms operative in the inhibition by cytochalasins of human peripheral blood T-lymphocytic rosette formation with sheep erythrocytes remain obscure in the light of the multiplicity of biologic effects of cytochalasins (Mookerjee and Jung, 1983).

6.13. Osmotic balance

Osmotic fragility and increased deformability of human erythrocytes have been reported after treatment with 20 to 40μg cytochalasin B per milliliter. Water movement across toad bladders was promoted by the action of vasopressin (antidiuretic hormone) or cyclic AMP. A reversible, dose-dependent inhibition of this response was obtained with cytochalasin B, producing a 50 to 100% reduction in water transport at 5 to 7.5μg cytochalasin B per milliliter. The profound disruption of fluid balance mechanisms in rodents at a toxic dose of Cytochalasin E or chaetoglobosin A was observed.

In rabbit gall bladder preparations 5 μg cytochalasin B per milliliter inhibited fluid transfer rate by approximately 50%. Cytochalasin B (20 μg/ml) also decreased fluid transport through rabbit corneal epithelium.

6.14. Electro-physiological effects

Inhibition of fluid transport by cytochalasin B was accompanied by alterations in the electrical properties of the membranes. Cytochalasin B reversibly eliminated the potential difference across rabbit corneal epithelium and produced a prompt, but reversible, reduction of the resistance (30%) and potential difference (40%) across rabbit gall bladder preparations.

Neurotransmission was also blocked through cat cervical sympathetic ganglia by high concentrations of cytochalasin B (50μg/ml). The electrophysiological and mechanical effects of cytochalasin D in the Langendorff perfused rabbit heart were observed (Kettlewell *et al.,* 2004). Cytochalasin B prolonged the falling phase of the action potential in frog muscle fibers, but did not alter resting potential or the amplitude of the action potential. Low concentration of cytochalasin B (0.8 to 4.0μg/ml) caused a rapid, nearly total loss of end-plate activity at nerve-muscle contacts in cultured chicken embryo muscle cells and spinal cord explants.

6.15. Axoplasmic transport

Axoplasmic transport of labelled protein was blocked by cytochalasin-B. Cytochalasin B at 200 to 400μm concentration blocked the fast axoplasmic transport of radio-labeled proteins in rat sciatic nerve. Isenberg *et al.* (1980) reported the evidence in favour of the actomyocin system driving axoplasmic transport with the injection of cytochalasin B. Half-maximal inhibition of serotonin binding to this protein was observed in the presence of cytochalasin B (50 μg) *in vitro*.

6.16. Metabolic cooperation

Metabolic cooperation is a form of cell communication requiring intimate cell to cell contact between cultured normal cells and mutant enzyme-deficient cells. In most studies mutant cells lacking the enzyme hypoxanthine-guanine phosphoribosyltransferase (HGPRT), showed a decreased or prevented under the influence of cytochalasin B (0.5 to 10 μg/ml).

Cytochalasin B (1 μg/ml) prevented the development of metabolic cooperation and abolished the preexisting cooperation between donor 3T3 cells and recipient BHK cells. Cooperation was reestablished by removal of cytochalasin B.

Cytochalasins dramatically enhanced DNA synthesis as mitogens, resulting in the increase of cyclic AMP, intracellular Ca+2 and phosphorylation of some kinases such as protein kinase A, protein kinase C and protein tyrosine kinase (Jasper *et al.,* 1995). Cytochalasins in certain lymphoid cells caused stimulation of intracellular signal transduction pathways for cell activation and /or differentiation (Kelleher *et al.*, 1995). Interleukin-8 production in the HeLa cells cultured with cytochalasin E blocked protein the kinase C inhibitors (Ikewaki *et al.,* 2003).

6.17. Ovulation

Cytochalasin B causes contraction of the follicle cells. The follicle contracts down around the oocyte, to lie collapsed against the ovarian wall, while the oocyte is free within the ovarian lumen (Smiley and Cloney, 1985). Trubnikova (2003) reported that the cytochalasin B, and aminoglutethimide suppressed ovulation most effectively.

In vitro inhibition of amphibian ovulation by cytochalasin B was reported. It prevents contraction of follicular constriction of the base of the follicle and ovulation (Martin and Talbot, 1981). It also completely inhibited the ovulation in ovary fragments of frogs and fish at concentration of 1 to 5μg. Luo *et al.* (2013) reported that chaetoglobosin K effectively inhibited angiogenesis through down regulation of VEGF-binding (Vascular epithelial growth factor) HIF-1 (hypoxia-inducible factor 1α) and could be an effective agent for cancer treatment. Ali *et al.* (2012) demonstrated the dual inhibitory effect of chaetoglobosin K on both the AKT and JNK signaling pathways in ras-transformed epithelial and human carcinoma cells.

6.18. Pigment granule dispersion

Cytochalasin B at 1 to 10μg/ml produced a reversible dose-depended blockage of (Melanocyte stimulation hormone) MSH-or cyclic AMP-induced darkening in frog skin in correlation with a drastic reduction in microfilaments of the dendritic processes of melanocytes. Cytochalasins B and D fully prevented pigment dispersion when administered to dark adapted animals immediately prior to their exposure to light and caused pigment aggregation to the RPE (Refinal pigment epithelial) cell base to fully light-adapted animals (Burnside *et al.,* 1983). It inhibits both aggregation and dispersion in chromato-phores of some invertebrate.

6.19. Interaction with contractile system

Cytochalasins causes a remarkable wide assortment of cytophysiological processes of cellular contractile mechanisms for the manifestation of a response. Cytochalasins causes change in microfilament structure and interact with various components of contractile system both *in vivo* and *in vitro*. However, the mechanisms by which cytochalasins alter microfilament-mediated process are very poorly understood. There is a continuing controversy over whether cellular microfilaments or the plasma membrane constitute the primary site of cytochalasin action. There are evidences to support both hypotheses. This situation is further complicated by the existence of atleast three classes of cellular microfilaments which differ in location, composition, and presumably function. Only indirect evidence implicating microfilament involvement in cytochalasin action has been forthcoming to date and usually consist of microscopical observations and visualization techniques correlating disappearance or alteration of microfilament structures with cytochalasin effects on cell functions. Despite the known effect a cytochalasins on cellular contractile processes, consistent evidence of toxin-induced disruption of contraction in muscle cells is lacking.

The effect of cytochalasins on alterations in the viscosity and ATPase activity and the binding of these toxins to the contractile proteins and their complexes has been studied. Many of the observation are contradictory. There are two hypotheses to explain this phenomenon. Cytochalasin B at concentrations 10 to 100-fold higher than the levels normally used to alter cell fractions in order to elicit a significant response.

Cytochalasin B (160 μg/ml) induced a rapid and marked decrease in viscosity of the actomyosin complex formed by the in vitro incubation of actin with myosin. A reduced viscosity was also noted when cytochalasin B was added prior to the initiation of actomyosin complex formation. An interaction of cytochalasin B with the actin component was indicated, since the viscosity of purified F-actin solutions was diminished, whereas that of myosin was not.

Phalloidin, a toxic cyclic peptide from the mushroom, *Amanita phalloides*, partially reversed the effect of cytochalasin B on the viscosity reduction of Filamentous actin (F-actin). Cytochalasin B (350 μM) did not alter the viscosity of rabbit muscle actin when complexed with the regulatory proteins, troponin-tropomyosin.

The polymerization of muscle Globular actin (G-actin) was effectively inhibited (80%) by 100 μM of cytochalasin A, whereas only a 15% inhibition was recorded with equimolar cytochalasin B. Cytochalasin A inhibited both the assembly and the colchicine binding to brain tubulin, presumably as a consequence of the reduction in tubulin sulfhydryl groups.

Cytochalasin D which is considerably more potent inhibitor of cell movement and cytokinesis than cytochalasin B, did not inhibit Ca^{+2} or Mg^{+2} – ATPase activity of muscle actomyosin or platelet actomyosin (thrombosthenin).

The binding of cytochalasins to platelet and muscle contractile components has been examined. Studies with ^{3}H-cytochalasin B demonstrated that most of the low affinity binding sites in platelets were associated with the actin-containing fraction. Purified rabbit skeletal muscle actin and myosin bound to ^{3}H- CB sites having a lower affinity myosin sites than the actin sites. Binding of ^{3}H- cytochalasin D by human platelets was preferentially localized in the myosin fraction.

Cytochalasin B (μM) reversibly prevented the gelation of rabbit pulmonary macrophage extracts and promoted the liquefaction of previously formed gels as well as the gelation of actin by the actin-binding protein. Cytoplasmic gelation in HeLa cell and *Acanthamoeba* extract was also inhibited by cytochalasin B concentrations ≥0.25 μM. Low concentration (2.5 μM) of cytochalasin A and D also inhibited cytoplasmic gelation in HeLa cell extracts.

The classically observed pattern of microfilament response to cytochalasin treatment has generally involved the cortical filament network. Within the several minutes, this network became dense and was retracted away from the cytoplasmic face of the plasma membrane, mouse 3T3 cells showed that actin was aggregated into star-shaped masses after a 1- to 5 min exposure to 20 μM cytochalasin B. Microfilament disruption by cytochalasin A (5 μg/ml) in mouse 3T3 cells produced actin-containing rod element approximately 5 μg long, in addition to the aggregate central actin masses usually observed with cytochalasin B treatment under conditions in which cell contraction had been induced.

6.20. Alterations in membrane structure and function

Cytochalasins mainly affect plasmamembrane at cellular level. These mainly cause alterations in the spatial arrangement and surface accessibility of specific membrane components including antigen, lectin and cell surface polypeptide and hormone receptors. These also inhibit membrane transport, particularly hexose transport system. These interact with high affinity membrane binding sites specially glucose carrier. Cytochalasin B altered the availability of surface sialic acid groups. It also blocked the ATP induced recovery of cell mobility in rat erythrocytes and interferes with contractile proteins in the membrane.

Other alterations in membrane properties may be a consequence of the well-established ability of the cytochalasins to disrupt microfilament function. Actin-containing microfilaments as well as microtubules are consistently found attached to plasmamembrane preparations in a variety of cell types, apparently representing a portion of the membrane-associated cytoskeletal system by which cell surface properties are mediated by cytoplasmic processes.

Ehrlich ascites tumor cell mobility was reversibly reduced by approximately 25% within 6 hr after incubation with as little as 0.2 μg cytochalasin B per milliliter. Cytochalasin B induced membrane vesicles (CIMVs) were successfully tested as a vector delivery of dye, nanoparticles and chemotherapeutic (Marina *et al.*, 2018).

6.21. Capping and agglutination

The role of cytochalasin B in altering membrane capping and agglutination phenomenon has been more extensively studied than other parameters of membrane topology.

Capping may be divided into two groups on the basis of the surface receptors involved. Immunoglobulin (Ig) capping induced cellular motility and was derived from an active primary contractile traction. Ig capping was readily inhibited by cytochalasin D, but was variably affected by relatively high concentrations of cytochalasin B (10 to 30 μg/ml). The dispersal of Ig caps was an energy-requiring process promoted by Ca+2 concentrations or cytochalasin D.

The second form of capping was induced by lectins such as concanavalin A. Lectin cap formation and maintenance was dependent upon cellular motility and was thought to be a passive aggregation of receptors secondary to the membrane movement involved in locomotion. Thus, low levels of cytochalasin B, which arrested locomotion, also inhibited the formation and promoted the spontaneous dispersal of lectin caps.

Cytochalasin B (10 to 30 μg/ml) was initially shown to partially and reversibly inhibit cap formation in spleen lymphocytes induced by rabbit antimouse Ig by up to 60% after a 30 min exposure. Different effects of cytochalasin B on capping were noted with a variety of Ig receptors in mouse thymus and spleen cells.

Concanavalin A- induced capping was particularly sensitive to cytochalasin B inhibition in mouse spleen cells, where surface Ig receptors were also incorporated into the cap along with other concanavalin A receptors.

Cytochalasin B (10μM) totally inhibited concanavalin A capping in Chinese hamster ovary cells, but was without effect in human polymorphonuclear leukocytes at cytochalasin B concentrations inhibitory to locomotion (5μg/ml).

Total inhibition of measles virus antigen-induced capping in human IgG-treated HeLa cells was reported with cytochalasin B (10μg/ml), whereas 1 μg cytochalasin B per milliliter produced only partial inhibition. High concentrations (38μg/ml) did not inhibit polyvalent Ig antisera capping of surface Ig in human lymphocytes.

Agglutination of rat ascites tumor cells by Concanavalin A and *Ricinus communis* agglutinin was inhibited by cytochalasin B, although no effect on the membrane binding of these lectins was observed. Cytochalasin B (10μg/ml) increased agglutinability by 60% after a 20 min exposure. Agglutination of trypsinized mouse 3T3 fibroblasts was totally inhibited by cytochalasin B, whereas agglutination of polyoma virus-transformed 3T3 cells was not affected.

Treatment of confluent hamster fibroblast cells with cytochalasin B (2.5 to 5 μg/ml) for 24 hrs promoted the loss of the LETS protein and other surface proteins to the medium. At 20 μg/ml cytochalasin B, one half of this protein was lost within 4 hrs.

Rapid, dose dependent release of the protein from human embryonic skin fibroblast was observed under the influence of cytochalasin B (0.3 to 3.0μg/ml) for 30 min.

Cytochalasin A, B and D (10μg/ml) induced a reduction of insulin and growth hormone surface receptors in cultured human lymphocytes.

Cytochalasin A, promoted 80% loss of human growth hormone receptor. However, the receptor loss was time-and dose-dependent without any noticeable alterations in the kinetics of the hormone-receptor interaction. Unlike most cytochalasin-induced cellular responses, the loss of surface receptors was neither reversible nor rapid onset. Cytochalasin A- induced loss of insulin receptors within 30 to 60 min.

Cytochalasin B blocked glucose transport and /or phosphorylation. It also inhibited the uptake of D-2-deoxyglucose, a nonmetabolizable analog of D-glucose, by horse leukocytes and mouse L-cell fibroblast with a concomitant decrease in glycolysis. Cytochalasin B inhibited deoxyglucose incorporation by 85% after exposure of leukocytes to 2μg/ml for 1 hr. Hexokinase activity and lactic acid production in cytochalasin B treated cell homogenates was unchanged. Therefore, alterations in the rate of glycolysis may be due to decreased glucose transport. Glucosamine incorporation has also been studied in relation to cytochalasin B inhibition of hexose transport, primarily as a probe for alterations in glycoprotein synthesis (Plagemann and Erbe, 1973).

Glucose uptake in HeLa cells and human erythrocytes was inhibited 50% by 8×10^{-8} M and 2 to 5×10^{-7} M cytochalasin B respectively. Glucosamine incorporation was inhibited 50% by 8×10^{-8} M cytochalasin B in mouse 3T3 fibroblasts and 1.0×10^{-7} M cytochalasin B in chicken embryo fibroblasts.

Dose-dependent inhibition of thymidin transport by cytochalasin B has been reported in mouse 3T3 fibroblast, Novikoff rat hepatoma cells, HeLa cells, mouse L and S180 cells and virally transformed mouse 3T3 fibroblasts. In these instances, transport inhibition ranged from 35 to 55% for cytochalasin B concentration of 4 to 20μM.

The transport of nucleic acid bases and other nucleosides was also competitively inhibited by cytochalasin B. Cytosine uptake was approximately 3 times more sensitive to cytochalasin B than L-glucose diffusional uptake with half-maximal inhibition being obtained at 2.8μM CB.

Calcium transport was not altered at cytochalasin B concentrations (3μM) which inhibited deoxyglucose uptake by over 90% in HeLa cells. Cytochalasin B was an effective reversible inhibitor of sodium absorption and potassium secretion in perfused rat salivary ducts. Sulfate and phosphate uptake studies have not shown any marked cytochalasin B-induced alterations. Leucine uptake was not inhibited in myoblasts, horse leukocytes or frog ganglionic cells at approximately 20μM cytochalasin B.

Chaetoglobosin A, cytochalasin A and B did not inhibit Na+-dependent glycine transport in rabbit reticulocytes at 100μg/ml. Choline transport was not affected by cytochalasin B in Novikoft hepatoma cells. The competitive binding affinity series CE>CA>CD>CB was obtained in human lymphocytes and leukocytes for the glucose-insensitive high affinity cytochalasin B binding sites.

6.22. Antimicrobial and antiparasitic activities

Betina *et al.* (1972) reported the antimicrobial activity of cytochalasin A and D. When cytochalasin A was active against *Bacillus subtilis* and *E. coli*, cytochalasin D was antimycotic. Cunninghan *et al.* (1979) recorded antibacterial activity against gram + ve bacteria where different physiological processes such as enzyme induction, transport and respiration were inhibited. The presence of α, β-unsaturated carbonyl group in the macrolide moiety and unsubstituted lactam functional groups proved to be a prerequisite for the antibacterial activity. Horn *et al.* (1995) feel that antimicrobial activity of phomopsichalasin isolated from endophytic *Phomopsis* sp was due to the 13-membered tricyclic in place of macrocycle. This substituted compound proved to be active against *Bacillus subtilis, Salmonella gallinarum* and *Staphylococcus aureus* as well as antifungal against yeast (*Candida tropicalis*). Pongcharoen *et al.* (2006) reported the scoparasins A and B isolated from endophytic fungus *Eutypella scoparia* PSU-D44, were active against the dermatophyte *Microsporum gypseum*. Makioka *et al.* (2004) reported that cytochalasins B, D, E and dihydrocytochalasin B caused significant decrease in amoebic infections (*Entamoeba histolytica*) at 10μM concentration. Cytochalasins E and K showed strong antifungal activity against *Eurotium repens* and *Mycotypha microspora*. Similarly cytochalasins E and K have good herbicidal activity against *Chlorella fusca* (Hidayat *et al.*, 2014). Cytochalasin Z28 isolated from the endophyte *Xylaria sp*. XC-16, showed strong antifungal activity against *Gibberella saubinetti* (MIC of 12.5 lM). Its fungitoxicity was more than a fungicide hymexazol (Zhang *et al.* 2014).

A considerable variation in the toxicity of different cytochalasins towards different microorganisms was observed. Cytochalasin A was more toxic to *E. coli,* while cytochalasin B and D were least toxic. Similarly cytochalasin A was potent fungicide, while protozoa and *Euglena* were unaffected by these toxins. Cytochalasins are reported to inhibit cell division or cytokinesis without affecting nuclear division which results in the formation of binucleate cells.

6.23. Phytotoxicity

Cytochalasin B inhibited Streaming of cytoplasmic and the growth of pollen tube of *Lilium longiflorum* which resulted in the death of tip of pollen tube. Several cytochalasins proved to be good biocontrol agents in the management of weeds (Scherlach *et al.,* 2010). Cimmino *et al.* (2008) and Berestetskiy *et al.* (2008) feel that a carbocyclic or a lactonic macrocycle fused to an unaltered perhydroisoindolyresidue as well as a secondary hydroxyl group at C-7 is essential for their biological activity. Evidente *et al.* (1998) reported that the cytochalasins produced by *Ascochyta* and *Phoma* exhibited phytotoxicity.

Cytochalasin H, a toxic cytochalasin of *Phomopsis* sp. isolated from weevil-damaged pecans, was markedly inhibitory to the growth of the floral development of tobacco plants. Pyrichalasin H, produced by *Pyricularia grisea,* and zygosporin D, isolated from the culture of *Metarrhizium anisopliae*, strongly inhibited the growth of rice seedlings (Fujii *et al.,* 2000). Cytochalasin Z6, an 24-oxa[14]-cytochalasin isolated from the culture of *Phoma exigua* var. *heteromorpha*, caused inhibition of rootlet elongation of tomato seedlings (Evidente *et al.,* 2003).

Three 24-oxa[14]cytochalasins, namely cytochalasins Z3, B, F and an [13] cytochalasin, deoxaphomin, from the culture of *Pyrenophora semeniperda,* caused remarkable inhibition of root elongation of wheat and tomato seedlings (Evidente *et al.,* 2002). Cytochalasins B, F, Z2 and Z3 and deoxaphomin, were potential mycoherbicide against the perennial weeds like *Sonchus arvensis* and *Cirsium arvense.* Deoxaphomin exhibited the highest level of toxicity on leaves of *S. arvensis,* whereas other cytochalasins showed low activity (Hseu *et al.,* 2011).

Several purified cytochalasins from a culture of *Phoma exigua* var. *heteromorpha* exhibited phytotoxic effects on germinating tomato seedlings (Evidente *et al.*, 2003). Cytochalasin E was toxic to brine shrimp larva and plants (Zhang *et al.,* 2014). Cytochalasin E killed the embryonic chicks (Doubravka *et al.,* 1983). Cytochalasin J and H were cytotoxic against human cervical cancer cells with low toxicity to non-cancer cells (Jouda *et al.,* 2016) and prevent infections caused by *Shigella flexneri* and *Vibrio cholerae*. Cytochalasin H, of *Phomopsis* sp. markedly inhibited the growth of the floral development in tobacco plants. Cytochalasin H isolated from *Xylaria sp.* A23 (Yu Li *et al.,* 2012), acts as a plant growth regulator in tobacco. It inhibits the floral development of tobacco plants at 10-2M concentration.

Cytochalasin E was identified as the most active component in brine shrimp toxicity and phytotoxicity potency in the fungal endophyte. Cytochalasin E was phytotoxic against *Lactuca sativa* and *Raphanus sativus* L. seedlings in a range of 10–80 lM, which is higher than that of glyphosate (Zhang *et al.,* 2014). Cytochalasin F and deoxaphomin showed a remarkable ability to inhibit root elongation of wheat and tomato seedlings (Evidente *et al.,* 2002).

Chaetoglobosin A and its 19-O-acetylchaetoglobosin A, isolated from the culture of *Calonectria morganii*, inhibited the root and shoot growth of cresson seedlings (von Wallbrunn *et al.,* 2001). Chaetoglobosins A, C and O, phytotoxic metabolites from *Cylindrocladium floridanum*, a causal fungus of alfalfa black rot disease, was potent growth inhibitor of alfalfa seedlings (Ichihara *et al.,* 1996). Similarly, chaetoglobosin K isolated from cultures of *Diplodia macrospora,* a corn pathogen causing ear and stalk rot showed growth-inhibiting effects on wheat coleoptiles (Cutler *et al.,* 1980).

6.24. Pharmacological properties

The cytochalasin B enhances tumor cell lysis and stasis are due to activated macrophages at a concentration of 10-7 moles. It does not act on the macrophage itself, but does exert its effect predominantly on the tumor cell, which may be due to the actin filament formation. It could be important for the destruction of tumor cells by activated macrophages (Martin *et al.,*1981).

Cytochalasin B combined with ultrasound therapy, proved to be efficient in destruction of neoplastic cells (Trendowski *et al.,* 2014). It proved to be a versatile chemotherapeutic agent that amplifies the damaged ultrasonic irradiation preferentially inflicts on malignant cells (Trendowski *et al.,* 2015).

Cytochalasin E which contains an epoxide group can be novel inhibitor of angiogenesis and tumor growth and can be a potential chemotherapeutic drug in the treatment of cancer and age related macular degeneration. Lipski *et al.* (1987) reported that cytochalasin E at its higher concentration caused unique histopathologic effects on the vasculature including hemorrhage and injury to the vascular walls accompanied by alterations in vascular permeability.

Cytochalasins proved to be crucial experimental agents for probing mechanisms of the cytoskeleton. Few clinicians would likely delineate cytochalasins as valid chemotherapeutic agents. In fact, these are yet to be a clinically approved microfilament-directed agents used in cancer therapy. There is a substantial evidence suggesting cytochalasins and other actin inhibitors as vital anticancer agents. Using chemotherapeutic agents to preferentially damage the cytoskeletons of malignant cells is a tactic long exploited in the clinical setting. Microtubule-directed agents such as taxanes and vinca alkaloids are commonly used as anticancer agents that primarily act as mitotic poisons, thereby inducing cell cycle arrest, and eventual apoptosis. Since microtubules are pivotal for successful mitosis, inhibiting normal function of these polymers can be particularly devastating for rapidly proliferating cells. Henceforth ideal for disrupting tumorigenic growths (Rubtsova *et al.,* 1998 and Mani *et al.,* 2004). Further, it has been shown that such agents also induce an apoptotic response in various cancer cell lines (Mollinedo and Gajate, 2003; Bergstralh and Ting, 2006 and Mooberry, 2007). Cytochalasins are also potent mitotic inhibitors, but act during cytokinesis at the end of the event. Therefore, it is likely that the concomitant use of microfilament and microtubule-directed agents would have a profound synergistic effect on disrupting cancer proliferation. It is hoped that the evidence provided for the efficacy of cytochalasins will need further investigation in the clinical setting.

According to Chao and Liu (2006) cytochalasin B decreases the Cell Survival, Inhibits Cell Growth, and Increases the Levels of G2/M Fractions in Human Lung Carcinoma Cells. The Blockage of Securin Protein increases cell death in both cytochalasin B treated Human Colorectal Carcinoma Cells (HCCC). Knockdown the Survivin Gene increases the cytochalasin B-induced cell death and cell growth inhibition in Human Lung Carcinoma Cells (HLCC).

Cytochalasin D inhibits the smooth muscle cell proliferation but not collar-induced neointima formation in rabbits (Bruijns and Bult, 2001). Cytochalasin D inhibits the neointima formation in a porcine drug-eluting stent model, where prolonged exposure was also ensured (Salu *et al.,* 2006). However, there was no significant effect of cytochalasin D in pig vein-grafts, despite an 80% inhibitory effect in organ cultures using the same pre-treatment regime.

Cytochalasin D can affect many cell functions, including adherence, cell motility, drug efflux, secretion, indicating that the cytochalasins might induce significant responses in experimental cancer chemotherapy model systems either as individual agents or, more likely, as agents amplifying responses induced by known antitumor drugs (Malecki *et al.,* 2010).

Cytochalasin D could significantly inhibit tumor growth and prolong survival in a manner similar to that of cisplatin. Anticancer properties of different cytochalasins have been reported sporadically (**Table 6.4**). Liposomal cytochalasin D induced significant tumour cell apoptosis. Significant inhibition of tumour angiogenesis was also observed in mice treated with liposomal cytochalasin D (Huang *et al.,* 2012). It also inhibited CT26 tumor growth potentially through inhibition of cell proliferation, induction of cell apoptosis and suppression of tumor angiogenesis (Huang *et al*., 2012). Cytochalasin H exhibited *in vitro* anti-angiogenic activity and suppressed cell growth and mobility in HUVEC (Human Umbilical Vein Endothelial Cells), and inhibited the pro-angiogenic protein-induced formation of new blood vessels *in vivo* (Lee *et al.,* 2014).

Hwang *et al.* (2013) suggested that cytochalasin B induced apoptosis via a mitochondrial-dependent pathway in HeLa cells. Huang *et al.* (2012) reported that cytochalasin D inhibit CT_{26} tumor growth potentially through inhibition of cell proliferation, induction of cell apoptosis and suppression of tumor angiogenesis. Cytochalasin B suppresses cancer cell proliferation or induces cancer cell apoptosis (Hwang *et al.,* 2013). Cytochalasin B acts on Lewis lung carcinoma cells in a similar manner to that of SV-40 transformed cells and efficiently kills the carcinoma cells.

Table 6.4. Anti-cancer properties of cytochalasins

Cytochalasins	**Mechanism of action**	**References**
Cytochalasin A	Actin targeting agent, reversible inhibitor of sugar uptake, poorly antiangiogenic	Yahara *et al.*(1982)
Cytochalasin B	Actin targeting agent, reversible inhibitor of sugar uptake (GLUT 1 binder), ion channel blockade	Choi *et al.* (2005)
Cytochalasin D	Actin targeting agent, protein synthesis inhibitor, antiangiogenic	Goddette *et al.* (1986)
Cytochalasin E	Actin targeting agent, antiangiogenic	Udagawa *et al.* (2000)
Cytochalasin H	Actin targeting agent, poorly antiangiogenic	Yahara *et al.*(1982) and Udagawa *et al.* (2000)
Deoxaphomin	Actin targeting agent	Yahara *et al.*(1982)
Chaetoglobosin K	Tumor angiogenesis	Haito *et al.* (2013)

Trendowski (2015) proved that cytochalasin B has potential therapentic value when combined with sonodynamic therapy in curing human leukemia. They further observed the synegstic effect of cytochalasin with ultrasonic irradiation in curing human leukemia (Trendowski *et al.,* 2014). Huang *et al.* (2013) reported that pegylated liposomal cytochalasin D could effectively inhibit the various kinds of tumor growth and prolong the survival time of tumor bearing mice. Cytochalasin D could effectively inhibit cell apotosis and caused tumor suppression and angiogenesis. Hence, cytochalasin D can be preferably used as a cytotoxic agent for cancer therapy (Huang *et al.,* 2012). Alienberg and Silverman (2003) reported the disruption of actin filaments in 3T3 cells due to antiapoptic activity in the presence of gelatinase A.

6.25. Mycotoxins and food spoilage

Mycotoxins are the low molecular weight, non-antigenic secondary metabolites of moulds which mostly live as saprophytes and sporulate heavily on variety of foods and feeds. Although the toxic effects of fungal metabolites have been known for centuries (St. Anthony's fire), the adverse effects of mycotoxins has gained dramatic importance due to the reduced application of antifungal treatment on food (Weidenborner, 2001).

These moulds infest foods, feed and fodders at different stages of harvest, storage and consumption cause not only huge loss due to their deterioration activity but also convert them into toxic (Muller *et al.,* 1998). The food poisoning and death of human beings is an age old problem. There are several instances of outbreak of diseases of unknown etiology which probably may be due to mould infestation and mycotoxin contamination. Creepy (2002) discussed the incidence and toxic effects of different mycotoxins in Europe. Similarly Bennet and Klich (2003) have discussed different aspects of mycotoxin problems in different parts of the world. Though several mycotoxins are unique to one species, but most mycotoxins are produced by more than one species. The most widespread mycotoxin roquefortine C, is produced by 25 species of *Penicillium* (Lugauskas, 2005).

Toxicological importance of cytochalasins in food spoilage and intoxication proved to be vital. Ohtsubo *et al.* (1978) demonstrated acute toxicity of chaetoglobosin A and C to rats and mice. Vesely *et al.* (1995) also reported chick embryo toxicity of Chaetoglobosin A. Similarly Sekita *et al.* (1982) observed cytotoxixity of different Chaetoglobosins to HeLa cells. Although toxicity of chaetoglobosins in mice is low, a potential health hazard might arise from additive or synergistic effects with co-occurring mycotoxins. *Penicillium expansum* which is most common in rotten fruits and vegetables, specially in apples is known to elaborate wide range of secondary metabolites including cytochalasins and are implicated in mycotoxicosis of humans.

6.26. Metabolism and macromolecular synthesis

Cytochalasin B causes inhibition of glucose incorporation in to glycogen in chicken embryo myoblasts. It also inhibited the glycolysis. At high concentration of cytochalasin B (100 μM) caused total inhibition of ATP synthesis in mitochondria probably due to rapid rate of ATP and Mg^{++} induced mitochondrial contraction. Cytochalasin D and E caused activation of superoxide generation. An inhibition of nucleic acid and mucopolysaccharide synthesis by cytochalasins was also reported. Cytochalasin B stimulates orotic acid and cause four fold incorporation into mouse 3T3 fibroblast RNA which may be attributed to increased transport of its precursor. Inspite of inhibition of variety of processes of cells like glucose and glucosamine uptake, morphological alterations, lectin induced agglutination, cell sorting, cytokinesis, antiviral interferon activity, lectin induced agglutination, virus replication, locomotion, ruffling membrane activity, cytoplasmic streaming, and protein synthesis remained unaffected. Cytochalasin B was potent inhibitor of glucosamine uptake. These may also interfere with cellular metabolism and function at points other than plasmamembrane and contractile microfilaments. Mitochondrial and nuclear processes also may be directly affected. Cytochalasin B alters some yet undefined intermediary steps in the hormone-receptor pathway. Cytochalasins affect microfilament structure *in vivo* and do interact with various components of contractile systems *in vitro.*

Cytoplasmic gelation in HeLa cell was also inhibited by cytochalasin A, B and D. The disaggregation of actin from a high molecular weight protein fraction was implicated in this process. Cytochalasin A disrupted microfilament in mouse

3T3 cells produced actin-containing rod elements in addition to the aggregate central actin masses. Variety of biological processes affected by cytochalasins are précised in **Table 6.5.**

Table 6.5. Biological effects of different cytochalasins

Name of cytochalasins	Biological activity	Reference
Cytochalasin A	Decrease of ATP, sterol content, sterol phosholipid and protein content	Albert (1984)
	Shedding of murine B cell membrane	Stephen *et al.* (1979)
	Inhibition of sugar uptake	Kletzien and Perdues (1972), Rampal *et al.* (1980)
	Reduction of insulin and growth of hormone in human lymphocytes	Sampath and Pollard (1991)
Cytochalasin B	Membrane change and fluidity	Marc *et al.* (1994)
	Fluidity of chicken erythrocytes	Nakajima *et al.* (1981)
	Fluidity of rat neutrophils	Masuda *et al.* (1989)
	Regulation of polymorph nuclear leukocyte membrane fluidity	Wiles *et al.* (1994)
	Virus and lectin agglutination of erythrocytes	Lyles and Landsberger (1976)
	Preimplantation and early post implantation of mouse embryos	Granholm (1979)
	Polymorphonuclear leukocytes stimulation	Ohno *et al.* (1982)
	Competitive inhibition of nucleoside transport	Peter *et al.* (1973)
	Insulin release and pancreatic islet cell metabolism	Jijakli *et al.* (2002)
	Antitumor activity against Lewis lung carcinoma, LA4, M109, B16F10, P388/S and P388/ADR leukemia cell lines	Tarman *et al.* (2011)
Cytochalasin C	Cytotoxic activity against tumour cell lines PC-3M, NCI-H460, SF-268, MCF-7 and MDA-MB-231	Wei *et al.* (2015)
Cytochalasin D	ATP and calmodulin aggregates	Edmonds and Koenig (1989)
	Cytoskeleton in exocytosis	Karine *et al.* (1999)

Name of cytochalasins	Biological activity	Reference
	Protein discharge induced by alpha and beta- adrenergic (or) cholinergic agonists in rat exorbital lacrimal glands	Muller *et al.* (1985)
	Acinar cell structure and secretion in rat parotid salivary glands	Tojyo *et al.* (1989)
	Phorbol myristate acetate on the apical endocytosis of ricin in polarised Caco-2 cells	Shurety *et al* (1996)
	Effect of actin and elongation of actin filament	Manfred (1982), Sampath and Pollard (1991)
	Cell motility process is actin	Ohmori *et al.* (1992)
	Effect of protein synthesis in *Xenopus laevis* oocytes	De Sousa and Masui (1990)
	Reversible inhibitor of protein synthesis in HeLa cells	Ornelles *et al.* (1986)
	Antibacterial activity against *S. sonnei, M. tubercylosis* and α-hemolytic *Streptococcus*	Wei *et al.*(2015)
Cytochalasin E	Blindness caused by an abnormal proliferation of blood vessels in the eye	Udagawa *et al.* (2000)
	Strong anti-angiogenic activities and potent and selective inhibitor of bovine capillary endothelial (BCE) cell proliferation	
Cytochalasin E	Insecticidal activity against brine shrimp. and Phytotoxic activity on *L. sativa* and *R. sativus*	Zhang *et al.*(2014)
Cytochalasin H	Highly toxic and possible teratogens.	Natarajan *et al.* (2000)
	Cytotoxic activity against K562 cell line	
	Particularly effective as an inhibitor of the cytoskeletal reorganisation that occurs in platelets following activation by adenosine diphosphate (ADP)	Li *et al.* (2012)
Cytochalasin H_2	Cytotoxic activity against HeLa and 293T cell lines	Li *et al.* (2012)
Cytochalasin G & H	Effects on Lymphocytes	Mookerjee *et al.* (1981)

Name of cytochalasins	Biological activity	Reference
Cytochalasin J	Blocks or slows chromosome motion and has a significant effect on spindle architecture	Snyder and Cohen (1995)
Cytochalasin Q	Cytotoxic activity against human epidermoid carcinoma cells of KB cell lines in mouth	Chinworrungsee *et al.* (2001)
	Antimalarial activity against *P. falciparum* K1. Cytotoxic activity against tumour cell lines KB, BC-1 and PC-3	Wei *et al.* (2015)
Cytochalasin U	Exhibit immunosuppressant activity in the MLCR with an IC50 of 5 μg ml[1]	Mann (2001)
Cytochalasin W	Toxic activity in the brine shrimp	Evidente *et al.*(1996)
Cytochalasin Z_1, Z_2, Z_3	Effect on root elongation of tomato and wheat seedlings.	Evidente *et al.* (2002)
Cytochalasin Z_{10}-Z_{15}	Cytotoxic effects on P388, A-549, HL-60, and BEL-7402 cell lines	Rui *et al.* (2008)
Cytochalasin Z_{17}	Cytotoxic assay against the human nasopharyngeal epidermoid carcinoma (KB) cell line	Zhang *et al.* (2010)
Cytochalasin Z_{21}-Z_{23}	Cytotoxic activities against A-549 cell lines	Wang *et al.* (2011)
Cytochalasin Z_{27}	Insecticidal activity against brine shrimp	Zhang *et al.*(2014)
Chaetoglobosin A	The nematicidal activity of ChA against *Meloidogyne incognita*	Yang Hu *et al.* (2013)
Chaetoglobosins B, D, F, G and J	Cytotoxicity against HeLa cells with IC50 values in the range 3.2-20 μg/mL	Sekita *et al.* (1982)
Chaetoglobosin C	Cytotoxicity against human HeLa cells Antimicrobial against *Mucor miehei* at 10μg/disk	Umeda *et al.* (1975)
Chaetoglobosin E	Cytotoxicity to HeLa cells.	Natori *et al.* (1977)
Chaetoglobosin G	Inhibits or prevents the proliferation of neoplasms.	Sekita *et al.* (1983)
Chaetoglobosin J	Decreased both the rate and the extent of actin polymerization	Maruyama *et al.* (1986)

Name of cytochalasins	Biological activity	Reference
Chaetoglobosin M	Blocks diplosporin biosynthesis	Christoph and Christoph (1988)
Chaetoglobosin U	Cytotoxic against the human nasopharyngeal epidermoid tumor KB cell line	Ding *et al.*(2006)
Chaetoglobosin V &W	Cytotoxic against four human cancer cell lines (KB,K562, MCF-7, and HepG2)	Zhang *et al.* (2010)
Chaetoglobosin Y	Cytotoxic against the KB cell line	Zhang *et al.* (2010)
Zygosporin D	Effective inhibitor of shoot elongation of rice seedlings	Fujii *et al.* (2000)
Zygosporin E	Effects on mammalian cell morphology and for their ability to influence transport phenomena at the cell membrane	Vedejs *et al.* (1988)
Sclerotionigrin A	Active against chronic lymphocytic leukemia	Petersen *et al.* (2014)
Sclerotionigrin A and B	Active against chronic lymphocytic leukemia	Petersen *et al.* (2014)
Multirostratin A	Cytotoxic active against HL-60, A-549, SMMC-7721, MCF-7 and SW-480 against five cancer cell lines	Chen *et al.* (2015)
Aspochalasin V & W	Cytotoxicity against the prostate cancer PC3 cell line and HCT116 cell line	Ying *et al.* (2014)
Alacahalasins A to G	Antimicrobial and antifungal activity	Zhang *et al.*(2008)
Aspochalasin F & G	Active against antimalerial agent	Fang *et al.* (1997)
Asperchalasin A	G1-phase cell cycle arrest by selectively inhibiting cyclin A, CDK2 and CDK6 in cancerous, but not normal, cells, highlighting it as a potentially selective cell cycle regulator against cancer cells.	Zhu *et al.* (2015)
Epicochalasines A and B	G2/M-phase cell-cycle arrest in HL60 And apoptosis in leukemia cells through the activation of caspase-3 and the degradation of poly (ADP-ribose) polymerase (PARP)	Zhu *et al.* (2016)
Penochalasins	Cytotoxic activity in the P388 lymphocytic leukaemia test system in cell cultures	Numata *et al.(*1996)

(i) Glycolysis

Rapid and dose dependent inhibition of glycolysis was recorded in animal and human leucocytes under the influence of cytochalasin. The inhibition of glycolysis was attributed to blocking of membranes to glucose transport rather than disruption of enzymatic mechanisms operative in glycolytic pathway. However, ATP synthesis in isolated rat mitochondria was totally inhibited by cytochalasin B. similarly glucose incorporation into glycogen was inhibited by cytochalasin B in chick embryo myoblasts. Inhibition of glycolysis was shown to be a direct consequence of the potent effect of cytochalasin B on blocking membrane glucose transport and was not a result of disruption of the enzymatic mechanisms operative in the glycolytic pathway. Hexokinase activity was unaffected in HeLa cells and leukocyte homogenates.

Since many of the diverse cellular functions are inhibited by cytochalasin B and other cytochalasins are energy-requiring processes. It is reasonable to suspect that disruption of cell activity may be secondary to suppression of energy production (ATP generation) via blocking glucose transport and consequently, glycolysis.

Reduced glucose uptake was not specifically implicated in the inhibitory effects of cytochalasin B on locomotion and phagocytosis in leukocytes, membrane ruffling activity and locomotion in chicken embryo glial cells, virus-induced fusion of Ehrlich ascites cells, lymphocyte-mediated cytolysis, lectin-induced agglutination of ascites cells, or salivary gland morphogenesis in mouse embryo. Glycolysis in human spermatozoa reversibly inhibited by cytochalasin B at 100μM concentration.

It is significant to note that treatment with 10 μg/ml cytochalasin B for 30 min did not alter cellular ATP levels of chicken embryo fibroblast, although morphological alterations were visible. Little decrease in ATP content of rat adipocytes and rat ascites tumor cells was observed after incubation with cytochalasin B (10 μg/ml) for 1 hr. However, ATP synthesis in isolated rat mitochondria was totally inhibited within 5 min by cytochalasin B at100 μM concentration.

Cytochalasin A, B and chaetoglobosin A did not alter leucine incorporation into rabbit reticulocytes at 10 μg/ml. However, higher concentrations of cytochalasin A (20 μg/ml) and cytochalasin B (50 μg/ml) depressed protein synthesis by 30%. Cytochalasin A (20 μg/ml) also did not affect general protein synthesis in the water mold *Achlya*, but it selectively inhibited the synthesis of the secretory protein, cellulase by approximately 80%.

(ii) Interaction with hormonal mechanisms

Cytochalasins may also interfere with cellular metabolism and functions at points other than the plasmamembrane and contractile microfilaments. Mitochondrial and nuclear processes may be directly affected.

a. Peptide hormones

Steroid synthesis was reversibly inhibited by cytochalasins B at concentrations as low as 10^{-8}M. Since cytochalasins B blocked the enzyme induction by cyclic AMP, the site of inhibition must necessarily be at point beyond the interaction of adrenocorticotropic hormone (ACTH) with plasmamembrane receptors. No effects on basal or unstimulated steroid synthesis was observed. This response was not attributable to glucose or ATP levels, protein synthesis inhibition, or alterations in cholesterol uptake by the tumor cells. Cytochalasins B decreased the ACTH-induced uptake of cholesterol by mitochondria, the major site of steroid synthesis.

A similar effect has been reported for the induction of lactate dehydrogenase by norepinephrine in rat glial cells. Maximal inhibition of lactate dehydrogenase nearly to basal enzyme levels, was obtained at the lowest concentration of cytochalasins B (5 μg/ml). The response was reversible and was not a consequence either of increased enzyme degradation or loss, or of generalized effects on protein synthesis. Cytochalasins B did not alter the hormone-induced increase in cyclic AMP and was still totally effective when added for 30 min, but not 6 hr, after hormone induction.

The induction of tyrosine aminotransferase (TAT) by insulin in Reuber hepatoma cells was partially inhibited by 10 μg/ml cytochalasins B. Although there was no alteration of insulin binding, cytochalasin B did not prevent the induction of TAT by cyclic AMP, suggesting that plasmamembrane effects were responsible for the inhibition.

Cytochalasins B also selectively dissociated the hydro-osmotic effect of vasopressin from its stimulatory activity on sodium transport in toad urinary bladder preparation. The site of inhibition again appeared to be distal to hormone-plasmamembrane interactions, since the hydro-osmotic effects of cyclic AMP was also blocked.

Cytochalasin B influences on thyroid hormone and growth hormone secretion (Scherlach *et al.,* 2010). Phosphatidylcholine and phosphatidylethanolamine biosynthesis are inhibited by cytochalasin B (George *et al.,* 1991).

(b) Steroid hormones

The basic mechanism of steroid hormone action is a sequential series of events which involve the diffusion of the hormone into the target cell, binding of the hormone to a specific high affinity cytoplasmic receptor protein, a temperature-dependent modification of the hormone-receptor complex, translocation of the activated complex to the nucleus and binding of the complex to specific chromatin sites and selective activation of gene transcription.

Steroid hormone activity is also significantly influenced by cytochalasins. Induction of glycerol phosphate dehydrogenase (GPDH) by hydrocortisone in rat glial cells was reversibly inhibited by cytochalasin B in a dose-dependent manner. The effects of cytochalasin B on the translocation of the estradiol-receptor complex from the cytoplasm to the nuclei of immature rat uteri was examined. A 15% reduction in nuclear binding was noted only under extreme conditions (5 x 10^{-4}M of cytochalasin B). Inhibition of hydrocortisone induction of tyrosine aminotransferase in Reuber hepatome cells was inhibited by cytochalasin B (10 to 50 μg/ml).

Cytochalasins B, D and E, inhibited actin polymerization that disrupt the formation of microfilaments, stimulate steroid hormone production in Y1 adrenocortical cells (Cortese and Wolff, 1978). Cytochalasin D effects basal and adrenocorticotropin (ACTH)-stimulated steroid production as well as stress fiber organization by bovine adrenocortical (BAC) cells (Rainey *et al.*, 1984). The stimulatory effect of cytochalasins on Y1 cells, but in conflict with the inhibitory effect of paclitaxel found on basal and cAMP-stimulated steroid hormone production in Y1 adrenocortical cells, but not in bovine adrenocortical cells (Rainey *et al.,* 1985). Cytochalasin B inhibits the steroidogenic response of mouse adrenal (Y-l) tumor cells to ACTH.

Cytochalasin B inhibits microfilament polymerization, as well as having other actions, such as inhibition of hexose transport, block tropic hormone-stimulated steroidogenesis. Cytochalasin D is a more effective inhibitor of microfilament polymerization than Cytochalasin B, but has no effect on hexose transport (Susan *et al.,* 1984). Cytochalasin B enhances hormone and Cholera toxin and its stimulated cyclic AMP accumulation in S49 lymphoma cells (Paul and Koachman, 1982).

■■■■

Chapter-7

Ecological Significance

In view of the production and distribution of our food shift from localized production and consumption toward an increasingly globalised network of distribution, mycotoxins in animal feed or human food now have the potential to pose serious health risk to consumers throughout the world.

The remarkable diversity of cytochalasins producing fungi and diverse biological activities may have significant ecological importance. One can speculate a possible ecological role and potential role of these substances in the maintenance of pathogenic or production of cytochalasins by the fungus may enable it to compete with other organisms in its natural environment and might play a major role in deterring predators or competitors.

There is likely to be caused relationship between exposure to mold contaminated indoor and outdoor environments and adverse effects due to mycotoxins has been difficult to prove. Despite extensive research there is still insufficient or inadequate epidemiological or toxicological evidence to allow us to determine whether an association exists between the presence of molds and/or mycotoxins in damp indoor environments and adverse health effects on otherwise healthy adults. Increased awareness of this will be critical to develop successful intervention and prevention strategies are to be acquired.

An understanding of the factors that favour cytochalasin producing fungi and cytochalasin accumulation in grain is helpful in devising management strategies for the disease and the mycotoxin. Strategies like integrated approach that combines the use of resistant varieties, fungicide application at antithesis, mycotoxin forecast models, practices that minimize to drying, crop residue management and crop rotation sequence that mycotoxin accumulation.

Production of chaetoglobosin by *Penicillium expansum* might be suited to provide mutational access and or to secure the food source (Scherlach *et al.*, 2010). Similarly, *Metarrhizum anisopliae* which occurs worldwide and invade variety of insect hosts and penetrate the insect cuticle, proliferate and finally cause death of insect probably by secreting toxin that either directly kill the host and/or suppress the host defense mechanism. Tan and Zou (2001) feel that the endophytes of certain plants obtain nutrients and protection and in return these fungi promote the health of plant by producing antimicrobials and protect the plant from the attack of pathogens and herbivores. These also help to maintain delicate equilibrium between the antagonism of host and fungus. Ecological function has been attributed to cytochalasin F of an endophyte of *Teuerinm scorosonia* which was shown to possess algicidal activity and inhibit photosynthesis (Konig *et al.,* 1999). Endophytic fungi are rich and reliable source of bioactive and chemically noval compounds with huge medicinal and agricultural potential research at molecular level in this field is needed for better understanding of host-endophyte interaction which paves the way for quicker recongnization of genetically particular and/or host gene carrying endophytes.

Vilcinskas *et al* (1997) reported that cytochalasins produced by two entomogenous fungi *Metarrhizium anisopliae* or *Beauveria bassiana* impair the cellular reorganization and defense reaction of their insect host. Cytochalasins produced by fungi during mycosis might prevent ingestion of fungal cells by plasmocytes.

Cytochalasin F produced by exhibited algicidal activity as well as it was found to be inhibitor of photosynthesis suggesting its ecological significance in plant symbiont relationship (Konig *et al.,* 1999).

Natural Incidence

Moisture damage, dampness and visible mould in buildings are consistently associated with adverse health onerous. Mendell *et al.* (2011) have discussed excellently epidemiological evidences and concluded that conditions of dampness and, mould are associated with multiple allergic and respiratory effects including asthma development and excrction, current asthma upper and lower respiratory tract symptom, respiratory infections, allergic rhinitis and eczema.

Excellent reviews such as WHO (2009), IOM (2004), Mendell (2011) and Tischer *et al.* (2011) reveal that all research on mycotoxins in indoor environments is placed dampness and visible mould are consistently associated with ill health but our knowledge as concerns the causal agents and diseases mechanisms is poor **(Fig. 7.1)**.

Most of our knowledge on the moulds of action of mycotoxins related to ingestion exposure such as inflammatory, immune-suppressive, cytotoxic and carcinogenic effects in various organs have been recorded. The natural incidence of cytochalasins in indoor environment has been recorded by many people **(Table 7.1).**

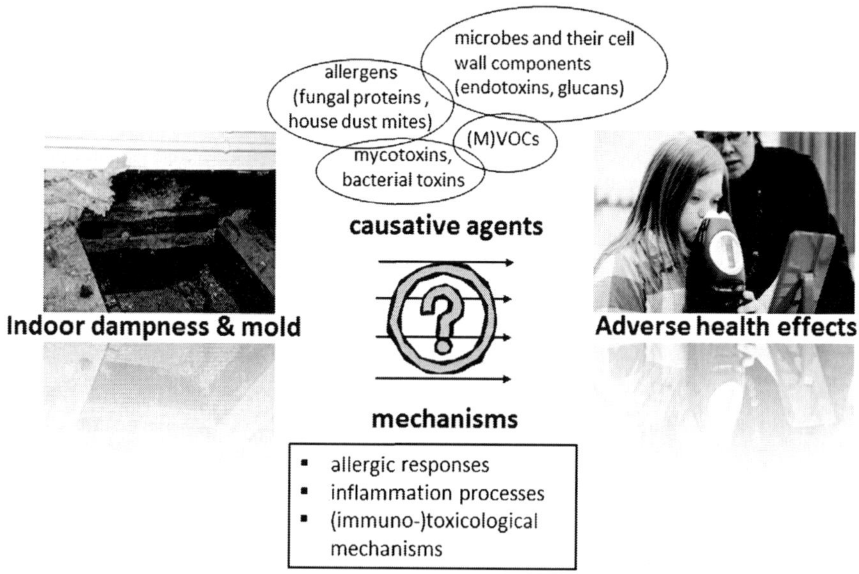

Fig 7.1: Moisture damage, dampness, and visible mold in indoor spaces and the related exposures and adverse health.

Table 7.1. Cytochalasins occurrence in indoor sample materials

Study sites and design	Sample materials	Target compounds	Analytical methodology	Main findings concerning indoor mycotoxin occurrence	Reference
Mould-infested domestic residence, several rooms/ location	Vacuumed samples or scraping from wall paper	Multiple mycotoxins	HPLC-DAD, TLC	Chaetoglobosin A and C	Nielsen *et al.* (1999)
Three schools with nonspecific indoor air problems	Dust swabs and filter materials from ventilations	40 mycotoxins	LC-MS	Chaetoglobosin A	Hintikka *et al.* (2009)
Seven water damaged buildings in Belgium	Samples of air, dust, wall paper, mycelium or silicone	20 mycotoxins	LC-MS	Chaetoglobosin A	Polizzi *et al.* (2009)

Study sites and design	Sample materials	Target compounds	Analytical methodology	Main findings concerning indoor mycotoxin occurrence	Reference
Damp buildings in Slovakia and Austria	Mould-infested material samples (wall scrapings, paper, soil/ wood etc)	186 fungal secondary metabolites	HPLC	Chaetoglobosin A, Chaetomin, Cytochalasin B and D	Vishwanath *et al.* (2009)
Five residential homes with moisture damage (Finland)	Vacuumed floor dust	186 fungal secondary metabolites	LC-MS	Cytochalasin D	Vishwanath *et al.* (2011)

Vishwanath *et al.* (2011) recorded cytochalasin D in the floor dust of finish homes with water damaged. Polizzi *et al.* (2009) have recorded chaetoglobosin A indoor air of moisture damaged homes in Belgium. Hintikka *et al.* (2009) have also recorded chaetoglobosin A in the intake and exhaust systems of indoor environment of seven schools of Finland. Chaetoglobosin A in the indoor environment of water damaged buildings of Belgium was recorded by (Polizzi *et al.,* 2009). Vishwanath *et al.* (2009) have recorded chaetoglobosin A, cytochalasin B and D in mold infested material of damp damaged buildings. Mc Mullin *et al.* (2013) recorded chaetoglobosin production by *Chaetomium globosum* isolated from indoor environment. Nielson *et al.* (1999) have recorded chaetoglobosin A and C in mold infested domestic samples. Polizzi *et al.* (2009) in their analysis of moisture damaged homes in Belgium detected six mycotoxins out of 20 air samples analyzed. Chaetoglobosin A is one of the toxin detected which was elaborated by *Chaetomium* sps. Mc Mullin *et al.* (2013) reported chaetoglobosins and azaphillones produced by canadians strains of *Chaetomium globosum* isolated from the indoor environment. Vishwanath *et al.* (2011) have detected variety of mycotoxins and volatile anthropogenic channels in indoor environments. Smoragiewicz *et al.* (1993) recoded variety of trichothecene mycotoxins in the dust of ventilation systems in office buildings.

The cytochalasins block micronucleus (CBMN) assay is a comprehensive system for measuring DNA damage. Cytosine and cytotoxicity-DNA damage events are scored specially in once-devided binucleated cells. End-points possible to be measured are micronuclei (MN) a biomarker of chromosome breakage and /or whole chromosome loss, nucleoplasmic bridge (NPB), a biomarker of DNA misrepair and /or telomere end-fusrions and nuclear buds (NBUD), biomarker of elimination of amplified DNA and /or DNA repair complexes. Cytostatic effects are measured via the proportion of mono-bi and multinucleated cells and cytotoxicity via necrotic and/or apoptotic cell ratios (Fenech *et al.*, 1999, Fenech, 2006 and Fenech, 2007).

The CBMN assay has become one of the most commonly used methods for assessing chromosome breakage and loss in human lymphocytes both *in vivo* and *ex vivo*. In their binucleated appearance after blocking cytokinesis with cytochalasin B, an inhibitor microfilament ring assembly required for causation of cytochalasins. The biological meaning of the presence of MN in mono-nucleated cells indicates DNA damage that was present in the cells before they were put into culture with cytochalasin B. whereas binucleated cells may contain preexisting MN as well as MN expressed during culture as a result of chromosome breaks accumulated during GO phase *in vivo* (Kirsch-Volders and Fench, 2001).

■■■■

Chapter-8

Structure and Activity Relationship

The number of cytochalasins is large, and there is a variety of reactive sites on the molecular structures to allow appropriate chemical modifications. In addition, the tremendously diverse array of biological processes affected by these toxins would permit the utilization of numerous biological models to promote the acquisition of correlative data.

Manito *et al.* (1973) have investigated series of cytochalasins and zygosporins and their derivatives for molecular relationships in cytotoxicity assays. Acylation of the C-hydroxyl group, reduction of the phenyl ring at C-10, or scission of the macrocyclic ring greatly reduced the cytotoxicity of these compounds.

The first report of differential biological potencies among cytochalasins stated that cytochalasin C and D were approximately ten times more active than Cytochalasin A and B, presumably in inhibiting cytokinesis and motility and promoting enucleation in mouse L-cell fibroblasts. No such clear-cut correlation or relationship could be recorded (**Table 8.1**). Cytochalasin B generally is less potent than the other congeners with the exception of glucose transport inhibition, for which all other cytochalasins are virtually inactive. Cytochalasin D except for its lack of inhibition of glucose transport, is more potent than cytochalasin B and toxin may also prove to be particularly active. The relatively low cytotoxicity of cytochalasin A belies its unexpectedly high activity towards a spectrum of cytophysiological functions (**Table 8.1**).

Table 8.1. Comparative toxicity of cytochalasin congeners

Biological effect of process	Biological system	Response[a]	Potency series[b]
Anti bacterial activity	*Escherichia coli:* *Bacillus subtilis*	En	CA>CD>(CB)
Antifungal activity	*Saccharomyces cerevisiae:* *Botrytis cinerea*	En	CA>(CB,CD)

Biological effect of process	Biological system	Response[a]	Potency series[b]
Acute toxicity	Mouse	En	CE ~ = CD>CC
Cytotoxicity	HeLa cells	En	CC ~ = CD>CB ~ = CA
Teratogenicity	Chicken embryo explants	En	CD>CA>CB>CE
Zeiosis and arborization	Mouse fibroblasts (3T3)	En	CE>CD>CB = dihydro-CB
Enucleation	Mouse mammary tumor cells (MCH-C18)	En	CB = dihydro-CB
Cytokinesis	Mouse embryo cell line (Rose)	In	CE>CA>CD>CB
	Mouse fibroblasts (3T3)	In	CB ~ = dihydro-CB
	Fertilized sea urchin eggs	In	CB = dihydro-CB
Membrane ruffling activity	Mouse fibroblasts (3T3)	In	CE>CD>CB = dihydro-CB
Cytoplasmic streaming	*Elodea* leaves	In	CB = dihydro-CB
Adhesion	Ehrilich ascites cells	In	CA>CB>CE
Axonal outgrowth	Chicken embryo dorsal root ganglia	In	CB = dihydro-CB
Histamine release	Rat peritoneal mast cells	In	CA>CB
	Human lung tissue	En	CA>CB
Insulin secretion	Rat pancreatic islets	En	CD = CB
Lymphcyte-mediated cytotoxicity	Immune mouse spleen Lymphocytes	In	CA = CB
Rosette formation	Rabbit alveolar macrophages and complement-coated human erythrocytes	In	CA>CE>CD>CB
Mitogen-induced lymphocyte activation	Human peripheral blood Lymphocytes	En	CA = CD = CE >>CB> (dihydro-CB, γ-lactone CB)
Viral infectivity	Poliovirus and Vero cells	En	CD>CB
	Herpes simplex virus and rabbit kidney cells	En	CD>CB
Blood clot retraction	Human platelet-rich plasma	In	CB = dihydro-CB

Biological effect of process	Biological system	Response[a]	Potency series[b]
Superoxide generation	Guinea pig polymorphonuclear Leukocytes	En	CE = CD>CC>(CA, CB)
Polymerization of G-actin	Rabbit muscle	In	CA > CB
Insulin binding to cell Surface	Human lymphocyte cell line (1M-9)	In	CA = CB = CD
Human growth hormone binding to cell surface	Human lymphocyte cell line (1M-9)	In	CA>CB
Glucose transport	Human erythrocyte ghosts	In	CB>> CA> (CC, CD,CE, dihydro-CB, γ-lactone CB)
	Mouse fibroblasts (3T3)	In	CB> (dihydro-CB)
Binding to high affinity membrane Site I (glucose carrier)	Human erythrocyte ghosts	En	CB>>CA> (CC,CD,CE dihydro-CB, γ-lactone CB)
Binding to high affinity membrane site II (nonglucose carrier site)	Human erythrocyte ghosts	En	CE>CA>CB>dihydro-CB> γ-lactone CB
Binding to high affinity membrane sites (nonglucose carrier sites)	Human erythrocyte ghosts	En	CE>CD> dihydro-CB ~ = CB
Binding to membrane receptors	Human peripheral blood lymphocytes and leukocytes	En	CE > CA > CD>CB
	Mouse fibroblasts (3T3)	En	CD>CB ~ = dihydro-CB

a. En = enhancement; In = inhibition.

b. Congeners listed in parentheses were inactive under the conditions employed

Chemical modification of the macrocyclic ring of cytochalasin B, either by reduction of a double bond (dihydro- cytochalasin B) or by lactonization and ring-opening (γ-lactone derivative of cytochalasin B) results in the non-toxicity in glucose transport (Murono *et al.*, 1982) and promotes lymphocyte activation, while, reduction of the double bonds of the macrocyclic rings of cytochalasin C and D did not alter their cytotoxicity.

At present, with the availability of a limited information on animal, toxicity and cytotoxicity information with zygosporins and chaetoglobosins, comparative structure and activity relationship has been obtained only with cytochalasin A to E in several *in vitro* systems for various biochemical and cytophysiological parameters (Table 8.1). In these cases, often only two or three congeners were compared. Hence, unequivocal statements of relative potency are often impossible to make. Unfortunately, there is no information at this time regarding the *in vivo* effects of any cytochalasin on specific biological processes.

Though much information on the architecture of natural cytochalasins and their biological activities, the chemical derivatization programmes are scarce. In 1973, Minato and co-workers tried to find the cytotoxicity of a broad range of cytochalasin D derivatives (Manito *et al.,* 1973). With reduced skin-irritating effects, while retaining cytotoxicity was negative. However, cytochalasin C, which was prepared from cytochalasin D by palladium-catalyzed hydrogenation resulted. Further they reported importance of a hydroxy function at C-7 and the C-3 benzyl group for biological activity.

The synthesis and biological activities of cytochalasin D derivatives were also examined by Patwardhan *et al.* (1982). Reaction of cytochalasin D with phenyl-trimethylammonium perbromide yielded 6,12-dibromocytochalasin D 38 (stereochemistry not given), which was selectively monodehydrohalogenated with lithium chloride. Nucleophilic substitution reactions with sodium iodide, cyanide and azide gave access to three novel cytochalasin derivatives 40–42, while cytochalasin C and D caused almost identical biological responses, derivatives with bromo 39, iodo 40 or azido42 functions significantly reduced their activity (30–50%).

■■■■

Chapter-9

Management of Cytochalasins Producing Fungi

Mycotoxins are considered to be natural contaminants of food and elementary raw materials is a worldwide problem. Let food be thy medicine and the medicine be thy food. The role of diet in keeping the health and well being of humans and animals has been scientifically proved today. Mycotoxins severely threaten the health of humans and animals. Statistics reveal that every year approximately 25% of world food grains are affected by variable levels of mycotoxins, which leads to considerable national economic loss (Muller *et al.*, 1998). Hence, the limitation of mycotoxin contamination is of utmost importance. Different countries have framed guidelines to regulate the maximum acceptable limits of mycotoxin levels in foods and feeds. Even though regulatory actions must be based on solid scientific knowledge, many aspects of their toxicological activity are still poorly understood. In particular, deepening knowledge on the primed molecular events triggering the toxic stimulus may be relevant to better understand the mechanism of action of mycotoxins. Cytochalasins, type of mycotoxins are produced by filamentous fungi such as *Chaetomium, Zygosporium* and *Phoma* etc, which infest food commodities during different stages of food production to consumption. The cytochalasin incidence mainly depends on environmental conditions both at the pre-harvest and post-harvest stages of food grain production. *Penicillium expansum* which is reported to be constantly associated with apple fruits besides elaborating patulin, citrinin and ochratoxin A, it was found to be constantly producing chaetoglobosin A. Chaetoglobosins are also being detected naturally infected apple juices and potato pulp and not patulin or citrinin (Andersen *et al.*, 2004). Hence, the control of cytochalasins problem has become a task of herculean proportion. Robens and Riley (2002) have elegantly discussed various aspects and strategies to be adapted to prevent mycotoxigenic moulds in foods and feeds. Keeping in mind that the full comprehension of mycotoxin toxicology is a major challenge for the scientific community and it requires interdisciplinary approach. The deeper the toxicological understanding the more appropriate, precise and effective the risk assessment of a molecular events can be made (Dellafiora *et al.*, 2018).

Indeed food and feed are among the most complex cocktails of low molecular weight xenobiotics to which the living organisms are exposed. This multiple, combined in some cases prolonged exposure may have severe impairing effects on humans and animals. It is important to quote that naturally occurring mycotoxins accumulating in food or feed produced by moulds are of major concern because they are typically hard to control and manage as they may severely threaten human and animal health.

Quality control and regulatory measures are most important in the context of wide movement of foods or feeds all over the world, in order to restrict mycotoxigenic moulds reaching different places.

A sound cooperation among interested parties drawn from scientists, consumers, industries and policy makers is needed to implement the regulation enacted for cytochalasins control in foods and feeds

The incidence of moulds depends on environmental conditions both at pre-harvest, post-harvest and even at consumption stage. It is a general experience that mycotoxin contamination is complex and unavoidable. Therefore, control of these toxins will be a herculean task. In general people in developing countries are victims of the first order. Hence, there is an urgent need to frame national guidelines to prevent cytochalasin producing fungi and cytochalasins contamination which include the following measures. Svingen and Vinggaard (2016) have excellently discussed risk of chemical cocktail including mycotoxin effects.

Cytochalasins, unique and structurally diverse metabolites elaborated by equally divergent fungi are reported to be increasing day by day in their incidence. Nearly more than 150 types of cytochalasins are produced by variety of fungi spread over all groups of fungi. These mycotoxins are unique in their mode of action which selectively inhibits cytokinesis without affecting karyokinesis as a result formation of binucleate and occasionally multinucleate cell are formed. Besides this various morphophysiological functions of cells are affected. Further, these mycotoxins are reported along with other mycotoxins produced either by the same fungus or other fungi growing in the same ecological nich may make their activity either get activated or inhibited. Thus the situation in nature becomes more complex and turned out to be health hazardous. However, reports of the natural incidence of these toxins and health hazards caused by them are limited which may be attributed either due attention was not paid or may be due to their limited incidence. Whatever may be the reasons, some precautions need to be taken such as

i. Harvesting fully matured grains/ears/cobs

ii. Drying to safe moisture levels (maximum 10%)

iii. Separating healthy sound grains/cobs/ears

iv. Drying the grains /cobs/ears on raised bamboo floors along with other appropriate drying methods

v. Using a water proof sheet (preferably non-plastic) to cover produce in order to protect them from rain or physical drying

vi. Storage ears/cobs etc with good ventilation

vii. Effective fumigation for the control of pest problems.

Methods for control of cytochalasins can be classified in to the following categories

1. Prevention of mould growth and cytochalasin elaboration
2. Decontamination and detoxification of cytochalasins
3. Rapid screening techniques
4. Control of environmental conditions

9.1. Prevention of mould growth and cytochalasin elaboration

Cytochalasins producing fungi may invade seeds during pre-harvest period in the field, during harvest, post-harvesting handling or during their storage and consumption. Though prevention is better than cure, integration of different methods is essential during different stages of crop production to consumption. A combination of methods appear to be essential in alleviating the cytochalasins problem. Some of the prevention methods include use of resistant crop varieties, crop rotation, short duration varieties, avoidance of over wintering, minimizing bird and insect damage and mechanical damage to seeds, quickly reducing grain moisture to 10-13%, increased sanitation in crop growth and inhibition of mould growth during storage. Prevention of cytochalasins producing fungi can be achieved in three stages.

i. Primary prevention: It should be initiated at an early stage of seed development. Some of the methods to be adapted are

i. Creating conditions unfavorable for mould growth

ii. Use of mould-resistant varieties of crop

iii. Proper planning of sowing, harvest and storage

iv. Reducing seed-moisture at post-harvest and during storage

v. Storing of seeds at low temperature

vi. Use of fungicides and insecticide to protect the seeds from moulds and insects.

ii. Secondary Prevention: Fungi that invade at an early stage or during storage should either be eliminated or their growth is to be restricted to avoid infestation by cytochalasins producing fungi. The following methods may be adapted to control cytochalasin contamination.

a. Prevention of mould infestation by drying the products to a minimum 10% moisture

b. Elimination of cytochalasin contaminated seeds

c. Protection of stored foods and feeds from conditions favorable for cytochalasins producing fungal growth.

iii. **Tertiary Prevention:** For highly infested foods and feeds where primary and secondary prevention will not be useful, only a few practices like destruction or detoxification of cytochalasins to a minimum level are recommended.

9.2 Decontamination and detoxification of cytochalasins

Although certain treatments have reduced the contamination level of particular cytochalasin, no single method proved to be effective in protection of different agricultural commodities. The detoxification process involves destruction, degradation or inactivation of elaborated cytochalasins by physical, chemical or biological methods. However, detoxification to non-toxic compound has its own limitations like deterioration in nutritive value and doubtful safety of chemical used or formed during the process.

9.3. Rapid screening techniques

By aggressive monitoring of cytochalasins and spending huge amount of money, the USA has been able to produce cytochalasin free food and feed for consumption. Contamination of food commodities by cytochalasins are very heterogeneous in nature and inadequate and/or inaccurate sampling are the major problems in checking cytochalasin contamination. The basic need in monitoring of cytochalasins and their sensitivity, reliability, repeatability and economy are important. Therefore, appropriate sampling and testing protocols that are less expensive and the compilation of regulatory guidelines are necessary. Many physical, chemical, biological, immunological and analytical techniques and procedures should be adapted to meet real situation in nature and effective control.

9.4 Control of environmental conditions

Storage environment plays a vital role in the safety of food grains with respect to mould infestation and cytochalasin contamination. Among various factors moisture, temperature, gaseous atmosphere, presence of contaminating micro-organisms of the commodity are important in minimizing cytochalasins contamination. Magan *et al.* (2004) elegantly discussed the effect of various factors on moulds growth and mycotoxins elaboration in stored food grains.

Several strategies such as introducing resistant varieties, control of insect population, use of natural products, practicing farm management techniques to minimize fungal contamination at the field level and post-harvest management of stored food grains have to be developed. Although significant research on different management strategies has been carried out, still there are many gaps in our understanding which require extensive basic and applied research. Fitzgerald *et al.* (1998) have used atoxigenic strains of *Pithomyces chartarum* to reduce sporodesmin producing strains in New Zealand pastures which resulted in the reduction of facial eczema in sheep and cattle. Such practices may be examined in the control of cytochalasins contamination.

For the successful control of cytochalasins contamination, an integrated strategy involving genetics, molecular biology, plant breeding, biochemistry, plant pathology and ecology is to be adopted. Genetic manipulation of both the host plant and the mycotoxigenic fungus through different molecular methods might provide the most promising results for achieving an effective control of cytochalasins in crops. Good farm management practices such as hygienic conditions, crop rotation, short duration varieties, raising resistant crop varieties and application of suitable and efficient bioagents could minimize the cytochalasins problem. Similarly, increased understanding of the metabolic pathways of both crop and cytochalasin producing fungi may help to modify crops genetically which could alleviate the cytochalasins problem. Recent attempts to raise resistant varieties have led to the isolation of genes involved in the biosynthesis of cytochalasins. The information gained through enzymology and molecular biology of these mycotoxigenic moulds can be exploited as detailed below:

1. To engineer a biological control agent;
2. To identify the compounds that block cytochalasins synthesis and to identify the compound that inhibits the growth of cytochalasin producing fungi.
3. In India, the problem of cytochalasins is being realized only recently and intensive surveys at geographical and climatic regions need to be undertaken. A strategy to control cytochalasins of that region has to be taken up accordingly. There is an urgent need for intensive and extensive survey on the natural incidence of these toxins. Intricacies involved in the elaboration of these mycotoxins need to be taken up at the molecular level. An integrated approach in the management of this problem is the need of hour.

Rigorous programmes for preventing human and animal exposure to contaminated food and feed have to be taken up. There are many reports of mycotoxicoses that need to be studied in depth. The idea of total avoidance of mycotoxins could be unrealistic, but exposure can be minimized through multiple approaches and good farm management practices. Factors affecting mycotoxin-induced apoptosis have enhanced our understanding of the mode of action of mycotoxins. Neverthless, the cytochalasin problem is complex and intricate, solutions are not likely to be easy to implement.

FAO, the World Health Organization (WHO) and the United Nations Environment Programme (UNEP) are engaged in providing essential information on various aspects of prevention and control of mycotoxins in all countries. Guidelines for international trade include: (i) procedures for sampling and analysis; (ii) surveillance and food control inspection systems; and (iii) protocols for detoxification and the quality control of the products. Low cost techniques for assessment, prevention and control of mycotoxins should be transferred from developed countries to developing ones.

Another control strategy is to identify compounds that specifically inhibit the transcription pathways of genes or inhibit the metabolic pathway or enzymes themselves may be more effective and useful. This would enable plant breeders to select the plant that contains inhibitory compounds or the plant that contains resistance inducing compounds. The intricacies involved in the elaboration of there cytochalasins need to be analysed at the molecular level. An integrated approach in the management of cyto chalasins problem needs special attention.

■■■■

Chapter-10

Prospects and Future

The diversity and complexcity of chemical structure of cytochalasins elaborated by equally diverse groups of moulds of different ecological niches makes these toxins quite interesting and intriguing. The only common, feature of these cytochalasins is inhibition of cytokinesis. In depth investigations certainly reveal interesting facts which may be far reaching significance. The elaboration of these toxins by entomogenous fungi, endophytes and soil fungi poses several eco-dynamic questions which need in depth analysis. Mycotoxic nature of these toxins are yet to be fully understood. These aspects certainly may reveal interesting facts of far reaching significance. Further, the relevance of moulds producing cytochalasins need to be investigated. Increasing amount of information on a anticancer nature of these toxins reveal their potential in the chemotherapy of cancer.

The risk assessment of mycotoxins and the advances in their toxicological understanding are closely related. Infact only a few of food related mycotoxins have been accounted in risk assessment studies and most of these are still lacking a satisfying toxicological characterization especially from a mechanistic point of view. Nevertheless, further insights on mycotoxins toxicology are needed to promote more risk assessment studies.

Investigating cytochalasins toxicology possess inherent challenges that are primarily depending on the poor availability of compounds keeping in mind that deciphering cytochalasins action inevitably requires inter-disciplinary approaches and both kinetic and dynamic aspects needs to be considered. Some of the possible applications in the frame work of hazard identification and hazard characterization are illustrated in Fig 10.1.

Molecular modeling approaches may be upstream and cost-effective analytical tools in the identification of cytochalasins competing for a specific receptor these by showing a a potential combined action. In addition the use of the so defined target fishing strategies may also concur to define and discover relevant receptors potentially targeted by a specific cytochalasins or cytochalasins group. This might be relevant especially for emerging cytochalasins as in many cases they have still unknown molecular initiating event.

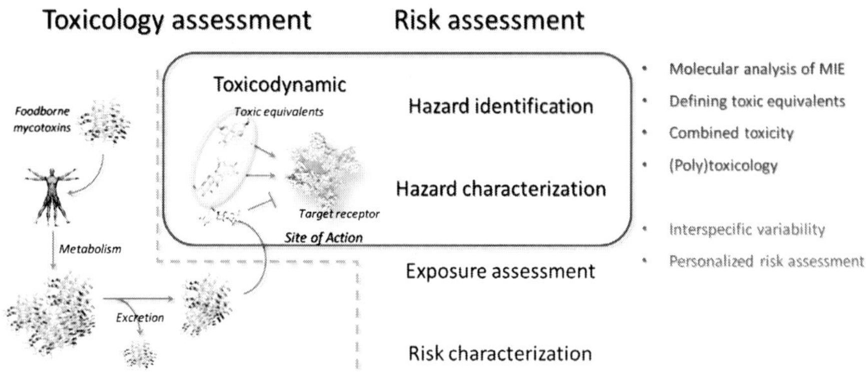

Fig 10.1. The interconnection between mycotoxins toxicology and the risk assessment process

Molecular modeling approaches may directly account the cytochalasin-receptor complex formation deepening the structural basis of cytochalasins interaction. Further, the computation of cytochalasins-receptor interaction may contribute to decipher the basis of the inter- and intra- specific variability.

Therefore, computational studies might be taken into consideration togather with experimental evidences for a more informal extrapolation of data from *In vivo* and *In vitro* animal studies or to support pioneering steps in the personalized risk assessment of cytochalasins.

In conclusion new methods for accounting the molecular initiating event of toxic action with particular reference to the key concepts and challenges of cytochalasins have to be investigated.

Abbreviations

μM= Micrometre

ACP= acyl carrier protein

ACTH=Adrenocorticotropic hormone

ADP =Adenosine diphosphate

A-MLV =Amphotropic Murine Leukemia Virus

AT= acyl transferase

ATP= Adenosine triphosphate

BAC=Bovine adrenocortical

BCE =Bovine capillary endothelial cells

BHK=Baby hamster kidney

BVMO = Baeyer-Villiger monooxygenase

CA=Cytochalasin A

CB=Cytochalasin B

CBMN = Cytochalasins block micronucleus

CC=Cytochalasin C

CD=Cytochalasin D

CE=Cytochalasin E

CheA = Chaetoglobosin A

CheB = Chaetoglobosin B

CheC = Chaetoglobosin C

CheD = Chaetoglobosin D

CheE = Chaetoglobosin E

CIMVs = Cytochalasin B induced membrane vesicles

CLL=Chronic lymphocytic leukemia

DH =dehydratase

ER= enoyl reductase

F-actin = Filamentous actin

fMLF = N-formyl-methionyl-leucyl phenylalanine

G-actin= Globular actin

GPDH=Glycerol phosphate dehydrogenase

HBC =Human breast cancer

HeLa= Human epithelial cells

HEp-2=Human epithelial type 2

HGPRT =Hypoxanthine-guanine phosphoribosyltransferase

HIF-1 =Hypoxia-inducible factor 1α

H-NMR =Proton nuclear magnetic resonance

HCCC = Human Colorectal carcinema cells

HLCC = Human Lung carcinema cells

HPLC= High pressure liquid chromatography

HSV =Herpes simplex virus

HUVEC =Human umbilical vein endothelial cells

IC = Inhibitory concentration

Ig=Immunoglobulin

ICG = Indocyanine green

ILT-4 =Immunoglobulin-like transcript -4

IMDA= Intramolecular Diels–Alder reaction

IOM = Institute of Medicine

IUPAC= International Union of Pure and Applied Chemistry

KR =ketoreductase

LC-MS = Liquid chromatography–mass spectrometry

MN = Micronuclei

NBUD = Nuclear buds

NPB = Nucleoplasmic bridge

PKS= polyketide synthase

PKS-NRPS= polyketide synthase-nonribosomal peptide synthetases

RNA= Ribonucleic acid

RPE = Retinal Pigment Epithelial

SMC =Smooth muscle cells

TAT=Tyrosine aminotransferase

TLC= Thin layer chromatography

TNF-α= Tumor necrosis factor alpha

UNEP =United Nations Environment Programme

UV= Ultra violet

VEGF= Vascular epithelial growth factor

WHO =World Health Organization

■■■■

Glossary

Acrogenous: Borne at the tip

Acrospore: An apical spore adpressed and in some cares fased along most of their length.

Anamorph: An asexual stage in the life cycle of the fungus

Ascocarp: Friut body covering asci and ascospores

Ascogenous: Non-spore bearing

ascospores or basidiospores

Aseptate: Lacking cross walls

Asexual: Reproducing without involving karyogamy and meiosis bearing conidia

Chlamydospora: A thick walled 1-2 celled sterile structure generally functions as a resting structure

Collulum: The neck of a phialide or annelide

Conidiamata: Asexual reproductive body producing conidia

conidiogenous cell

Conidiogenous cells: A cell producing conidia Conidiophores

Conidium: Any asexual spore which rohen mature in liberated from a conidiophores or

Corniform: Shaped like horn

Cruciate: A form of cross

Crustose: Conidial masses in the form of crust

Cuneiform: Wedge shaped

Cupulate: Cup like

Demataceous: Darkly pigmented

Denticulate (Conidiogenous cell, conidiophores): Bearing deuticle

Determinate (Conidiogenous cell, conidiophores): Ceasing to grow in length on production of terminal conidium or chains of conidia.

Dictyoseptate: A spore with both transverse and longitudinal septa

Discrete (Conidiogenous cells): Separate from the main axis or branches of a conidiophore and often having distinct shape.

Doliform: Barrel shaped

Endogenous: Produced entirely within a phialide

Enteroblastic (Conidogenous cells): Blastic with inner wall only or neither wall contributing towards the formation of conidial wall.

Exogenous: Formed on the surface of a protoplasmic mass protruding from the end of a

Extrolites : Secondary metabolites produced by fungi

Exudates (Extrolites): Droplets excreted by the mycelium

Falaform: Formed like a bill hook

Fasicle: Aggregation of conidiophores

Floccose: Loose aerial mycelium

Metulae : A sporephore branch having phialides

Micronematous (Conidiophore): Morphologically resembling a vegetative hyphae but

Monoblastic (Conidiogenous cell): Holoblastic and blowing out at only one point

Mononematous (Conidiophore) : Composed of a single thread or filament secondary

Mycotoxin: Toxins produced by fungi which are toxic to warm blooded animals including man.

Obovoid: Egg shaped with the broad end at the base

Penicillate: A paint brush like arrangement

Peridium: The wall of an ascocarp or other fruiting body

Phialide (Conidiogenous cells): A discrete or integrated phialidic conidiogenous cells phialide.

Phialides: Conidiogenous cells

Polyblastic: Holoblastic and blowing out at several points

Pulvinate: Cushion shaped

Setulae: A delicate appendage arising from the surface of a conidium

Solitary: A rising singly at one point.

Sporodochia: A pulvinate stroma with closely packed relatively short conidiophores.

Stellate: Star shaped

Stipe (Conidiophore): The main, stont axis or unbranched lower part.

Stipe: A stalk, in the present context a specialised hypha arising from a fertile hypha and supporting a pcnicillus

Stipe: Fertile hyphae

Synnemata (Conidiophore): Composed of numerous threads or filaments tightly

Teleomorph: Refers to the state of a fungus which bears sexual spores such as

Unitunicate: Refers to an ascus with a fused or single wall

Verruculose: Finely warted

Verticil: A cluster of metulae or phialides with a common origin.

■■■■

References

Albert SG. 1984. Cytochalasin B does not serve as a marker of glucose transporter in rabbit erythrocytes. *Biochem. Int.* **9**: 93-103.

Aldridge DC and WB. Turner. 1969. Structures of cytochalasins C and D. *Journal of the Chemical Society C: Organic.* **6**: 923-928.

Aldridge DC, BF. Burrows and WB. Turner. 1972. The structures of the fungal metabolites cytochalasins E and F. *Journal of the Chemical Society, Chemical Communications.* **3**: 148-149.

Aldridge DC, JJ. Armstrong, BRN. Speake and WB. Turner. 1967. The structures of cytochalasins A and B. *J. Chem. Soc. Commun.* **0**: 1667-1676.

Ali A, SS. Tatyana and DF. Matesic. 2012. Dual Modulation of JNK and Akt Signaling Pathways by Chaetoglobosin K in Human Lung Carcinoma and *ras*-Transformed Epithelial Cells. *Invest New Drugs.* **31**: 525–534.

Alienberg M and M. Silverman. 2003. Cytochalasin D disruption of actin filaments in 3T3 cells produces an anti-apoptotic response by activating gelatinase A extracellularly and initiating intracellular survival signals. *Biochim Biophys Acta.* **1593**:249-58.

Amnuaykanjanasin A, J. Punya, P. Paungmoung, A. Rungrod, A. Tachaleat, S. Pongpattanakitshote, S. Cheevadhanarak and M. Tanticharoen. 2005. Diversity of type I polyketide synthase genes in the wood-decay fungus *Xylaria* sp. BCC 1067. *FEMS Microbiol. Lett.* **251**: 125–136.

Araki T and JG. Williams. 2012. Perturbations of the actin cytoskeleton activate a Dictyostelium STAT signalling pathway. *Eur J Cell Biol.* **91:**420-5.

Andersen B, J. Smedsgaard and JC. Frisvad. 2004. *Penicillium expansum*: Consistent production of patulin, chaetoglobosins, and other secondary metabolites in culture and their natural occurrence in fruit products. *J. Agric.Food.Chem.***52**: 2421-2428.

Auclair K, A. Sutherland, J. Kennedy, DJ. Witter, JP. Van den Heever, CR. Hutchinson and JC. Vederas. 2000. Lovastatin nonaketide synthase catalyzes an intramolecular Diels-Alder reaction of a substrate analogue. *J. Am. Chem. Soc.* **122**:11519–11520.

Bennett JW and M Klich. 2003. Mycotoxins. *Clin. Microbiol. Rev.* **16**:497-516.

Berestetskiy A, A. Dmitriev, G. Mitina, I. Lisker, A. Andolfi and A. Evident. 2008. Nonenolides and cytochalasins with phytotoxic activity against *Cirsium arvense* and

Sonchus arvensis: A structure–activity relationships study. *Phytochemistry*. **69**: 953-960.

Bergmann S, J. Schümann, K. Scherlach, C. Lange, AA. Brakhage and C. Hertweck. 2007. Genomics-driven discovery of PKS-NRPS hybrid metabolites from *Aspergillus nidulans*. *Nat. Chem. Biol.* **3:** 213–217.

Bergstralh DT and JP. Ting. 2006. Microtubule stabilizing agents: Their molecular signaling consequences and the potential for enhancement by drug combination. *Cancer Treat. Rev*. **32**:166–179.

Betina V, D. Micekova and P. Nemec.1972. Antimicrobial Properties of Cytochalasins and their Alteration of Fungal Morphology. *Journal of General Microbiology*. **71**:343-349.

Binder M and C. Tamm. 1973. The Cytochalasans: A New Class of Biologically Active Microbial Metabolites. *Angewandte Chemie International Edition*. **12**: 370–380.

Binder M, CH. Tamm and H. Minato. 1973. Nomenclature of a class of biologically active mold metabolites. Cytochalasins, phomins, and zygosporins. *J. Chem. Soc. Perkin. Trans*.**11:** 1146-1147.

Binder M, JR. Kiechel and C. Tamm. 1970. Zur biogenese des antibioticums phomin. 1 Teil: die grundbausteine. *Helv Chim Acta*. **53**:1797–1812.

Bonder EM and M. Mooseker.1986. Cytochalasin B slows but does not prevent monomer addition at the barbed end of the actin filament. *J. Cell Biol*. **102**:282-288.

Brett JG and GC. Godman. 1984. Microvacuolation induced by cytochalasin: its relation to the cytoskeleton; morphological and cytochemical observations. *Tissue Cell*. **16**: 311-324.

Bruijns RHJ and H. Bult. 2001. Effects of local cytochalasin D delivery on smooth muscle cell migration and on collar-induced intimal hyperplasia in the rabbit carotid artery. *Br J Pharmacol*. **134**: 473-483.

Buchanan MS, T. Hashimoto and Y. Asakawa. 1996. Cytochalasans a *Daldinia* sp from *fungus*. *Phytochemistry*. **41**: 821-828.

Buchanan MS, T. Hashimoto and Y. Asakawa. 1995. Five 10-phenyl-[11]-cytochalasans from a *Daldinia* fungal species. *Phytochemistry*. **40:** 135-140.

Burlot L, JC. Cherton, O. Convert, I. Correia and B. Dennetiere. 2003. New chaeto-globosins from maize infested by *Phomopsis leptostromiformis* fungi. Production, identification, and semi-synthesis. *Spectroscopy*. **17** : 725–734.

Burnside B, R. Adler and PO' Connor. 1983. Retinomotor Pigment Migration in the Teleost Retinal Pigment Epithelium/Roles for Actin and Microtubules in Pigment Gronule Tronsporr ond Cone Movement. *A Journal of Dosic and Clinical Researc*. **24**:1-15.

Bylund J, S. Pellmé, H. Fu, UH. Mellqvist, K. Hellstrand, A. Karlsson and C. Dahlgren. 2004. Cytochalasin B triggers a novel pertussis toxin sensitive pathway in TNF-alpha primed neutrophils. *BMC Cell Biology*. **5**:2, 1-14.

Cameron AF, AA. Freer, B. Hesp and CJ. Strawson. 1974. Isolation, and crystal and molecular structure of cytochalasin G: an [11]cytochalasan containing an indole group

Capasso R, A. Evidente and A. Ritieni. 1988. Ascochalasin, A new cytochalasin from *Ascochyta heteromorpha* . *J. Nat. Prod.* **51**: 567-71.

Capasso R, A. Evidente and M. Vurro. 1991. Cytochalasins from *Phoma exigua* var. *heteromorpha*. *Phytochemistry*. **30** :3945-3950

Carlier MF, D. Pantaloni and ED. Korn. 1986. The effects of Mg^{2+}at the high-affinity and low-affinity sites on the polymerization of actin and associated ATP hydrolysis. *J. Biol Chem.* **261**:10785–10792.

Carrieri C, R. Galasso, F. Semeraro, CT. Ammollo, N. Semeraro and M. Colucci. 2010. The role of thrombin activatable fibrinolysis inhibitor and factor XI in platelet-mediated fibrinolysis resistance: a thromboelastographic study in whole blood. *Journal of Thrombosis and Haemostasis*. **8**: 154-162.

Carter SB. 1967. Effects of cytochalasins on mammalian cells. *Nature*. **213**: 261-264.

Casella JF, DF. Michael and S. Lin. 1981. Cytochalasin D inhibits actin polymerization and induces depolymerization of actin filaments formed during platelet shape change. *Nature*. **293**: 302–305.

Chao JI and HF. Liu. 2006. The Blockage of survivin and securin expression increases the cytochalasin B-Induced cell death and growth inhibition in human cancer cells. *Mol Pharmacol*. **69**:154-164.

Chen C, Q.Tong, H. Zhu, D. Tan, J. Zhang, Y. Xue, G. Yao, Z. Luo, J. Wang, Y. Wang and Y. Zhang. 2016. Nine new cytochalasan alkaloids from *Chaetomium globosum* TW1-1 (Ascomycota, Sordariales). *Scientific Reports*. **6:**18711,1-8.

Chen XM, NF. La Russo. 2000. Mechanisms of attachment and internalization of *Cryptosporidium parvum* to biliary and intestinal epithelial cells. *Gastroenterology*. **118**:368-79.

Chen ZM, HP. Chen, Y. Li, T. Feng and JK. Liu. 2015. Cytochalasins from cultures of endophytic fungus *Phoma multirostrata* EA-12. *The Journal of Antibiotics*. **68**: 23-26.

Chinworrungsee M, P. Kittakoop, M. Isaka, A. Rungrod, M. Tanticharoen and Y. Thebtaranonth. 2001. Antimalarial halorosellinic acid from the marine fungus *Halorosellinia oceanica. Bioorg Med Chem Lett*. **11**: 1965-1969.

Choi BH, JA. Park, KR. Kim, GI. Lee, YT. Lee, H. Choe, SH. Ko, MH. Kim, YH. Seo, YG. Kwak. 2005. Direct block of cloned hKv1.5 channel by cytochalasins, actindisrupting agents. *Am J Physiol Cell Physiol*. **289**: 425–436

Choo SJ, BS. Yun, IJ. Ryoo, YH. Kim, KH. Bae ID. and Yoo. 2009. Aspochalasin I, a melanogenesis inhibitor from *Aspergillus* sp. *J. Microbiol. Biotechnol.* **19:** 368–371.

Christoph S and T. Christoph . 1988. Chaetoglobosin M, a New Metabolite of a Mutant of *Diplodia macrospora*, belonging to the family of (1*H*-indol-3-yl)-substituted 10,11-diethyl-10,11-dinorcytochalasans. *Helvetica Chimica Acta*. **71**: 8, 1881–1884.

Cimmino A, A. Andolfi, A. Berestetskiy and A. Evidente A. 2008. Production of phytotoxins by *Phoma exigua var.exigua*, A potential mycoherbicide against perennial thistles. *J. Agric. Food Chem.* **56**: 6304–6309.

Clark DA and PL. Fuchs. 1979. Cytochalasin support studies. Macrocycle synthesis via enolate-assisted, intraannular 1,4-fragmentation reactions. *J. Am. Chem. Soc.* **101**: 3567–3576

Collet JP, G. Montalescot, C. Lesty, J. Soria, Z. Mishal, D. Thomas and C. Soria. 2001. Disaggregation of in vitro preformed platelet-rich clots by abciximab increases fibrin exposure and promotes fibrinolysis. *Arterioscler Thromb Vasc Biol.* **21**: 142–8.

Collings DA, BES. Gunning, PK. Hepler and D. Menzel.1996. Actin in living and fixed characean internodal cells: identification of a cortical array of fine actin strands and chloroplast actin rings. *Protoplasma.* **190**:25-38.

Collings DA, GO. Wasteneys and RE. Williamson. 1995. Cytochalasin rearranges cortical actin of the alga *Nitella* into short, stable rods. *Plant Cell Physiol.* **36**: 765-772.

Cooper JA. 1987. Effects of cytochalasin and Phalloidin on Actin. *The Journal of Cell Biology.* **105**: 1473-1478.

Cortese F and J. Wolff. 1978.Cytochalasin-stimulated steroidogenesis from high density lipoproteins. *J Cell Biol.* **77**:507–516.

Cox RH, HG. Cutler, RE. Hurd and RJ. Cole.1983. Proton and C-13 nuclear magnetic resonance studies of the conformation of cytochalasin-H derivatives and plant growth regulating effects of cytochalasins. *J Agr Food Chem.***31**:405–408.

Creppy EE. 2002. Update of survey, regulation and toxic effects of mycotoxins in Europe. *Toxicol Lett.* **127**:19-28.

Cunningham D, D. Schafer, S. W. Tanenbaum and M. Flashner. 1979. Physiological responses of bacteria to cytochalasin A: effects on growth, transport, and enzyme induction. *J. Bacteriol.***137:** 925–932.

Cutler HG, FG. Crumley, RH. Cox, RJ. Cole, JW. Dorner, JP. Springer, FM. Laterell, JE. Thean and AE. Rossi. 1980. Chaetoglobosin K: a new plant growth inhibitor and toxin from Diplodia macrospora. *Journal of Agricultural and Food Chemistry*. **28:** 139e142.

Dagne E, AAL. Gunatilaka, S. Asmellash, D. Abate, D G I. Kingston, GA. Hofmann, RK. Johnson. 1994. Two new cytotoxic cytochalasins from *Xylaria obovata. Tetrahedron.* **50**: 5615-5620.

De Sousa PA and Y. Masui. 1990. The effect of cytochalasin D on protein synthesis in Xenopus laevis oocytes. *Mol Reprod Dev.* **26**:248-52.

Dellafiora L, C. Dall'Asta and G.Galaverna. 2018. Toxicodynamics of Mycotoxins in the Framework of Food Risk Assessment-An In Silico Perspective. *Toxins.* **52**: 1-23.

Demain AL, NA. Hunt, V. Malik, B. Kobbe, H. Hawkins, K. Matsuo, GN. Wogan. 1976. Improved procedure for production of cytochalasin E and tremorgenic mycotoxins by *Aspergillus clavatus. Appl Environ Microbiol.* **31**:138-140.

Ding G, H. Wang , L. Li, B. Song, H. Chen, H. Zhang, X. Liu and Z. Zou. 2014. Trichodermone, a spiro-cytochalasan with a tetracyclic nucleus (7/5/6/5) skeleton from the plant endophytic fungus *Trichoderma gamsii. J. Nat. Prod.* 77**:** 164–167.

Ding G, YC. Song, JR. Chen, C. Xu, HM. Ge, XT. Wang and RX. Tan. 2006. Chaetoglobosin U, a cytochalasin alkaloid from endophytic *Chaetomium globosum* IFBE019. *J. Nat. Prod.* **69**: 302-304.

Ding GH, W Li Li and L. Chen. 2012. Trichoderones A and B: Two Pentacyclic Cytochalasans from the Plant Endophytic Fungus *Trichoderma gamsii. Fitoterapia.* **83**: 541–544

Donoso R, A. R. Sagredo, J.A. Hueso-Rodriguez and S.W. Elson. 1997. A new chaetoglobosin isolated from a fungus of the genus *Discosia. Natural Product Letters.***10**: 49-54.

Doubravka V, D. Vesely and R. Jelink. 1983. Cytochalasin E: production by strain *Aspergillus clavatus* and toxic effects upon embryonic chicks. *Microbiol. Aliment Nutr.* **4**: 393-398.

Durackova Z , V. Bekina and P. Nemec. 1976. *J. Chromatogr.* **116**: 155.

Edmonds BT and E. Koenig. 1989. ATP and calmodulin dependent actomyosin aggregates induced by cytochalasin D in goldfish retinal ganglion cell axons in vitro. *J. Neurobiol.* **21**: 555-566.

Edwards RL and DJ. Maitland .1989. Metabolites of the higher fungi. Part 24. Cytochalasin N, O, P, Q, and R. New cytochalasins from the fungus *Hypoxylon terricola* Mill. *J. Chem Soc. Perkin Trans.* I. **1**: 57-65

El-Kady IA and ME Mostafa. 1995. Production of cytochalasins C, D & E from dematiaceous hyphomycetes. *Folia Microbiologica.* **40:** 301–303.

Espada A, A. Rivera-Sagredo, JM. De La Fuente, JA. Hueso-Rodriguez and SW. Elson. 1997. New cytochalasins from the fungus *Xylaria hypoxylon. Tetrahedron* **53**: 6485–6492.

Evidente A, A .Andolfi, M. Vurro, MC. Zonno and A. Motta A. 2003. Cytochalasins Z4, Z5, and Z6, Three new 24-oxa[14]cytochalasins produced by *Phoma exigua* var. *heteromorpha. J. Nat. Prod.* **66:**1540-1544.

Evidente A, A. Andolfi, M. Vurro, MC. Zonno and A. Motta A. 2002. Cytochalasins Z1, Z2 and Z3, Three 24-oxal[14]cytochalasans produced by *Pyrenophora semeniperda. Phytochemistry.* **60:** 45-53.

Evidente A, R. Capasso, A. Andolfi, M. Vurro and MC. Zonno. 1998. Structure activity relationships studies of putaminoxins and pinolidoxins: phytotoxic nonenolides produced by phytopathogenic Phoma and Ascochyta species. *Natural Toxins.* **6:** 183-188.

Evidente A, R. Capasso, M. Vurro and A. Bottalico. 1996. Cytochalasin W, a New 24i0xa[14]cytochalasan From *Phoma exigua* Var. *heteromorpha. Natural Toxins.* 453-457.

Evidente A, R. Lanzetta, R. Capasso Vurro and A. Bottalico. 1992. Cytochalasins U and V, two new cytochalasans from *Phomu exigua* var. *hereromorpha. Tetrahedron.* **48:**63 17-6324.

Fang F, Ui. Hideaki, K. Shiomi, R. Masuma, Y. Yamaguchi, C. Gang Zhang, X. Wu Zhang, Y. Tanaka and S. Omur. 1997. Two New Components of the Aspochalasins Produced by *Aspergillus* sp. *The journal of antibiotics.* **50:** 919-925.

Fenech M, J. Crott, J. Turner and S. Brown. 1999. Necrosis, apoptosis, cytostasis and DNA damage in human lymphocytes measured simultaneously within the cytokinesis-block micronucleus assay:description of the method and results for hydrogen peroxide. *Mutagenesis.* **14**:605-12.

Fenech M. 2006. Cytokinesis-block micronucleus assay evolves into a "cytome" assay of chromosomal instability, mitotic dysfunction and cell death. *Mutat Res* . **600**:58-66.

Fenech M. 2007. Cytokinesis-block micronucleus cytome assay. *Nat Protoc.* **2**:1084-104.

Feng YJ, JW. Blunt, ALJ. Cole and MHG. Munro. 2002. Three novel cytochalasins X, Y, and Z from *Pseudeurotium zonatum. J. Nat. Prod.* **65**: 1274-1277.

Fitzgerald JM, R. Collin and NR. Towers. 1998. Biological control of sporidesmin-producing strains of *Pithomyces chartarum* by biocompetitive exclusion. *Letters in Applied Microbiology.* 26:17-21.

Foissner I and GO. Wasteneys. 2007. Wide-ranging effects of eight cytochalasins and latrunculin A and B on intracellular motility and actin filament reorganization in characean internodal cells. *Plant & cell physiology.* **48** :585-597.

Foissner I and GO. Wasteneys. 2000. Microtubule disassembly enhances reversible cytochalasin-dependent disruption of actin bundles in characean internodes. *Protoplasma.* **214** :233-44.

Forscher P and SJ. Smith. 1988. Actions of Cytochalasins on the Organization of Actin Filaments and Microtubules in a Neuronal Growth Cone. *The Journal of Cell Biology.* **107**: 1505-1516.

Frisvad JC, RA. Samson, RR. Birgitte, MI. Van der Horst, FTJ van Rijn and J. Stark. 1997. *Penicillium discolor,* a new species from cheese, nuts and vegetables. *Antonie van Leeuwenhoek.* **72**: 119–126.

Fujii Y, H. Tani, M. Ichinoe and H. Nakajima. 2000. Zygosporin D and two new cytochalasins produced by the fungus *Metarrhizium anisopliae. J. Nat. Prod.* **63**: 132–135.

Gebhardt K, S. Judith, H. Alexandra, D. Konrad, D. Siegfried, B. Winfried, R. Joachim and F. Hanspeter. 2004. Aspochalamins A-D and Aspochalasin Z Produced by the Endosymbiotic Fungus *Aspergillus niveus* LU 9575. *The Journal of Antibiotics.* **11**: 707-714

George TP, WH. Cook, DM. Byers, Frederick B.St.C. Palmer Matthew W. Spence. 1991. Inhibition of phosphatidylcholine and phosphatidylethanolamine biosynthesis by cytochalasin B in cultured glioma cells: Potential regulation of biosynthesis by Ca^{2+}-dependent mechanisms. *Lipids and Lipid Metabolism.* **1084**: 185-193

Goddette DW and C. Frieden. 1986. Actin polymerization. The mechanism of action of cytochalasin D. *J Biol Chem.* **261**: 15974–15 980.

Goldman RD and DL. Spector. 2004. Live cell imaging: a laboratory manual. Cold Spring Harbor Laboratory Press. Cold Spring Harbor. NY.

Granholm NH. 1979. Effects of cytochalasin B on preimplantation and early postimplantation mouse embryos in vitro. *Abnormal Embryogenesis.* Chapter-**5**: 71-94.

Haidle A and AG. Myers. 2004. An enantioselective, modular, and general route to the cytochalasins: Synthesis of L-696,474 and cytochalasin B. *Proc. Nat. Acad. Sci. U.S.A.* **101**:12048-12053.

Haitao L, B. Li, Z. Li, SJ. Cutler, GO. Rankin and YC. Chen. 2013. Chaetoglobosin K inhibits tumor angiogenesis through down regulation of vascular epithelial growth factor-binding hypoxia-inducible factor 1α. *Anticancer Drugs.* **24:**1-20.

Halo LM, JW. Marshall, AA. Yakasai, Z. Song, CP. Butts, MP. Crump, M. Heneghan, AM. Bailey, TJ. Simpson, CM. Lazarus and RJ. Cox. 2008. Authentic heterologous expression of the tenellin iterative polyketide synthase nonribosomal peptide synthetase requires coexpression with an enoyl reductase.*Chem Bio Chem*. **9**: 585–594.

Hayakawa S, T. Malisushima and M. Kimura. 1973. *Annu. Rep. Shinogi Res. Lab.* **23**: 2-3.

Hayakawa S, T. Matsushima, T. Kimura, T. Minato and K. Katagiri. 1968. Zygosporin A. A new antibiotic from *Z. masonii. J. Antibiol. (Yokyo).* **21**: 523-534.

Hazel D, P G. Steel and EJ. Thomas. 1989. Cytochalasan synthesis: total synthesis of cytochalasin G. *J. Chem. Soc., Perkin Trans.* **1**: 525-528

Hertweck C. 2009. The biosynthetic logic of polyketide diversity. *Angew. Chem., Int. Ed.***48**: 4688-4716.

Hidayat H , C. Kliche-Spory, A. Al-Harrasi, A. Al-Rawahi, G. Abbas, I. Robert Green, B. Schulz, K. Krohn, A. Shah. 2014. Antimicrobial constituents from three endophytic fungi. *Asian Pacific Journal of Tropical Medicine.* **7**: 224-227.

Hintikka EL, R. Holopainen, A. Asola, M. Jestoi, M. Peitzsch and S. Kalso. 2009. Mycotoxins in the ventilation systems of four schools in Finland. *World Mycotox J.* **2**:369-79.

Hirose T, Y. Izawa, K. Koyama, S. Natori, K. Iida, I. Yahara, S. Shimaoka and K. Maruyama. 1990. The effects of new cytochalasins from *Phomopsis* sp. and the derivatives on cellular structure and actin polymerization. *Chem. Pharm. Bull.* **38:** 971-974.

Horn WS, MSJ. Simmonds, RE. Schwartz and WM. Blaney. 1995. Phomopsichalasin, a novel antimicrobial agent from an endophytic *Phomopsis* sp. *Tetrahedron.* **51**: 3969-3978.

Howard TH, J. Casella and S. Lin. 1981. Correlation of the Biologic Effects and Binding of Cytochalasins to Human Polymorphonuclear Leukocytes. *Blood.* **57**: 399-405.

Hseu YC, SC. Chen, WH. Lin, DZ. Hung, MK. Lin, YH. Kuo, MT. Wang, HJ. Cho, L. Wang and HL. Yang. 2011. *Toona sinensis* (leaf extracts) inhibit vascular endothelial growth factor (VEGF)-induced angiogenesis in vascular endothelial cells. *J. Ethnopharmacol.* **134**: 111–121.

Hu Y, D. Dietrich, W. Xu, A. Patel, Justin A. J. Thuss, J. Wang, W. Yin, K. Qiao, KN. Houk, J C. Vederas and Y. Tang1.2014. A Carbonate-Forming Baeyer-Villiger Monooxygenase. *Nat Chem Biol.* **10**: 552–554.

Huang FY, Y. Nan Li, W. Mei, H. Fu Dai, P. Zhou, G. Hong Tan. 2012. Cytochalasin D, a tropical fungal metabolite, inhibits CT26 tumor growth and angiogenesis. *Asian Pacific Journal of Tropical Medicin.* **5:** 169-174.

Huang J, SK. Das, P. Jha, W. Al Zoughbi, S. Schauer, T. Claudel, V. Sel, P. Vesely, R. Birner-Gruenberger and D. Kratky. 2013. The PPARα agonist fenofibrate suppresses B-cell lymphoma in mice by modulating lipid metabolism. *Biochim Biophys Acta.* **1831**:1555–1565.

Hungate RW, JL. Chen, KE. Starbuck, SA. Macaluso and RS. Rubino. 1996. New cytochalasins: Synthetic studies of a novel HIV-1 protease inhibitor. *Tetrahedron Lett.* **37 :** 4113-4116.

Hwang J, M. Yi, X. Zhang, Y. Xu, JH. Jung and DK. Kim. 2013. Cytochalasin B induces apoptosis through the mitochondrial apoptotic pathway in HeLa human cervical carcinoma cells. *Oncol Rep.* **30**:1929-35.

Ichihara A, K. Katayama, H. Teshima, H. Oikawa and S. Sakamura. 1996. Chaetoglobosin O and Other Phytotoxic Metabolites from *Cylindrocladium floridanum,* a Causal Fungus of Alfalfa Black Rot Disease, *Bioscience, Biotechnology, and Biochemistry.* **60**: 360-361.

Ikewaki N, A. Yamada and H. Inoko. 2003. Depolymerization of Actin Filament by Cytochalasin E Induces Interleukin-8 Production and Up-Regulates CD54 in the HeLa Epithelial Cell Line. *Microbiol. Immunol.* **47**: 775–783.

Institute of Medicine (IOM). 2004. Damp Indoor Spaces and Health, Washington, DC, National Academy Press.

Isenberg G, P. Schubert and GW. Kreutzberg.1980. Experimental approach to test the role of actin in axonal transport. *Brain Res* **194**:588–593.

Isowa N, SH. Keshavjee M. and Liu. 2000. Role of microtubules in LPS-induced macrophage inflammatory protein-2 production from rat pneumocytes. *Am. J. Physiol. Lung Cell. Mol. Physiol.* **279***:* L1075–L1082.

IUPAC. 1974. *Inform. Bull Tech. Dep.* **8**: 1.

Iwamota C, T. Yamada, Y. Ito, K. Minoura and A. Numata. 2001. Cytotoxic cycochalasins from Penicillium species separated from a marine alga. *Tetrahedron.* **57**: 2997-3004.

Izawa Y, T. Hirose, T. Shimizu, K. Koyama and S. Natori. 1989. Six new 10-phenyl-[11] cytochalasans, cytochalasins N–**S** from *Phomopsis* sp.*Tetrahedron.***45**: 2323-2335.

Janet T and CG. Gabriel. 1983. Cytochalasin D Induces Increased Actin Synthesis in HEp-2 Cells. *Molecular and Cellular Biology,* **3:** 132-142.

Jasper JR, SR. Post, KH. Desai, PA. Insel and D. Bernstein. 1995. Colchicine and cytochalasin B enhance cyclic AMP accumulation via post receptor actions. *J. Pharmacol. Exp. Ther.* **274***:* 937–942.

Jeremy BB, J.Migrainea, VL. Faivrec, BL. Loumagnea, AB. Claire Lukaszewiczc, DB. Payenc and B. Favier. 2013. Exocytosis acts as a modulator of the ILT4-mediated inhibition of neutrophil functions. *PNAS* 110. **44 :** 17957–17962.

Jiao W, Y. Feng, JW. Blunt, ALJ. Cole and MHG. Munro. 2004. Chaetoglobosins Q, R, and T, Three further new metabolites from *Chaetomium globosum J. Nat. Prod.* **67:** 1722– 1725.

Jijakli H, Zhang HX, Dura E, Ramirez R, Sener A and Malaisse WJ.2002. Effects of cytochalasin B and D upon insulin release and pancreatic islet cell metabolism. *Int J Mol Med.* **9**:165-72.

Jouda JB, JD. Tamokou, CD. Mbazoa, C. Douala-Meli, P. Sarkar, PK. Bag, J. Wandji. 2016. Antibacterial and cytotoxic cytochalasins from the endophytic fungus Phomopsis sp. harbored in Garcinia kola (Heckel) nut. *BMC Complement Altern Med.* **16**:462.

Junliang L, Z. Hu, H. Huang and Q. Xu. 2012. Aspochalasin U, a moderate TNF-inhibitor from Aspergillus sp. *The Journal of Antibiotics.* **65**: 49-52.

Kakeya H, M. Morishita, C. Onozawa, R. Usami, K. Horikoshi, K. Kimura, M. Yoshihama and H. Osada. 1997, RKS-1778, its preparation, a cell cycle inhibitor and an antitumour agent produced by *Phoma* sp. *J Nat Prod.* **60**: 669–672.

Karine MV, FD. Gumkowski and JD. Jamieson. 1999. The subapical actin cytoskeleton regulates secretion and membrane retrieval in pancreatic acinar cells. *Journal of Cell Science.* **112:** 81-96.

Kelleher D, A. Murphy, C. Feighery and EB. Casey. 1995. Leukocyte function-associated antigen-1 (LFA-1) and CD44 are signaling molecules for cytoskeleton-dependent morphological changes in activated T cells. *J. Leukoc. Biol.* **58:** 539-546.

Keller-Schierlein W and E. Kupfer.1979. Stoffwechselprodukte von Mikroorganismen. 186. Mitteilung. Über die Aspochalasine A, B, C und D. *Helvetica Chimica Acta.* **62**: 1501-1524

Kelly WL. 2008. Intramolecular cyclizations of polyketide biosynthesis: mining for a "Diels-Alderase. *Org. Biomol. Chem.* **6**:4483-4493.

Kelly SE. 1991. Comprehensive Organic Synthesis 1; Trost, B. M. and Flemming, I., Eds.; Pergamon: Oxford, U.K. p 729.

Kennedy J, K. Auclair, SG. Kendrew, C. Park, JC. Vederas and CR. Hutchinson.1999. Modulation of polyketide synthase activity by accessory proteins during lovastatin biosynthesis. *Hutchinson, Science.* **284**: 1368-1372.

Kettlewell S, NL.Walker, SM. Cobbe, FL. Burton and GL. Smith. 2004. The electrophysiological and mechanical effects of 2,3-butane-dione monoxime and cytochalasin-D in the Langendorff perfused rabbit heart. *Experimental Physiology.* **89**: 163–172.

Kimura Y, H. Nakajima and T. Hamasaki T. 1989. Structure of Rosellichalasin, a New Metabolite Produced by *Rosellinia necatrix. Agr Biol Chem Tokyo.* **53**:1699–1701.

Kirsch-Volders M and M. Fenech. 2001. Inclusion of micronuclei in non-divided mononuclear lymphocytes and necrosis/apoptosis may provide a more comprehensive cytokinesis block micronucleus assay for biomonitoring purposes. *Mutagenesis.* **16**:51-8.

Kletzien R and J. Perdues. 1972. The Inhibition of Sugar Transport in Chick Embryo Fibroblasts by Cytochalasin B. *The Journal of Biological Chemistry.* **248:** 711-719.

Knudsen PB, B. Hanna, S. Oh, L. Sellner, T. Zenz, H. Dohner, S. Stilgenbauer, TO Larsen, P. Lichter and M. Seiffert. 2014. Chaetoglobosin A preferentially induces apoptosis in chronic lymphocytic leukemia cells by targeting the cytoskeleton. *Leukemia*, 1-10.

Kohno J, N. Nonaka, M. Nishio, T. Ohnuki, K. Kawano, T. Okuda and S. Komatsubara. 1999. TMC-169, a new antibiotic of the aspochalasin group produced by *Aspergillus flavipes. J. Antibiot.* (Tokyo). **52**: 575–577.

Konig GM, AD. Wright, HJ. Aust, S. Draeger and B. Schulz. 1999. Geniculol, a new biologically active diterpene from the endophytic fungus *Geniculosporium* sp. *Nat. Prod.* **62:** 155–157.

Lang G, JW. Blunt, NJ. Cummings, AL. Cole and MH. Munro . 2005. Hirsutide, a cyclic tetrapeptide from a spider-derived entomopathogenic fungus, *Hirsutella* sp. *J Nat Prod.* **68**:1303-5.

Ledu C and H. Mc Comble. 2003. Effects of cytochalasin B on fertilization and ploidy in the Pacifie oyster *Crassostrea gigas. Invertebrate Reproduction and Development.* **44**:131-137

Lee J, JM. Yi , H. Kim, YJ. Lee, JS. Park, OS. Bang and NS. Kim. 2014. Cytochalasin H, an active anti-angiogenic constituent of the ethanol extracts of *Gleditsia sinensis* thorns. *Biol Pharm Bull.* **37**:6-12.

Lei CW, ZQ. Yang, Y. Ping Zeng, Y. Zhou, Y. Huang, X. Sheng He, G. You Li and X. Hong Yuan. 2018. Xylastriasan A, a new cytochalasan from the fungus *Xylaria striata, Natural Product Research.* **32**: 7-13.

Lewitt MS and RC. Baxter. 1991. Cytochalasin B stimulates insulin-like growth factor-binding protein-1 production by HepG2 cells. *Mol. Cell. Endocrinol.* **77:** 149–157.

Lin ZJ, TJ. Zhu, HJ. Wei, GJ. Zhang, H. Wang, QQ. Gu. 2009. Spicochalasin A and New Aspochalasins from the Marine-Derived Fungus *Spicaria elegans. Eur. J. Org. Chem.* **18:** 3045–3051.

Lipski KM, JD. McQuiggan, KJ. Loucy and TP. Fondy. 1987. Cytochalasin B: Preparation, analysis in tissue extracts, and pharmacokinetics after intraperitoneal bolus administration in mice. *Anal Biochem.* **161**:332–340.

Liu J, Z. Hu, H. Huang, Z. Zheng and Q. Xu. 2012. Aspochalasin U, a moderate TNF-alpha inhibitor from *Aspergillus* sp. *J. Antibiot. (Tokyo).* **65**:49-52.

Liu JT, B. Hu, Y. Gao, JP. Zhang, BH. Jiao, XL. Lu and XY. Liu. 2014. Bioactive tyrosine-derived cytochalasins from fungus *Eutypella* sp. D-1. *Chem Biodivers.* **11**:800-6.

Liu R, Q. Gu, W. Zhu, C. Cui, G. Fan, Y. Fang, T. Zhu and H. Liu. 2006. 10-Phenyl-[12]-cytochalasins Z_7, Z_8, and Z_9 from the marine-derived fungus *Spicaria elegans J. Nat. Prod.* **69:**871– 875.

Lombardo YB, S. Drago, A. Chicco, P. Fainstein, R. Gutman, JJ. Gagliardino and CL. Gomez Dumm. 1996. Long-term administration of a sucrose-rich diet to normal rats: Relationship between metabolic and hormonal profiles and morphological changes in the endocrine pancreas. *Metabolism.* **45**: 1527-1532

Lugauskas, A. 2005. Potential toxin producing micromycetes on food raw material and products of plant origin. *Bot. Lithuanica.* **7**:3–16.

Luo H, B. Li, Z. Li, JC. Stephen, O. Gary Rankin and YC. Chen. 2013. Chaetoglobosin K inhibits tumor angiogenesis through downregulation of vascular epithelial growth factor-binding hypoxia-inducible factor 1α. *Anticancer Drugs.* **24:**1-20.

Lyles DS and FR. Landsberger. 1976. Virus and lectin agglutination of erythrocytes: Spin label study of membrane lipid-protein interactions. *Proc. Nati. Acad. Sci.* USA. **73**: 3497-501.

Magan N, V. Sanchis and D. Aldred. 2004. Role of spoilage fungi in seed deterioration. Chapter 28, In fungal biotechnology in agricultural, food and environmental applications. (eds: D.K.Aurora Marcell Dekker). Pp. 311-323.

Magdline SHS. 2015. The Involvement of Microtubules and Actin during the Infection of Japanese Encephalitis Virus in Neuroblastoma Cell Line, IMR32. *BioMed Research International,* Article ID 695283, 1-9.

Makioka A, M. Kumagai, S. Kobayashi, T. Takeuchi. 2004. Different effects of cytochalasins on the growth and differentiation of *Entamoeba invadens. Parasitology Research.* **93**: 68–71.

Malecki JM, A. Bentke, B. Ostrowska and P. Laidler P. 2010. Cytochalasin D, LY294002 and olomoucine synergize in promoting death of melanoma cells through activation of caspase-3 and apoptosis. *Melanoma Res.* **20**: 52-58.

Manabu N and N. Tsuneo. 1991. Productivity of Pyrichalasin H, a Phytotoxic Metabolite, from Different Isolates of *Pyricularia grisea* and from Other Isolates of *Pyricularia* spp. *Agricultural and Biological Chemistry.* **55**: 1899-1900.

Manfred S. 1982. Action of Cytochalasin D on Cytoskeletal Networks. *The Journal of Cell Biology.* **92**: 79-91.

Mani S, M. Macapinlac Jr, S. Goel, D. Verdier-Pinard, T. Fojo, M. Rothenberg, D. Colevas. 2004. The clinical development of new mitotic inhibitors that stabilize the microtubule. *Anticancer Drugs.* **15**: 553-8.

Manito H, Matsumoto M and Katayama T. 1973. *J. Chem. Soc. Perkin Trans.* 1.p: 1819.

Mann J. 2001. Natural products as immunosuppressive agents. *Nat. Prod. Rep.***18**: 417-430.

Marc E. Wiles, JA. Dykens and CD. Wright. 1994. Regulation of polymorphonuclear leukocyte membrane fluidity: effect of cytoskeletal modification. *Journal of Leukocyte Biology* . **56:** 192-199.

Marina G, S. Kletukhina, S. Kurbangaleeva and A. Rizvanov. 2018. Evaluation of cytochalasin B induces membrane vesicles fusion specificity with target cells. *BioMed Research International.* **10**:1-6.

Martin F, NO Olsson and JF Jeannin. 1981. Effects of Four Agents that Modify Microtubules and Microfilaments (Vinblastine, Colchicine, Lidocaine, and Cytochalasin B) on Macrophage-mediated Cytotoxicity to Tumor Cells. *Cancer Immunol. Immunother.* **10**: 113-119.

Martin GG and P. Talbot .1981. Drugs that block smooth muscle contraction inhibit in vivo ovulation in hamsters. *J Exp Zool.* **216**: 483-91.

Maruyama K, M. Oosawa, A. Tashiro, T. Suzuki, M. Tanikawa, M. Kikuchi and S. Natori. 1986. Effects of chaetoglobosin J on the G-F transformation of actin. *Biochim Biophys Acta.* **874:** 137-143.

Maslow DE and E. Mayhew. 1972. Cytochalasin B Prevents Specific Sorting of Reaggregating Embryonic Cells. *Science.* **177**:281-282.

Masuda M, Komiyama Y, Nishikado H, Kuriki H, Egawa H, Murata K.1989. Change of membrane fluidity of rat neutrophils accompanying *Escherichia coli* inoculation. *J Leukoc Biol.* **46(2):**169-74.

May JA, H. Ratan, JR. Glenn, W. Lösche, P. Spangenberg, S. Heptinstall. 1998. GPIIb-IIIa antagonists cause rapid disaggregation of platelets pre-treated with cytochalasin D. Evidence that the stability of platelet aggregates depends on normal cytoskeletal assembly. *Platelets.* **9**: 227-32.

McMullin DR, MW. Sumarah and JD. Miller. 2013. Chaetoglobosins and azaphilones produced by Canadian strains of *Chaetomium globosum* isolated from the indoor environment. *Mycotoxin Res.* **29**:47-54.

Melissa MW, J. Corwin, G. Strobel and J. Clardy.2000. Three New Cytochalasins Produced by an Endophytic Fungus in the Genus *Rhinocladiella. J. Nat. Prod.* **63**: 1692-1695.

Mendell MJ, AG. Mirer, K. Cheung, M. Tong and J. Douwes. 2011. Respiratory and allergic health effects of dampness, mold, and dampness-related agents: a review of the epidemiologic evidence. *Environ. Health Perspect.***119**: 748–756.

Merifield E and EJ .Thomas. 1990. Cytochalasan synthesis: total synthesis of cytochalasin D *J. Chem. Soc., Chem. Commun.* **6**: 464-466.

Merifield E and EJ .Thomas. 1999. Total synthesis of cytochalasin D: total synthesis and full structural assignment of cytochalasin O. *J. Chem. Soc., Perkin Trans. 1*:**22**: 3269-3283.

Minato H and M. Matsumoto. 1970. Studies on the Metabolites of *Zygosporium Masonii.* I. Structure of Zygosporin A. *J Chem Soc Perkin.* **11**: 38-45.

Mitchell T, A. Lo, MR. Logan, P. Lacy and G. Eitzen. 2008. Primary granule exocytosis in human neutrophils is regulated by Rac-dependent actin remodeling. *Am. J. Physiol. Cell Physiol.* **295:**1354-C1365.

Mollinedo F and C. Gajate. 2003. Microtubules, microtubule-interfering agents and apoptosis. *Apoptosis*. **8**: 413–450.

Mooberry SL. 2007. Strategies for the development of novel Taxol-like agents. *Methods Mol. Med.* **137:**289–302.

Mookerjee BK and CY Jung. 1983. Effects of cytochalasins on lymphocytes: Mechanism of inhibition of rosette formation. *J. Immunol.* **131**(3):1126-30.

Mookerjee BK, Cuppoletti J, Rampal AL, Jung CY. 1981. The effects of cytochalasins on lymphocytes. Identification of distinct cytochalasin-binding sites in relation to mitogenic response and hexose transport. *J Biol Chem.* **256**:1290-1300.

Morris A and Tannenbaum J. 1980. Cytochalasin D does not produce net depolymerization of actin filaments in HEp-2 cells. *Nature*. 287(5783):637-9.

Mortensen K and LI. Larsson. 2003. Effects of cytochalasin D on the actin cytoskeleton: association of neoformed actin aggregates with proteins involved in signalling and endocytosis. *Cell. Mol. Life Sci.* **60:** 1007-1012.

Muller AM, JE. Reimann, U. Schunacher and K. Schwadorf. 1998. Natural occurrence of *Fusarium* toxins in oats harvested during five years in an area of southwest Germany. *Food. Addi. Contam.* **15**: 801-806.

Muller P, AM. Chambaut-Guerin and B. Rossignol. 1985. Comparative effects of cytochalasin D on the protein discharge induced by alpha- and beta-adrenergic or cholinergic agonists in rat exorbital lacrimal glands. *Biochim. Biophys. Acta.* **844**: 158-166.

Murono EP, T. Lin, J. Osterman and HR Nankin. 1982. Relationship between inhibition of interstitial cell testosterone synthesis by cytochalasin B and glucose. *Biochemical and Biophysical Research Communications.* **104**: 1: 299-306

Nakajima M, E. Tamura, T. Irimura, S. Toyoshima and H. Hirano. 1981. Mechanism of the Concanavalin A-Induced Change of Membrane Fluidity of Chicken Erythrocytes. *J. Biochem.* **89:** 665-675.

Namatame I, H. Tmoda and M. Arai. 2000. Effect of Fungal Metabolites Cytochalasans on Lipid Droplet Formation in Mouse Macrophages. Omura, S. *J. Antibiot.* **53**: 19-25.

Natarajan JA .May, HM. Sanderson, M. Zabe , P. Spangenberg, S. Heptinstall.2000. Effects of Cytochalasin H, a Potent Inhibitor of Cytoskeletal Reorganisation, on Platelet Function. *Platelets* **11** : 467-476.

Natori S. 1997. In Mycotoxins in human health and animal health. (J.V. Rodricks, C.W. Hesseltine and M.A. Mehlman eds.). Pathotox Publ., Park Forest South. Illinois.

Nessa B, J. Fallon, K. McLoughlin and K. Kavanagh . 2011. Disruption of haemocyte function by exposure to cytochalasin b or nocodazole increases the susceptibility of Galleria mellonella larvae to infection. *Microbes Infect.* **14**:1191-1198.

Nicolaou KC, MW. Härter, JL Gunzner and A. Nadin A. 1997. The Wittig and Related Reactions in Natural Product Synthesis . *Liebigs Ann.Recueil.* **7**: 1283-1301

Nielsen KF, S. Gravesen, PA. Nielsen, B. Andersen, U. Thrane and JC. Frisvad. 1999. Production of mycotoxins on artificially and naturally infested building materials. *Mycopathologia.* **145**:43–56.

Nobuaki N, Y. Haruaki , M. Shinji, S. Yosuke and O. Toshikazu. 1993. Aspochalasin E, a new antibiotic isolated from a fungus. *The Journal of Antibiotics.* **46**: 679-681.

Nukina M, 1987. Pyrichalasin H, a New Phytotoxic Metabolite Belonging to the Cytochalasins from *Pyricularia grisea* (Cooke) Saccardo. *Agric. Biol. Chem.*51: 2625–2628.

Numata A, C. Takahashi, Y. Ito, K. Minoura, T. Yamada, C. Matsuda and K. Nomoto. 1996. Penochalasins, a novel class of cytotoxic cytochalasans from a *Penicillium* species separated from a marine alga: structure determination and solution conformation. *J. Chem. Soc., Perkin Trans.* **3:** 239-245.

Ohmori H, S. Toyama and S. Toyama. 1992. Direct proof that the primary site of action of cytochalasin on cell motility processes is actin. *J. CellBiol.* **116**:933-941.

Ohno YI, KI. Hirai, T. Kanoh, H. Uchino and K. Ogawa .1982. Sub cellular localization of hydrogen peroxide production in human polymorphonuclear leukocytes stimulated with lectins, phorbol myristate acetate, and digitonin: an electron microscopic study using $CeCl_3$. *Blood.* **60**:1195-202.

Ohtsubo K, M. Saito, S. Sekita, K. Yoshihira and S. Natori. 1978. Acute toxic effects of chaetoglobosin A, a new cytochalasin compound produced by *Chaetomium globosum,* on mice and rats. *Jpn J Exp Med.* **48**: 105-10.

Oikawa H and T. Tokiwano. 2004. Enzymatic catalysis of the Diels–Alder reaction in the biosynthesis of natural products. *Nat. Prod. Rep.* **21**:321-352.

Oikawa H, Y. Murakami and A. Ichihara . 1992. Biosynthetic study of chaetoglobosin A: origins of the oxygen and hydrogen atoms, and indirect evidence for biological Diels–Alder reaction. *J. Chem. Soc., Perkin Trans.***1**: 29-55.

Olorundare OE, SR Simmons and RM Albrecht.1992. Cytochalasin D and E: effects on fibrinogen receptor movement and cytoskeletal reorganization in fully spread, surface-activated platelets: a correlative light and electron microscopic investigation. *Blood.* **79**:99-109.

Oppermann U, C. Filling, M. Hult, N. Shafqat, X. Wu, M. Lindh, J. Shafqat, E. Nordling, Y. Kallberg, B. Persson and H. Jörnvall. 2003. Short-chain dehydrogenases/ reductases (SDR): The 2002 update. *Chem. Biol. Interact.* **143**:247–253.

Ornelles DA, GF. Edward and P. Sheldon. 1986. Cytochalasin Releases mRNA from the Cytoskeletal Framework and Inhibits Protein Synthesis. *Molecular and Cellular Biology*. **6**: 1650-1662.

Padhye SB, CR. Joshi and GS. Pendse. 1975. Thin-layer chromatography of the cytochalasins. *J. Chromatogr*. **108:** 405-408.

Patwardhan BH, M. Flashner, CA. Miller and SW. Tanenbaum. 1982. Structure activity co relationship of cytochalasins. Novel halogenated and related CC and CD derivatives. *J. Med. Chem.* **25**: 663–666.

Patwardhan SA, RC. Pandey and S. Dev.1974. Toxic cytochalasins of *Phomopsis paspalli,* a pathogen of Kodo millets. *Phytochemistry*. **13**:1985–1988.

Paul AI and AM. Koachman. 1982. Cytochalasin B Enhances Hormone and Cholera Toxin-stimulated Cyclic AMP Accumulation in S49 Lymphoma Cell. *Journal of Biolocical Chemistry*. **257**: 9717-9723.

Peter G, W. Plagemenn and RD Estensen. 1973. Cytochalasin B. VI. Competitive inhibition of nucleoside transport by cultured Novikoff rat hepatoma cells. *The Journal of Cell Biology*. **55**: 179-185.

Petersen LM, TB. Tanja, C. Dürr , M. Seiffert, JC. Frisvad, CH. Gotfredsen and TO. Larsen. 2014. Isolation, Structural Analyses and Biological Activity Assays against Chronic Lymphocytic Leukemia of Two Novel Cytochalasins - Sclerotionigrin A and B. *Molecules*. **19**: 9786-9797

Peterson JR and TJ. Mitchison. 2002. Small molecules, big impact: a history of chemical inhibitors and the cytoskeleton. *Chemistry & Biology*. **9**: 1275-1285.

Plagemann PGW and J. Erbe. 1973. Transport and metabolism of glucosamine by cultured Novikoff rat hepatoma cells and effects on nucleotide polls. *Cancer Res.* **33**: 482-492.

Polizzi V, B. Delmulle, A. Adams, A. Moretti, A. Susca, AM. Picco, Y. Rosseel, R. Kindt, J. Van Bocxlaer, N. De Kimpe, C. Van Peteghem and S. De Saeger. 2009. JEM Spotlight: fungi, mycotoxins and microbial volatile organic compounds in mouldy interiors from water damaged buildings. *J. Environ. Monit.***11**: 1849–1858.

Pongcharoen W, V. Rukachaisirikul, M. Isaka and K. Sriklung . 2007. Cytotoxic metabolites from the wood-decayed fungus Xylaria sp. BCC 9653. *Chem Pharm Bull (Tokyo)*. **55**:1647-8.

Pongcharoen W, V.Rukachaisirikul, S. Phongpaichit, N. Rungjindamai and J. Sakayaroj. 2006. Pimarane Diterpene and Cytochalasin Derivatives from the Endophytic Fungus *Eutypella scoparia* PSU-D44. *J. Nat. Prod.* **69** : 856-858.

Prentki M, C. Chaponnier, B. Jeanrenaud, and G. Gabbiani. 1979. Effect of Phalloidin and Cytochalasin D. *J. Cell Biolog*. **81**: 592-607.

Probst A and C. Tamm. 1981. 1982.Chaetoglobosin L, a New Metabolite of *Diplodia macrospora*. HCA. **65**: 1543-1546.

Probst A and C. Tamm. 1981. Biosynthesis of the Cytochalasans. Biosynthetic studies on chaetoglobosin A and 19-*O*-acetylchaetoglobosin A. *Helv. Chim. Acta*. **64**: 2065-2077.

Qiao KJ, YH. Chooi and Y. Tang. 2011. Identification and engineering of the cytochalasin gene cluster from *Aspergillus clavatus* NRRL 1. *Met ab. Eng.* **13**:723-32

Qin JC, YM. Zhang, JM. Gao, MS. Bai, SX. Yang, L. Hand and AL. Zhang. 2009. Bioactive metabolites produced by *Chaetomium globosum*, an endophytic fungus isolated from *Ginkgo biloba Bioorg. Med. Chem. Lett.* **19**:1572-1574.

Rainey WE, JW. Shay and JI. Mason.1984. The effect of cytochalasin D on steroid production and stress fiber organization in cultured bovine adrenocortical cells. *Mol Cell Endocrinol.***35**:189-97.

Rainey WE, RE. Kramer, JI. Mason and JW. Shay. 1985. The effects of taxol, a microtubulestabilizing drug, on steroidogenic cells. *J Cell Physiol.* **123**:17-24.

Rampal AL, HB. Pinokofsky and CY Jung. 1980. Structure of cytochalasins and cytochalasin B binding site in human erythrocyte membranes. *Biochemistry.* **19:**679-683.

Rasmussen I and V. Frederik. 2015. Macropinocytosis Is the Entry Mechanism of Amphotropic *Murine Leukemia* Virus. *J Virol.* **89**: 1851-1866.

Ressom HW, JF. Xiao, L. Tuli, RS. Varghese, B. Zhou, TH. Tsai, MR. Ranjbar, Y. Zhao, J. Wang, C. Di Poto, AK. Cheema, MG. Tadesse, R. Goldman and K. Shetty. 2012. Utilization of metabolomics to identify serum biomarkers for hepatocellular carcinoma in patients with liver cirrhosis. *Analytica chimica acta.* **743:** 90-100.

Richard HJB and B. Hidde. 2001. Effects of local cytochalasin D delivery on smooth muscle cell migration and on collar-induced intimal hyperplasia in the rabbit carotid artery. *Br J Pharmacol.* **134**: 473-483.

Robens J and RT. Riley. 2002. Introduction: afl atoxin/fumonisin elimination and fungal genomics workshops, Phoenix, Arizona. *Mycopathologia* .**155:** 1-3.

Robert JL and Tamm C. 1975. Biosynthesis of cytochalasans. Part 5. The incorporation of deoxaphomin into cytochalasin B (phomin). *Helv Chim Acta.* **58**:2501-2504.

Rochfort S, J. Ford, S. Ovenden,SS. Wan, S. George, H. Wildman, RM. Tait, B. Meurer-Grimes B, S. Cox, J. Coates and D. Rhodes. 2005. A novel aspochalasin with HIV-1 integrase inhibitory activity from *Aspergillus flavipes*. *J. Antibiot. (Tokyo)*, **58**: 279-283.

Rothweiler W and CH. Tamm. 1966. Isolation and structure of Phomin. *Experientia.* **22**: 750-752.

RothweilerW and Tamm C. 1970. Isolierung und Struktur der Antibiotica Phomin und 5-Dehydrophomin . *Helv. Chim. Acta.* **53**: 696-724.

Rubtsova SN, RV. Kondratov, PB. Kopnin, PM. Chumakov, BP. Kopnin and JM Vasiliev. 1998. Disruption of actin microfilaments by cytochalasin D leads to activation of p53. *FEBS Lett.* **430**:353-7.

Rui L, Q. Gu, W. Zhu, C. Cui, G. Fan, Y. Fang, T. Zhu and H. Liu 2006. 10-Phenyl-[12]-cytochalasins Z_7, Z_8, and Z_9 from the Marine-Derived Fungus *Spicaria elegans*. *J. Nat. Prod.* **69** : 871–875

Rui L, Z. Lin, T. Zhu, Y. Fang, Q. Gu and W. Zhu. 2008. Novel Open-Chain Cytochalsins from the Marine-Derived Fungus *Spicaria elegans. J. Nat. Prod.* **71** : 1127-1132

Salu KJ, JM. Bosmans, YM. Huang, M. Hendriks, M. Verhoeven and A. Levels. 2006. Effects of cytochalasin D-eluting stents on intimal hyperplasia in a porcine coronary artery model. *Cardiovasc Res.* **69:** 536-544.

Sampath P and TD. Pollard. 1991. Effects of cytochalasin, phalloidin and pH on the elongation of actin filaments. *Biochemistry.* **30**: 1973-1980.

Sampson K, JD. Pickett-Heaps and A. Forer.1996. Cytochalasin D blocks chromosomal attachment to the spindle in the green alga Oedogonium. *Protoplasma.* **192**: 130-144.

Satoru Y, K. Tomita, T. Kudo, S. Ando and M. Yamazaki. 2005. Induction of multicellular 3-D spheroids of MCF-7 breast carcinoma cells by neutrophil-derived cathepsin G and elastase. *Cancer Sci.* **96**: 560–570.

Sauter R, EJ. Thomas and JP. Watts. 1989. Cytochalasan Synthesis: An Alternative Approach to Cytochalasin H. *J. Chem. Soc., Chem. Commun.* **18**: 1449-1450 .

Scherlach K, D. Boettger, N. Remme and C. Hertweck. 2010. The chemistry and biology of cytochalasins. *Nat. Prod. Rep.* 27:869– 886.

Schmidlin T, P. E. Burckhardt, N. Waespe-Sarcevic and C. Tamm. 1983. Approaches to the Synthesis of Cytochalasans. Part 4. Improved Synthesis of the Tetrahydroisoindoline Subunits Related to the Cytochalasins and Aspochalasins. *Helv. Chim. Acta.* **66**: 450-465.

Schuemann J and C. Hertweck. 2006. Advances in cloning, functional analysis and heterologous expression of fungal polyketide synthase genes. *J. Biotechnol.* **124**: 690–703.

Schuemann J and C. Hertweck. 2007. Molecular Basis of Cytochalasan Biosynthesis in Fungi: Gene Cluster Analysis and Evidence for the Involvement of a PKS-NRPS Hybrid Synthase by RNA Silencing. *J. Am. Chem. Soc.* **129 :** 9564–9565.

Sekita S, K. Yoshihira and S. Natori. 1983. Chaetoglobosins, cytotoxic 10-(indol-3-yl)-[13]cytochalasans from *Chaetomium* spp. IV. Carbon-13 nuclear magnetic resonance spectra and their application to a biosynthetic study. *Chem Pharm Bull.* **31**:490-498.

Sekita S, K. Yoshihira, S. Natori and H. Kuwano. 1982. Chaetoglobosins, cytotoxic 10-(indol-3-yl)-[13]cytochalasans from *Chaetomium* spp. III. Structures of chaetoglobosins C, E, F, G, and J *Chem. Pharm. Bull.* **30**: 1629- 1638.

Sekita S, K.Yoshihira, S. Natori, S. Udagawa, F. Sakabe, H. Kurata, M. Umeda, 1982. Chaetoglobosins, cytotoxic 10-(indol-3-yl)-[13]cytochalasans from Chaetomium spp. I. Production, isolation and some cytological effects of chaetoglobosins A–J *Chem. Pharm. Bull.* **30**:1609-1617.

Seshime Y, PR. Juvvadi, M. Tokuoka, Y. Koyama, K. Kitamoto, Y. Ebizuka and I. Fujii. 2009. Functional expression of the *Aspergillus flavus* PKS-NRPS hybrid CpaA involved in the biosynthesis of cyclopiazonic acid. *Med. Chem. Lett.* **19**: 3288-3292.

Sheikh S and Nash GB. 1998. Treatment of neutrophils with cytochalasins converts rolling to stationary adhesion on P-selectin. *J Cell Physiol.* **174** :206-16.

Sheng G, Y. Chen, L. Han, Y. Huang, X. Liu, L.Z. Mao. 2016. Encapsulation of indocyanine green into cell membrane capsules for photothermal cancer therapy. *Acta Biomaterialia*. **43**: 251–261

Shu Xu, HM. Ge, Y. Chun Song, Y. Shen and RX. Tan. 2009. Cytotoxic cytochalasin metabolites of endophytic *Endothia gyrosa. Chem Biodivers.* **6** :739-45

Shurety W, NA. Bright and JP. Luzio.1996. The effects of cytochalasin D and phorbol myristate acetate on the apical endocytosis of ricin in polarised Caco-2 cells. *J Cell Sci.* **109**: 2927-35.

Sims JW, JP. Fillmore, DD. Warner and EW. Schmidt EW. 2005. Equisetin biosynthesis in *Fusarium heterosporum. Chem. Commun.* **14**: 186–188.

Smiley S and RA. Cloney. 1985. Ovulation and the fine structure of the stichopus californicus (Echinodermata: Holothuroidea) fecund ovarian tubules. *Biol Bull.* **169**: 342-364.

Smith GF, MA. Ridler and JA. Faunch. 1967. Action of Cytochalasin B on Cultured Human Lymphocytes. *Nature.* **216** : 1134-1135.

Smoragiewicz W, B. Cossette, A. Boutard and K. Krystyniak. 1993. Trichothecene mycotoxins in the dust of ventilation systems in office buildings. *Int Arch Occup Environ Health* . **65**:113-7.

Snyder JA and L. Cohen. 1995. Cytochalasin J affects chromosome congression and spindle microtubule organization in PtK_1 cells. *Cell Motil. Cytoskeleton.* **32**: 245–257.

Song Z, RJ. Cox, CM. Lazarus and TJ. Simpson. 2004, Fusarin C biosynthesis in Fusarium moniliforme and *Fusarium venenatum. Chem BioChem* **5**:1196–1203.

Springer JP, J. Clardy, JM. Wells, RJ. Cole, JW. Kirksey, RD. MacFarlane and DJ. Torgerson. 1976. Isolation and structure determination of the mycotoxin chaetoglobosin C, a new [13] cytochalasan. *Tetrahedron Lett.* **17**: 1355–1358.

Staunton J and KJ Weissman. 2001. Polyketide biosynthesis: a millennium review. *Nat. Prod. Rep.* **18:** 380–416.

Stefan B, A. Encinas, J. Keck and CF. Nising. 2009. Chemistry and Biology of Mycotoxins and Related Fungal Metabolites. *Chem. Rev.* **109**: 3903-3990.

Stephen L. S. Brenner and DK. Edward. 1979. Substoichiometric Concentrations of Cytochalasin D Inhibit Actin Polymerization. *The Journal of Biological Chemistry.* **254**: 9982-9985.

Steyn PS, FR. van Heerden and CJ Rabie. 1982. Cytochalasins E and K, Toxic metabolites from *Aspergillus clavatus. J. Chem. Soc., Perkin Trans.* **1**: 541– 544.

Stocking EM and RM. Williams. 2003. Chemistry and biology of biosynthetic Diels-Alder reactions. *Angew Chem Int Ed Engl.* **14**:3078-115.

Stork G and E. Nakamura. 1983. A simplified total synthesis of cytochalasins via an intramolecular Diels-Alder reaction. *J. Am. Chem. Soc.* **105**:5510-5512.

Stork G, Y. Nakahara, Y. Nakahara and WJ Greenlee. 1978. Total synthesis of cytochalasin B. *J. Am. Chem. Soc.* **100**: 7775- 7777.

Susan LS, B. Norman and JF. Strauss. 1984. Re-evaluation of the Effects of Cytochalasins on Steroidogenesis: Studies on Hamster Granulosa Cells. *Endocrinology.* **115**: 1511-1516.

Svingen T and AM Vinggaard. 2016. The risk of chemical cocktail effects and how to deal with the issue. *J. Epidemiol. Commun. Health.* **70**: 322-323.

Tamm C. 1978. *Front. Biol.* **46:**15–51.

Tan RX and WX. Zou. 2001. Endophytes: a rich source of functional metabolites. *Nat. Prod. Rep.* **18:** 448-459.

Tanenbaum SW. In Tanenbaum, SW, Ed. 2010. Cytochalasins, Biochemical and Cell Biological Aspects; North-Holland Publishing: Amsterdam, 1978 (b) Scherlach K, Boettger D, Remme N and Hertweck C. *Nat. Prod. Rep.* 27: 869.

Tao J, M.Wang, L. Li, J. Si, B. Song, C. Zhou, M. Yu, X. Wang, Y. Zhang, G.Ding and Z. Zou· 2016. Over expression of the Global Regulator LaeA in *Chaetomium globosum* Leads to the Biosynthesis of Chaetoglobosin Z. *J. Nat. Prod.* **79** : 2487-02494.

Tarman K, GJ. Palm, K. Wende, U. Lindequist. 2011. Biological and chemical study of two Indonesian marine endophytic fungi. *Planta Med.* **77** - SL71.

Thohinung S, S. Kanokmedhakul, K. Kanokmedhakul, V. Kukongviriyapan, O. Tusskorn and K. Soytong. 2010. Cytotoxic 10-(indol-3-yl)-[13]cytochalasans from the fungus *Chaetomium elatum* ChE01 *Arch. Pharm. Res.* **33**: 1135-1141.

Thomas EJ and JWF Whitehead. 1986. Cytochalasan synthesis: total synthesis of cytochalasin H. *J. Chem. Soc., Chem. Commun.* **9**: 727-728.

Thomas EJ. 1991. Cytochalasan synthesis: macrocycle formation via intramolecular Diels-Alder reactions. *Acc. Chem. Res.* **24**:229-235.

Tian Q, S. Pahlavan, K. Oleinikow, J. Jung, S.Ruppenthal, A. Scholz, C. Schumann, A. Kraegeloh, M. Oberhofer, P. Lipp and L. Kaestner. 2012. Functional and morphological preservation of adult ventricular myocytes in culture by sub-micromolar cytochalasin D supplement. *Journal of Molecular and Cellular Cardiology.* **52**: 113-124.

Tischer C, CM. Chen and J. Heinrich. 2011. Association between domestic mould and mould components, and asthma and allergy in children: a systematic review. *Eur. Respir. J.* **38**: 812–824.

Tojyo Y, K. Okumura, M. Kanazawa and Y. Matsumoto. 1989. Effect of cytochalasin D on acinar cell structure and secretion in rat parotid salivary glands *in vitro. Archives of Oral Biology.* **34** : 847-855

Tomas F. 1981. Structures of cytochalasin k, l and m, isolated from *Chalara microspora. Tetrahedron Letters.* **22**: 2703-2706.

Tomikawa T, K. Shin-Ya, H. Seto, N. Okusa, T. Kajiura and Y. Hayakawa. 2002. Structure of aspochalasin H, a new member of the aspochalasin family. *J. Antibiot. (Tokyo).* **55**: 666-668.

Tomikawa T, K. Shin-Ya, H. Seto, T. Kinoshita, A. Miyajima, H. Seto and Y. Hayakawa. 2001. Selective cytotoxicity and stereochemistry of aspochalasin D. *J. Antibiot. (Tokyo).* **54:** 379-381.

Tominaga M, K. Morita, E. Sonobe, E. Yokata and T. Shimmen. 1997. Microtubules regulate the organization of actin filaments at the cortical region in root hair cells of Hydrocharis. *Protoplasma.* **199**: 83-92.

Tomioka T, Y. Izawa, K. Koyama and S. Natori. 1987. Three New 10-Phenyl[11] cytochalasins, Cytochalasins N, O, and P from *Phomopsis* sp. *Chem. Pharm. Bull.* **35**:902-905.

Tomoda H, I. Namatame, N. Tabata, K. Kawaguchi, S. Si and S. Omura. 1999. Phenochalasins, Inhibitors of Lipid Droplet Formation in Mouse Macrophages, Produced by *Phomopsis* sp. FT-0211. *J. Antibiot.* **52**: 851–856.

Trendowski M, G. Yu, V. Wong, C. Acquafondata, T. Christen, TP. Fondy. 2014. The real deal: Using cytochalasin B in sonodynamic therapy to preferentially damage leukemia cells. *Anticancer Res.* **34:**2195–2202.

Trendowski M, V. Wong, G. Yu, TP. Fondy. 2015. Enlargement and multinucleation of U937 leukemia and MCF7 breast carcinoma cells by antineoplastic agents to enhance sensitivity to low frequency ultrasound and to DNA-directed anticancer agents. *Anticancer Research.* **35**:65-76.

Trendowski M. 2014. The promise of sonodynamic therapy. *Cancer Metastasis Rev.* **33:** 143–160.

Trendowski M. 2015. Using Cytochalasins to Improve Current Chemotherapeutic Approaches. *Anticancer Agents Med Chem.* **15**: 327–335.

Trendowski M. 2015. The inherent metastasis of leukaemia and its exploitation by sonodynamic therapy. *Critical Reviews in Oncology/Hematology.* **94**:149-163.

Trost BM and FW Gowland. 1979. An approach to enolonium equivalents. Application to a total synthesis of (.+-.)-pyrenophorin. *J. Org. Chem.* **44**: 3448–3450.

Trost BM and SJ. Brickner. 1983. Palladium-assisted macrocyclization approach to cytochalasins: a synthesis of antibiotic A26771B. *J. Am. Chem. Soc.* **105 :** 568–575.

Trost BM, M. Ohmori, SA. Boyd, H. Okawara and SJ. Brickner. 1989. Palladium-catalyzed synthesis of macrocycles. A total synthesis of (-)-aspochalasin B. *J. Am. Chem. Soc.***111**: 8281-8284.

Trubnikova OB. 2003. Effects of Cycloheximide, Aminoglutethimide, Cytochalasin B, and Colchicine on Ovulation and the Ultrastructure of the Ovarian Follicle Wall in the Stellate Sturgeon *Acipenser stellatus* Pall. *Russian Journal of Developmental Biology.* **34**: 110–120

Udagawa S, T. Muroi, H. Kurata, S. Sekita, H. Yoshihira, S. Natori and M. Umeda. 1979. The production of chaetoglobosins, sterigmatocystins, O-methyl sterigmatocystin and chaetocin by *Chaetomium* spp. and related fungi. *Can J Microbiol.* **25**: 170-177.

Udagawa T, J. Yuan, D. Panigrahy, YH. Chang, J. Shah, RJ. D'Amato. 2000. Cytochalasin E, an epoxide containing Aspergillus-derived fungal metabolite, inhibits angiogenesis and tumor growth. *J Pharmacol Exp Ther.* **294**: 421–427.

Umeda M, K. Mastumoto, M. Siato, S.sekita, K. Yoshilira, S. Natori, S.S. Adgaura, F.Sakabe and H. Kurata. 1975. *Exprrientia.* **31**: 435-438.

Vassiliev I, A.Tsui, W. Xian Kang, S. McIlfatrick and MB. Nottle. 2014. Effect of cytochalasin types on the production of heterozygous parthenogenetic porcine embryos and the isolation of putative parthenogenetic embryonic stem cells. *Stem Cell Biol Res.* **1:**1-7.

Vedejs E and S. Ahmad. 1988. Synthesis of the 11-membered cytochalasin ring system by modified reformatsky cyclization. *Tetrahedron Letters.* 29: 2291-2294.

Vedejs E and SJ Wittenberger. 1990. The Total Syntheses of dl-Zygosporin E and dl-C 8-Desmethylcytochalasin D. *J. Am. Chem. SOC.* **112**: 4357-4364.

Vedejs E and RC. Gadwood.1978. An approach to cytochalasins: Diels-Alder addition of alpha, beta - unsaturated imides. *J. Org. Chem.* **43:** 376-378.

Vedejs E, JD. Rodgers and SJ. Wittenberger. 1988. A sulfur-mediated total synthesis of Zygosporin E. *J. Am. Chem. Soc.* **110** : 4822-4823.

Vedejs E, JG. Reid, JD Rodgers and SJ Wittenberg.1990. Synthesis of cytochalasins: the route to sulfur-bridged [11]cytochalasins. *J. Am. Chem. Soc.* **112**:4351- 4357

Vedejs E and Reid JG. 1984. Total synthesis of carbocyclic cytochalasins. *J. Am. Chem. Soc.* 1984, 106, 4617-4618.

Vederas JC, W. Graf, L. David and C. Tamm 1975. Biosynthesis of cytochalasans. Part 4. The mode of incorporation of common naturally-occurring carboxylic acids into cytochalasin D1 *Helv. Chim. Acta.* **58**: 1886-1898.

Vesely D, D. Vesela and R. Jelinek. 1995. *Penicillium aurantiogriseum* Dierckx produces chaetoglobosin A toxic to embryonic chickens. *Mycopathologia.* **132:** 31-33.

Vilcinskas A, V. Matha and P. Got. 1997. Effects of the entomopathogenic fungus *Metarhizium anisopliae* and its secondary metabolites on morphology and cytoskeleton of plasmatocytes isolated from the greater wax moth, *Galleria mellonella. J. Insect Physiol.* **43**:1149–1159.

Vishwanath V, M. Sulyok, G. Weingart, B. Kluger, M. Teaubel, S. Mayer, R. Schuhmacher and R. Krska. 2011. Evaluation of settled floor dust for the presence of microbial metabolites and volatile anthropogenic chemicals in indoor environments by LC-MS/MS and GC-MS methods. *Talanta.* **85**: 2027-2038.

Vishwanath V, M. Sulyok, R. Labuda, W. Bicker and R. Krska. 2009. Simultaneous determination of 186 fungal and bacterial metabolites in indoor matrices by liquid chromatography/tandem mass spectrometry, *Anal. Bioanal. Chem.* **395**:1355-1372.

VonWallbrunn C, Luftmann H, Bergander K, Meinhardt F. 2001. Phytotoxic chaetoglobosins are produced by the plant pathogen *Calonectria morganii* (anamorph *Cylindrocladium scoparium*) *J. Gen. Appl. Microbiol.* **47**: 33- 38.

Vurro M, A. Bottalico, R. Capasso and A. Evidente. 1997. Cytochalasins from phytopatogenic *Ascochyta* and *Phoma species.* Upadahyay RK, Mukerji KG Toxins in Plants Disease Development and Evolving Biotechnology. New Delhi; Oxford & IBH Publishing Co. 127-147.

Wang FZ, HJ. Wei, TJ. Zhu, DH. Li, ZJ.Lin and QQ. Gu. 2011. Three New Cytochalasins from the Marine-Derived Fungus *Spicaria elegans* KLA03 by Supplementing the Cultures with L- and D-Tryptophan. *Chemistry & Biodiversity.* **8:** 887-894.

Wei G, D. Tan, C. Chen, Q.Tong, X. Li, J. Huang, J. Liu, Y. Xue, J. Wang, Z. Luo, H. Zhu and Y. Zhang. 2017. Flavichalasines A–M, cytochalasan alkaloids from *Aspergillus flavipes. Scientific Reports.* **7**: 1-12.

Wei H, Y. Xu, PE. Artiles, MX. Liu, JG. Luo, J M. U'Ren, AE Arnold and AAL Gunatilaka. 2015. Sesquiterpenes and other constituents of *Xylaria* sp. NC1214, a fungal endophyte of the moss *Hypnum* sp. *Phytochemistry*. **118**: 102–108.

Weidenbörner M. 2001. Encyclopedia of food mycotoxins. Springer. Berlin, Germany.

Wells JM, HG Cutler and RJ Cole. 1976. Toxicity and plant growth regulator effects of cytochalasin H isolated from *Phomopsis* sp. *Can. J.Microbiol.* **22**: 1137–1143.

Wells JM, RJ. Cole, HG Cutler and DH Spalding. 1981. *Curvularia lunata,* a new source of cytochalasin B. *Appl Environ. Microbial.* **41**: 967-971.

WHO. 2009. Dampness and Mould -WHO Guidelines for Indoor Air Quality, Geneva, World Health Organization.

Wieland T and I. Low. 1980. Concentration-Dependent Influence of various Cytochalasins and Chaetoglobosins on the Phalloidin-Induced Polymerization of G-Actin in 0.6 M Potassium Iodide. *Biochemistry*. **19**: 3363-3367.

Wiles ME, JA. Dykens and CD. Wright. 1994. Regulation of polymorphonuclear leukocyte membrane fluidity: effect of cytoskeletal modification. *Journal of Leukocyte Biology*. **56**: 192-199.

Xue M, Q. Zhang, JM. Gao, H. Li, JM. Tian and G. Pescitelli. 2012. Chaetoglobosin Vb from endophytic *Chaetomium globosum*: absolute configuration of chaetoglobosins. *Chirality*. **24:**668-74.

Yahara I, F. Harada, S. Sekita, K. Yoshihira and S. Natori. 1982. Correlation between effects of 24 different cytochalasins on cellular structures and cellular events and those on actin in vitro. *J Cell Biol*. **92**: 69–78.

Yang Hu, Weipu Zhang, Ping Zhang, Weibin Ruan,and Xudong Zhu. 2013. Nematicidal Activity of Chaetoglobosin A Poduced by Chaetomium globosum NK102 against *Meloidogyne incognita*. *J. Agric. Food Chem*. **61**: 41–46.

Ying L, S. Zhao, W. Ding, P.i Wang, X.n Yang and J. Xu. 2014. Methylthio-Aspochalasins from a Marine-Derived Fungus *Aspergillus* sp. *Mar. Drugs*. **12**: 5124-5131.

Yu Li, C. Lu, Y. Huang, Y. Li and Y. Shen. 2012. Cytochalasin H2, a New Cytochalasin, Isolated from the Endophytic Fungus *Xylaria* sp. A23. *Rec. Nat. Prod*. **6**:2121-126.

Zhang D, H. Ge, D. Xie, R. Chen, J. Zou, X. Tao and J. Dai. 2013. Periconiasins A–C, New Cytotoxic Cytochalasans with an Unprecedented 9/6/5 Tricyclic Ring System from Endophytic Fungus *Periconia* sp. *Org. Lett.* **15** :1674–1677.

Zhang GH, Q. Wang, JJ. Chen, XM. Zhang, SC. Tam and YT. Zheng. 2005. The anti-HIV-1 effect of scutellarin. *Biochem. Biophys. Res. Commun*. **334**:812–816.

Zhang HW, J. Zhang, S. Hu, Z. Jian Zhang, Cheng-Jian Zhu, S. Weng Ng, R. Xiang Tan. 2010. Ardeemins and Cytochalasins from *Aspergillus terreus* residing in *Artemisia annua*. *Planta Med.* **76:** 1616-1621.

Zhang HW, Y. Chun Song and R. Xiang Tan. 2006. Biology and chemistry of endophytes *Nat. Prod. Rep.* **23:** 753–771

Zhang J, HM. Ge, RH. Jiao, J. Li, H. Peng, YR. Wang, JH. Wu, YC. Song and RX. Tan. 2010. Cytotoxic chaetoglobosins from the endophyte *Chaetomium globosum. Planta. Med.* **76**: 1910–1914.

Zhang Q, HQ. Li, SC. Zong, JM. Gao and AL. Zhang. 2012. Chemical and bioactive diversities of the genus *Chaetomium* secondary metabolites. *Mini-Rev. Med. Chem.* **12**:127–148.

Zhang Q, J. Xiao, QQ. Sun, JC. Qin, G. Pescitelli and JM Gao. 2014. Characterization of cytochalasins from the endophytic *Xylaria* sp. and their biological functions. *J Agric Food Chem.* **62**:10962-10969.

Zhang Y, R. Tian, S. Liu, X. Chen, X. Liua and Y. Che. 2008. Alachalasins A-G, new cytochalasins from the fungus *Stachybotrys charatum. Bioorganic & Medicinal Chemistry.* **16**: 2627-2634.

Zhang J, GM. Ge, RH. Jiao, J. Li , H. Peng, YR. Wang , JH. Wu, YC. Song and RX. Tan. 2010. Cytotoxic chaetoglobosins from the endophyte *Chaetomium globosum Planta. Med.* **76:** 1910– 1914.

Zhen JL, GJ. Zhang, TJ. Zhu, R. Liu, HJ. Wei and QQ. Gu. 2009. Bioactive Cytochalasins from *Aspergillus flavipes*, an Endophytic Fungus Associated with the Mangrove Plant *Acanthus ilicifolius. Helvetica Chimica Acta.* **92:**1538-1544.

Zheng CJ, CL. Shao, L. Wu, M. Chen, K.Wang, D. Zhao, X. Sun, G. Chen and C. Wang. 2013. Bioactive Phenylalanine Derivatives and Cytochalasins from the Soft Coral-Derived Fungus, *Aspergillus elegans. Mar. Drugs.* **11:** 2054-2068.

Zheng, Ming-Zhu Kong, Qin Zhao , Guo-Dong Chen, Hai-Yan Tian, Xiao-Xia Li, Liang-Dong Guo, Jia Li, Yi-Zhi Zheng and Hao Gao. 2014. Chaetoglobosin Y, a new cytochalasin from *Chaetomium globosum. Fitoterapia.* **93:** 126–131.

Zhenjian L, T. Zhu, H. Wei, G. Zhang, H. Wang and Q. Gu. 2009. Spicochalasin A and New Aspochalasins from the Marine-Derived Fungus *Spicaria elegans. Eur. J. Org. Chem.* **18**: 3045–3051.

Zhou GX, EM.Wijeratne, D. Bigelow, LS. Pierson, HD. VanEtten and AA. Gunatilaka .2004. Aspochalasins I, J, and K: three new cytotoxic cytochalasins of *Aspergillus flavipes* from the rhizosphere of *Ericameria laricifolia* of the Sonoran. *Desert. J Nat Prod.* **67**:328-32.

Zhu H, C. Chunmei, T. Qingyi, L. XiaoNian, Y. Jing, X. Yongbo, L. Zengwei, W. Jianping, Y. Guangmin and Z. Yonghui. 2016. Epicochalasines A and B: Two Bioactive Merocytochalasans Bearing Caged Epicoccine Dimer Units from *Aspergillus flavipes. Angewandte Chemie.* **128**:3547-3551.

Zhu H, C. Chen, Y. Xue, Q. Tong, XN. Li, X. Chen, J. Wang, G. Yao, Z. Luo and Y. Zhang. 2015. Asperchalasine A, a Cytochalasan Dimer with an Unprecedented Decacyclic Ring System, from Aspergillus flavipes. *Angew Chem Int Ed Engl.* **54**:13374-8.

Zhu X, D. Zhou, F. Liang, Z. Wu, Z. She and C. Li. 2017. Penochalasin K, a new unusual chaetoglobosin from the mangrove endophytic fungus *Penicillium chrysogenum* V11 and its effective semi-synthesis. *Fitoterapia.* **123**: 23-28.

■■■■